Membranes and Ion Transport

Volume 2

Membranes and Ion Transport

Volume 2

Edited by

E. Edward Bittar
Department of Physiology,
The University of Wisconsin,
Madison, Wisconsin, U.S.A.

WILEY-INTERSCIENCE
a division of John Wiley & Sons Ltd.
LONDON NEW YORK SYDNEY TORONTO

Library of Congress Catalog card No. 71–110649

ISBN 0 471 077119

Printed in Great Britain by
Dawson & Goodall Ltd.
The Mendip Press, Bath.

This work is dedicated to
the late
EDWARD J. CONWAY
1894–1968

Preface

Thirty years have now elapsed since the publication of the celebrated paper by Boyle and Conway. During these years much has happened in the field of membrane metabolism and ion transport. The following work is an attempt to deal with the notable advances that have already been made and to treat the subject both systematically and critically. Membrane transport may be said to be an integral part of the life sciences but it does not yet seem to cover a comprehensive field. To undertake a general survey of it is therefore to invite criticism, in particular at a time when research is being conducted more intensively than ever. Written primarily for the novice and professed student of the field, the work aims at providing an intelligible view of the wide scope and importance of the subject, and of the different lines of research that may lead to a more unified discipline.

The work is divided into three books. Volume 1 contains a section on the structure, chemistry and behaviour of both artificial and natural membranes, and a section on the theoretical aspects of transport. A useful glossary has been appended at the beginning of this volume. Volume 2 deals with ion movements in symmetrical cells and sub-cellular organelles. And Volume 3 begins with an account of ion movements in asymmetrical cells, as well as water movements in cells in general, and ends with a collection of chapters dealing with ion regulatory mechanisms and cellular interaction and continuity.

The field of ion transport owes a great deal to the late Professor E. J. Conway. It therefore seemed more than fitting to dedicate this work to him.

Madison, Wisconsin
April, 1970

E. EDWARD BITTAR

Contributors to volume 2

C. C. Ashley — *Department of Zoology, University of Bristol, Bristol, England.*

R. Cereijo-Santalo — *The New Mount Sinai Hospital, Toronto, Ontario, Canada.*

D. J. Fry — *Medical Professorial Unit, Saint Bartholomew's Hospital, London, E.C.1, England.*

P. J. Garrahan — *Departmento de Quimica Biologica, Facultad de Farmacia y Bioquimica, Universidad de Buenos Aires, Argentina.*

P. J. Goodford — *The Biophysics and Biochemistry Department, The Wellcome Research Laboratory, Beckenham, Kent, England.*

W. P. Hurlbut — *The Rockefeller University, New York, N.Y.* 10021, *U.S.A.*

R. M. Marchbanks — *Department of Biochemistry, University of Cambridge, Cambridge, England.*

Winifred G. Nayler — *Baker Medical Research Institute, Melbourne, Australia.*

I. Seidman — *Department of Pathology, New York University Medical Center, New York, N.Y.* 10016, *U.S.A.*

Contents

I

Ion Movements in Symmetrical Cells

CHAPTER 1

Ion movements in skeletal muscle

C. C. Ashley
Department of Zoology,
University of Bristol,
Bristol, England

I. GENERAL INTRODUCTION

In this chapter little mention is made of the nature of active transport processes, in particular the sodium–potassium coupled system. Current speculation as to the molecular mechanisms involved in this and other transport systems have been discussed by Caldwell (1970) in Volume I. Attention has been drawn to the use of single muscle cells as a preparation for investigating ionic movements. Not only do they represent intact functional units, but in certain cases their large size permits the internal medium to be altered in a defined way.

II. SODIUM AND POTASSIUM

A. Sodium and Potassium Activity in Muscle

1. *Introduction*

Interpretations of resting and action potentials in muscle in terms of the Nernst and Constant Field equations imply that the majority of the sodium and potassium in muscle and nerve is free and able to diffuse readily. Theories which suggest that these ions form a bound or immobile fraction by their interaction with charged molecules in the cytoplasm (Ling, 1960, 1962) have not received a great deal of experimental support. Although not directly pertinent to this article, the work of Baker, Hodgkin and Shaw (1961, 1962), Oikawa, Spyropoulos, Tasaki and Teorell (1961) and Tasaki, Watanabe and Takanaka (1962) using giant axons from the squid *Loligo*, should be mentioned. In these experiments the major part of the axoplasm was removed and replaced or the axon perfused with simple salt solutions. Axons with replaced 'axoplasm' showed normal resting and action potentials which suggests that if sodium and potassium ions are bound, the fraction must be small and would appear to play only a minor role in the ionic basis for the electrical events of the axon.

2. *Evidence from Ion-selective Glass Electrodes*

Small ion-selective electrodes have been developed which permit the activities of sodium and potassium ions to be measured directly within the cytoplasm of isolated muscle and nerve cells. (Hinke, 1959; Lev, 1964).

(a) The activity of potassium ions in muscle fibres from both the lobster *Homarus* as well as from the giant axon of the squid *Loligo* was similar in value and was close to the potassium activity observed in potassium chloride solutions of similar ionic strength. Hinke observed an activity coefficient of 0·55 for *Homarus* fibres, whilst Lev, using much smaller electrodes on single frog fibres, reported values of 0·75–0·79. This latter figure is similar to the activity coefficient observed in solutions of potassium chloride having a similar ionic strength to that of frog muscle. In summary, the experiments using potassium-selective electrodes suggests that the bulk of the potassium in both vertebrate and invertebrate muscle is free and able to diffuse readily throughout the cells. Theoretical calculations carried out by Nanninga (1961) for frog predict an activity coefficient of 0·87, which is reasonably close to the observed value. More recently, the experiments of Kushmerick and Podolsky (1969) using single frog and barnacle fibres have indicated that the longitudinal diffusion of potassium ions is close to that of potassium chloride in free solution at similar ionic strengths. The diffusion coefficient in the fibre was reduced by only a factor of two, relative to free solution, and similar reductions were observed for inert solutes such as sorbitol or sucrose. Kushmerick and Podolsky suggest that the reason for the similar reduction in all the diffusion coefficients may involve physical factors common to all the solutes, such as the higher viscosity of the cell cytoplasm compared to free solution, rather than a specific interaction of certain of the ionic solutes with charged protein molecules in the cytoplasm.

(b) The activity of sodium, in contrast to that of potassium, has been estimated by Hinke to be considerably lower than the total concentration of sodium inside the fibre. In *Homarus* and in *Carcinus* fibers, Hinke observed sodium activity in the range 0·012–0·016 M/kg, whilst the sodium concentration was close to 0·052–0·055 M/kg for both species. These figures indicate an activity coefficient for sodium of 0·22–0·29, whilst McLaughlin and Hinke (1966) have reported an even smaller activity coefficient of 0·13–0·19 for sodium in the large fibres from the barnacle *Balanus nubilus*. Similar low values for sodium have also been indicated by Lev (1964) for frog muscle. However the activity coefficient of sodium in squid axon is about 0·46, which is close to the value for sodium chloride solutions of similar ionic strength and similar to the value for potassium ions in the cell. One obvious explanation for the finding in muscle would be, according to the 'fixed-charge' theory of Ling, that the major part of the sodium is bound on immobile structures in the cytoplasm and is unavailable to the cation-selective electrode. In the absence of preferential binding sites for sodium in muscle compared to nerve, it is not easy to see why most of the muscle sodium should apparently be bound yet only a small fraction of the cell potassium. In addition, the experiments of Kushmerick and Podolsky already mentioned indicated that potassium and sodium behaved in similar ways in both frog and barnacle muscle, where the values of the diffusion coefficients were only a factor of two smaller than in free solution. An explanation of this apparently low activity coefficient for sodium in muscle might involve a compartmentalization of the sodium, rather than binding to cytoplasmic structures. There are several compartments, present in both vertebrate and invertebrate muscle which penetrate the fibre, yet are still in direct communication with the bathing saline. In frog the external saline contains about 100 mM and in most crustacea 450–500 mM sodium. The volume occupied by one such compartment, the transverse or T-tubular system of the sarcoplasmic reticulum, has been estimated as between 0·3–1 per cent in frog (Huxley, 1964; Endo, 1964) and so contributes only a small fraction to the total fibre volume (see later section). In crustacean fibres however the situation is more complex for although the homologue of the T-tubules is present, there are also numerous invaginations of the sarcolemma, termed 'clefts', extending considerable distances into the fibre. Typical cross-sections of muscle fibres from the barnacle *Balanus* as well as from the crabs *Paralithoides* and *Carcinus* illustrate the extensive nature of the cleft system (Selverston, 1967) and estimates from such cross sections of fibres fixed under isosmotic conditions suggest a cleft volume of 15–20 per cent. However, Maclaughlin and Hinke (1968) and Hinke (1969) using ^{14}C-inulin as a marker for this intermediate space report figures that are somewhat lower, in the region of 5–6 per cent.

If an average value of 0·65 is taken for the activity coefficient for sodium in aqueous solution at equivalent ionic strength, then the sodium activity reported by Hinke for barnacle would predict a concentration in the range 15–20 mM/kg. The remaining 30–60 mM sodium would be in an extracellular compartment, and at an extracellular sodium concentration of 450 mM/kg the estimate of the cleft volume has only to be 10 per cent of the fibre volume to give a fibre sodium concentration within the observed range of values. In addition, Brinley (1968) quotes a considerably smaller value (21 mM/kg) for the internal sodium concentration in barnacle than that reported by Maclaughlin and Hinke (1968) and this was based on a series of timed rinses in ONa, isotonic sucrose salines. Values in the range of 20–30 mM/kg have also been reported for barnacle fibres by Hagiwara and Naka (1964) and by Beaugé and Sjodin (1967). Brinley suggests that perhaps the higher figures quoted by Maclaughlin and Hinke could be explained by inadequate rinsing of the fibres before analysis permitting a large amount of sodium to remain in the clefts. It is also worth noting the extracellular space values determined in the taenia coli of the guinea pig using ^{14}C-inulin were some 10–20 per cent lower than the values estimates by smaller molecular weight markers such as ^{14}C-sorbitol or ^{60}CO-EDTA (Brading and Jones, 1969). Moreover, the clefts in both barnacle and crab fibres appear to be filled with electron dense material, which if present in the living fibre could well slow diffusion and perhaps hinder the entry of large molecules such as inulin.

Other cellular compartments may well contribute to the removal of sodium from the sarcoplasm, for example the mitochondria, nuclei and possibly the longitudinal elements of the sarcoplasmic reticulum (SR). Estimates place the volume of the longitudinal SR in the range of 12–13 per cent of the cell volume (Peachey, 1965) and although histochemical reactions indicate a high sodium concentration in the T-system and on the outer borders of the surface membrane, there was a negative reaction within the sarcoplasm or in the longitudinal SR (Zadunaisky, 1966). It is possible that the sodium-specific reagent, potassium pyroantimonate, gave a negative result intracellularly because it was unable to penetrate the surface membrane; however, other cells appear to be permeable to the reagent (Zadunaisky, Gennaro, Bashirelahi and Hilton, 1968). It appears therefore in muscle that the main site for sodium is within the cleft and T-system tubules which form an intermediate compartment, whose lumina are in direct contact with the bathing saline.

The results obtained by Lev (1964) on single frog fibres suggest that about 70 per cent of the estimated fibre sodium is not available to the sodium-sensitive electrode. This implies that of the total sodium some 6–9 mM is available to the electrode and the remaining 20–23 mM is contained in some other compartment that might tentatively be associated with part of the sarcoplasmic reticulum. The only likely possibility in frog are the T-tubules which, as has been mentioned, seem to occupy only about 1 per cent of the total fibre volume. This space, which is in contact with the extracellular fluid, can swell considerably under conditions where there is a loss of potassium and chloride from the fibre (Foulkes, Pacey and Perry, (1965) in frog; Giradier, Reuben, Brandt and Grunfest (1963) in crayfish and Selverston (1967) in *Carcinus*). A gradual gain of sodium and loss of potassium appears to occur in normal saline after muscles have been isolated (Simon, Shaw, Bennett and Muller, 1957).

Thus the gradual swelling of the T-system in frog fibres which have been isolated could well explain the rather high figures for the total fibre sodium reported by Lev as 0·029 M/kg and by Hodgkin and Horowicz (1959a) as 0·026 M/kg. The sodium concentration in freshly dissected frog fibres, as Hodgkin and Horowicz point out, was considerably smaller (0·01–0·018 M/kg). It seems therefore that part of the reason for the discrepancy between the activity of sodium, measured by the sodium electrode, and the total sodium concentration may be explained by the gradual swelling of the T-system and retention of extracellular sodium within the lumen of this intermediate compartment. This receives some support from the work of Kostyuk, Sorokina and Kholodova (1969) where the sodium concentration and activity were examined in isolated and non-isolated muscles. Non-isolated muscles had lower total sodium than isolated muscles but had similar values for the activity of sodium, as measured by a sodium electrode. There is also evidence from the study of sodium and potassium fluxes across muscle cell membranes to support the idea of an intermediate compartment, in contact with the external saline, which increases in size gradually from time of isolation. This evidence, together with the flux experiments on single muscle fibres which strongly suggests that sodium ions are normally free to diffuse within the cell, will be mentioned in the next section.

B. Sodium and Potassium Movements

1. *Sodium and Potassium Fluxes*

Harris and Burn (1949), Keynes (1951, 1954) and Carey and Conway (1954) were amongst the first to use radioactive tracers for following sodium and potassium fluxes across muscle membranes. These early experiments revealed that for the most accurate kinetic analyses of the membrane flux, muscles that were thin and had small diffusion distances were the most suitable. Keynes (1954) used the small frog toe muscle and obtained a value of 4·5 pmole/cm^2/sec for the potassium flux across the membrane. The influx being slightly smaller than the efflux, whilst that for sodium was more difficult to measure but was estimated as about 10 pmole/cm^2/sec. The situation was simplified by the use of single frog fibres where the problems of extracellular space and extraneous binding of the isotope were considerably reduced (Hodgkin and Horowicz, 1959a). They observed that the loss of both sodium and potassium ions followed single exponential kinetics as would be expected if the surface membrane was acting as the main barrier and the ions were essentially in freely diffusible forms on either side. The potassium flux at rest was about 5·4 pmole/cm^2/sec with the efflux slightly greater than the influx, as was observed by Keynes (1954). The sodium flux was 3·5 pmole/cm^2/sec and the influx was slightly greater than the efflux. The time constants for the loss of sodium was 1·5 hours and for potassium 9 hours. Occasionally, a rapid loss of sodium preceded the slower loss and this was attributed to imperfect cleaning of the muscle fibre during the dissection. Hodgkin and Horowicz point out however that they could not detect sodium or potassium losses with time constants of less than 1–2 minutes. In addition, they confirmed the early findings of Fenn and Cobb (1936) on whole rat sartorius, who reported a net increase in sodium and loss of potassium following electrical stimulation. Hodgkin and Horowicz were able to estimate in single frog fibres that the net gain of sodium per action potential was 15·6 pmole/cm^2 and loss of potassium of about 9·6 pmole/cm^2 at 21°C. These values for sodium gain and potassium loss during an impulse are somewhat larger than have been observed in squid axon (sodium gain = 3·5 pmole/cm^2 and potassium loss = 3·0 pmole/cm^2 at 22°C, data taken from Hodgkin (1958, 1964)). Hodgkin and Horowicz found no evidence for an intermediate extracellular compartment during the sodium efflux experiments from single frog fibres. However, Carey and Conway (1954) using whole frog sartorius muscle and Shaw (1958), using single muscle fibres from the crab *Carcinus*, did find evidence that suggested an intermediate space and Carey and Conway estimated it as about 10 per cent of the fibre volume. It seems likely, in view of the evidence outlined so far, that part of the intermediate extracellular compartment observed in frog muscle by Conway and his collaborators could be explained in terms of extraneous binding of the isotope to damaged fibres and part to the swelling of the T-system in muscles that were not freshly dissected. Evidence for the compartment, located within the cleft system, as reported by Shaw for *Carcinus* has since received extensive support from the electron microscopy of this and other crustacean muscles.

More recently, the use of the large invertebrate muscle fibres has permitted the isotope to be injected directly into the fibre sarcoplasm making the interpretation of the efflux experiments even simpler. The work on large crab muscle fibres by Bittar (1966), Bittar, Caldwell and Lowe (1967) and on the large fibres from the pacific barnacle *Balanus nubilus* (Brinley, 1968) indicated that the fluxes followed simple exponential kinetics. The surface membrane acted as the main barrier to the loss of the tracer and thus agreed with the experiments of single vertebrate fibres by Hodgkin and Horowicz (1959a). The rate constant for the loss of the isotope rose rapidly in these fibres after axial injection and reached a steady value representing the exponential phase within 5–10 minutes. The initial phase, when the rate constant was increasing, presumably represents the diffusion of the ^{22}Na from the region of the injection throughout the fibre. The flux of sodium across the membrane of these fibres is considerably higher than for frog, and values of 49 and 39 pmoles/cm^2/sec respectively for the influx and efflux have been reported by Brinley (1968) for *Balanus* and an efflux of 221 pmole/cm^2/sec for *Maia* fibres (Bittar, Caldwell and Lowe, 1967). If account is taken of the large infoldings of the sarcolemma (the cleft system) which appear to increase the surface area by 10–20 times (Selverston, 1967), then the sodium efflux can be reduced to values which are not too dissimilar to those reported for frog fibres. The sodium efflux from these large cells seemed to behave in a similar way to that of frog (and squid nerve) and was reduced, although not completely eliminated by the external application of the cardiac glycoside ouabain, or by removal of external potassium. The potassium influx and efflux from *Balanus* fibres has been reported by Brinley (1968) to be 28 and 60 pmole/cm^2/sec respectively and the influx was reduced by cardiac glycosides, the efflux being unaffected. Bittar (1966) also observed that the sodium efflux from *Maia* fibres was not reduced to the levels of passive exchange by cardiac glycosides and from the results of a variety of metabolic inhibitors on the efflux and the resting potential has suggested that part of the efflux may be electrogenic and driven directly by respiration (see also Cross, Keynes and Rybova (1965); Adrian and Slayman (1964)). Additionally, in frog muscle Keynes and Swan (1959) have shown that in frog sartorius the sodium efflux is reduced by half if the external sodium is replaced by choline or lithium. This suggests that about half the efflux is an exchange diffusion mechanism, whilst the remaining efflux was reduced further by the application of ouabain; this fraction is presumably the coupled sodium–potassium pump.

The experiments on large single muscle fibres give no direct evidence for the part played by the cleft system as an intermediate space. Certainly however the application of ouabain or the application of OK salines resulted in a slower fall in the sodium efflux (Bittar, Caldwell and Lowe, 1967) than was observed in squid axon (Caldwell and Keynes, 1959). It seems likely that this is due to the presence in *Maia* of the clefts which increase diffusion times, and the fall in the efflux implies that appreciable sodium extrusion can occur across the cleft membranes.

2. *Net Changes in Sodium and Potassium*

There is an appreciable amount of evidence to suggest that in a normal muscle cell as in squid axon, there is a coupling of the efflux of sodium from the cell to the influx of potassium. This ionic pump mechanism, whose activity is in part reflected by the flux experiments already mentioned, requires the participation of metabolic energy by the cell. This active process maintains the ionic gradients across the cell membrane which are necessary for the functioning of the action potential mechanism. Keynes and Maisel (1954) found no evidence for the inhibition of the sodium efflux from whole frog muscle with either 0·2 mM dinitrophenol or a mixture of 0·5 mM iodoacetate and 3 mM cyanide applied for 3–3·5 hours. The work on squid axon (Caldwell, 1960) suggests that a drop in the sodium efflux would only be expected when all the phosphagen (in this case arginine phosphate) has been split resulting in a fall in the ATP concentration. Since the distribution of high energy phosphates was not examined in the experiments of Keynes and Maisel and the fibres were not in rigor, an appreciable concentration of ATP must have been present during the period of time in the metabolic inhibitors (Bendall, 1951).

However, Conway and his associates (Carey, Conway and Kernan, 1959) were able to demonstrate the effect of metabolic inhibitors on the net extrusion of sodium from frog muscle under certain conditions. Paired sartorius muscles were used and were initially loaded with sodium by immersion in a potassium-free saline at about 0°C for about 24 hours. At the end of this time, one sartorius was removed and the total sodium content estimated. If the second muscle was placed in a recovery saline containing the same amount of sodium together with the normal amount of potassium (10 mM) at 18°C, no net loss of sodium occurred. If, however, the second muscle was placed in a medium containing 10 mM potassium but a lower concentration of sodium than the loading saline, then the muscles extruded sodium. The loss of sodium at 18°C over a 2 hour period was about 20 mM, a decrease from 59 mM/kg in the control to about 39 mM/kg in the second muscle. Similar results were obtained by Frazier and Keynes (1959) and by the earlier work of Desmedt (1953) who was also able to show that the peak of the overshoot of the action potential was reduced in the sodium-loaded muscles. This interesting finding implies that the extra sodium is within the cell sarcoplasm in a free form, rather than bound to protein molecules or within an intermediate compartment of the type already discussed. Frazier and Keynes (1959) observed an increase in the rate constant for the loss of tracer sodium in loaded fibres as would also be predicted if the sodium was free in the cell sarcoplasm. Brinley (1968) has recently reported a similar finding in single barnacle muscle fibres micro-injected with sodium.

Although Carey, Conway and Kernan (1959) were able to inhibit sodium extrusion from loaded muscles by either 2mM cyanide, ouabain, low temperatures or iodoacetate, the work of Frazier and Keynes (1959) indicated that cyanide alone was ineffective unless mixed with iodoacetate. Later work by Conway (1960) suggested that the difference in results in cyanide-containing saline depended very much on the metabolic state of the muscles. Conway and his associates interpreted the overall results from the sodium-extrusion experiments in terms of a 'critical energy barrier' or, alternatively, a free energy which has to be supplied before net extrusion of sodium will take place (Conway, Kernan and Zadunaisky, 1961). Certain experiments strongly suggested such a concept, in particular the finding that a decrease in the resting potential, by an increase in the potassium concentration in the external saline, rendered the sodium extrusion insensitive to cyanide. The decrease in the resting potential decreased the energy barrier required for the extrusion of a sodium ion (see, for example, Conway (1960); Kernan (1962, 1968); Caldwell (1968)). The relationships between the free energy needed to extrude a sodium ion, the resting potential and the sodium concentration can be expressed in the following way:

$$\frac{\mathrm{d}G}{\mathrm{d}n} = \mathrm{R}T \ln \frac{[\mathrm{Na}]_\mathrm{o}}{[\mathrm{Na}]_\mathrm{i}} + V_\mathrm{m}\mathrm{F}$$

which represents the sum of the osmotic and electrical gradients. Where $\mathrm{d}G/\mathrm{d}n$ is the change in free energy for an equivalent of sodium extruded, Na_o and Na_i are the sodium activities in the outside saline and inside the muscle respectively and V_m is the resting potential of the muscle. Thus depolarization of the fibres, which affects the term $V_m\mathrm{F}$, will decrease the free energy required for the extrusion of a sodium ion. It seems likely from the work on squid axon, where a coupled sodium–potassium pump was only operative in the presence of both ATP and phosphagen (Caldwell, Hodgkin, Keynes and Shaw (1959); Caldwell (1960); Caldwell, Hodgkin, Keynes and Shaw (1960)), that the extent of sodium extrusion in frog muscle will also depend on the posphagen content. This would help to explain the reported effects of inhibitors on the extrusion of sodium, in that they might not be effective in muscles that had high posphagen contents. This tends to be confirmed by the findings of Conway (1960), where cyanide was apparently ineffective in muscles with higher amounts of high energy phosphate compounds (i.e. ATP+creatine phosphate). Kernan (1962) has also shown that the presence of lactate or insulin in the recovery medium stimulates sodium extrusion under conditions where it is not normally observed. It seems possible that both compounds by stimulating metabolism may well increase the phosphagen content of the muscles, so providing extra free energy in the form of high energy phosphate compounds for the extrusion of sodium.

III. CALCIUM

A. Internal Calcium Changes

1. *Introduction*

Calcium ions appear to be required not only for the functional integrity of muscle membranes but are also required if the normal processes of mechanical activation are to occur. The work of Ringer (1883), and Mines (1913) and, more recently, Ware, Bennett and McIntyre (1955) indicated that in the absence of calcium ions the mechanical activity of cardiac muscle was abolished, despite the presence of electrical activity on the cell membrane. Heilbrunn and Wiercinski (1947) were amongst the first to suggest that calcium was required for the activation of skeletal muscle. Of the major physiologically occurring ions, only calcium was effective when injected directly into the sarcoplasm of single frog fibres.

The theory that the entry of calcium across the outer cell membrane was sufficient to activate the contractile systems was opposed on theoretical grounds by Hill (1948, 1949). He pointed out that the entry of calcium followed by diffusion would not be sufficiently rapid to account for the fast mechanical activation observed in frog fibres (see also Sandow, 1952). It was not until the re-discovery of the sarcoplasmic reticulum (SR) by Bennett and Porter (1953), as a series of tubules ramifying throughout the cell, that it did appear that Hill's fundamental objection might be by-passed. The SR surrounded each myofibril and had certain elements that originated at the surface membrane of the cell. These were described more clearly by Porter and Palade (1957) as transverse or T-tubules which made close associations with the longitudinal SR surrounding the myofibrils. The physiological role of the T-tubules in the activation of contraction in both vertebrate and invertebrate muscle was confirmed by the elegant local activation experiments of Huxley and Taylor (1958). Much evidence now indicates that in striated muscle the contractile system is activated by a reduction of the potential on the outer fibre membrane which is transmitted along the T-system and which results in the release of intracellular calcium from the longitudinal elements of the SR. Experimental evidence also indicates that in the resting state the ionized calcium concentration in the sarcoplasm is very low (probably $<1 \times 10^{-7}$ M, Portzehl, Caldwell and Rüegg (1964)) and that during activation this increases considerably. It has been suggested that during maximum fibre activation as much as $0{\cdot}5\text{–}1 \times 10^{-4}$M calcium could be released so that every cross-bridge on the myosin would be activated (Davies, 1963). Calcium can then be returned to the SR from the sarcoplasm and contractile apparatus by an active process (Hasselbach and Makinose, 1961, 1962, 1963) requiring the participation of ATP. Surface binding of the calcium to the membrane of the SR could equally well account for the rapid removal of a large fraction of calcium from the sarcoplasm (Ohnishi and Ebashi, 1964). During this process the sarcoplasmic calcium is lowered to its original resting level, at which concentration the calcium-activated ATPase of the myofibrils has a low activity (Weber, Herz and Reiss, 1964, 1969).

2. *Tracer Studies*

There are several ways in which the movement of calcium ions within the sarcoplasm of whole muscles or single fibres can be followed, both at rest and during contraction. Woodward (1949) used ^{45}Ca as a tracer for calcium movements within frog sartorius muscle. The muscle was initially loaded with the tracer by soaking in an 'active' calcium solution and the efflux subsequently followed into 'inactive' saline. Initially there was an extremely rapid efflux which probably represented loss of the tracer from extracellular spaces and from extraneous binding sites. The efflux decreased to a low level, characteristic of the resting state and was increased some 30–200 per cent by electrical stimulation. Analysis of the efflux curves during stimulation revealed that this extra calcium was most certainly derived from inside the muscle fibres, rather than from interfibre spaces. In this case the efflux of calcium across the outer membrane acted as an index of the mobilization of intracellular calcium, although the time course of the efflux bore no simple relationship to the actual sarcoplasmic events. The tracer technique has been used extensively by Bianchi and Shanes (1959) and Shanes and Bianchi (1960), using whole sartorius muscle, who showed that associated with the increased efflux of calcium during contraction there was an equal influx so that the total fibre calcium level would be expected to remain reasonably constant. Contractions induced by raised potassium or caffeine salines or by electrical stimulation all produced an increase in the flux of calcium across the membrane.

As with the early experiments of Woodward, the efflux experiments from whole sartorius muscle were complicated by the presence of a rapid extra-cellular component. This inherent difficulty can be overcome if the isotope is introduced directly inside the fibre sarcoplasm. The discovery of striated muscle fibres large enough to permit cannulation and subsequent axial injection (Caldwell and Walster, 1963; Hoyle and Smyth, 1963), provided an ideal experimental system in which to investigate the efflux independent of the problems of extracellular space. Direct injection of ^{45}Ca inside the large single fibres from the crab *Maia* led to a rapid loss of the isotope within the first 5–10 minutes after injection, and this was associated with a small contraction. This rapid loss was followed by a fall in the rate constant to a low resting value after about 60–90 minutes. Application of salines containing a raised amount of potassium or caffeine, or electrical stimulation at this point led to an increase in the rate constant for the loss of the isotope from the fibre (Lowe and Caldwell, 1964). More interestingly, levels of potassium insufficient to produce a detectable contraction led to an increase in the efflux of the isotope from the fibre. A similar response in the efflux was observed when subcontraction doses of caffeine were applied to the fibre (Lowe, 1964).

In both types of experiment, whether the tracer is initially introduced across the outer membrane or by direct axial injection, a fraction of the tracer calcium is removed from the fibre sarcoplasm by the calcium-sequestering mechanism that is, the SR. Stimulation of the fibre or partial excitation by subcontractile doses of potassium or caffeine are sufficient to release some of this calcium into the sarcoplasm and it is this mobilization which is eventually observed by an increased efflux of the isotope across the outer membrane. Although the efflux acts as a useful index of the changes in intracellular calcium during a contractile response, it is not able at present to predict either the time course or the absolute magnitude of the response with any certainty.

3. *Complexiometric Methods*

Estimates of the calcium concentrations at rest and during contraction have been made by using the compound EGTA (ethylene glycol-bis(β-aminoethyl ether)*N*, N^1-tetraacetic acid), which under physiological conditions chelates calcium much more strongly than magnesium ions. When EGTA is mixed with proportions of calcium, solutions containing stabilized, ionized calcium concentrations can be produced in the range of about $0{\cdot}01–2{\cdot}0 \times 10^{-6}$M. Injection of such solutions into single muscle fibres suggested that at rest the ionized calcium concentration was $<1 \times 10^{-7}$M and that detectable contraction occurred if the calcium concentration was raised to $3–25 \times 10^{-7}$M. (Portzehl, Caldwell and Rüegg, 1964). Similar values for detectable contraction have been reported for other large fibres from the barnacle *Balanus nubilus* (Hagiwara and Nakajima, 1966; Ashley, 1967) and for the giant fibres from the king crab *Paralithoides* (Ashley, 1967). These figures are much lower than the earlier estimates made by Heilbrunn and Wiercinski (1947) as to the threshold calcium concentration for shortening. The more recent calcium concentrations agree reasonably well with the threshold values needed to activate the myofibrillar ATPase in frog (Weber, Herz and Reiss 1964), crab (Portzehl, Zaoralek and Grieder, 1965) and in barnacle (Hasselbach, 1966).

Injection of EGTA alone into single fibres is able to suppress almost completely the contraction brought about by either potassium or caffeine salines or by electrical stimulation (Ashley, Caldwell, Lowe, Richards and Schirmer, 1965; Ashley, 1967). From the amount of EGTA required to suppress the contractile responses to just threshold levels, the amount of calcium released into the sarcoplasm by the electrical stimulation can be estimated as $<2\times10^{-4}$M. Whilst vigorous stimulation induced by raised potassium or caffeine salines released $<2\times10^{-3}$M. Because the presence of EGTA internally may well disturb the delicate calcium releasing and calcium reaccumulating mechanism, the calcium concentrations estimated by this method are likely to be a good deal higher than in the uninjected fibre (see next section).

4. *Optical Methods*

More recently however, optical methods have been described which for the first time permit the rapid changes in intracellular calcium to be followed during the time course of a single contraction (Ashley and Ridgway, 1968, 1969; Ridgway and Ashley, 1967). The method uses a bioluminescent protein which, under physiological conditions, emits light solely in the presence of calcium ions (Shimomura, Johnson and Saiga, 1962, 1963; Shimomura and Johnson, 1969). The protein aequorin, extracted from the luminous jellyfish *Aequorea*, is dissolved in buffer solution and injected into the sarcoplasm of an isolated muscle fibre. If the fibre is now made to contract either by electrical stimulation or spontaneously when the preparation is cooled rapidly, the fibre emits light in response to the changes in intracellular calcium. These changes in light emission can be observed by the dark-adapted eye but are more conveniently followed using a photomultiplier tube which is positioned close to the fibre. The variations in internal calcium concentration can be readily related to the changes in the membrane response and tension development which are recorded simultaneously. In addition, the fibre in the resting state emits a steady low level light emission which reflects the resting calcium concentration in the sarcoplasm.

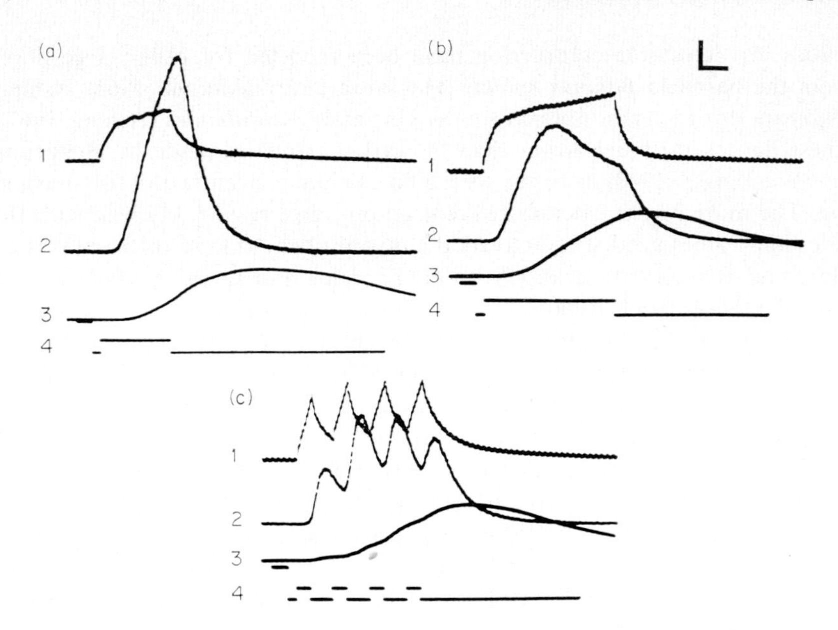

Figure 1. Illustrates the effect of (a) short, (b) long and (c) brief tetanic stimuli on the internal calcium changes in single barnacle muscle fibres. Trace 1, membrane response, (a) 20 mV/cm, (b) and (c) 10 mV/cm; trace 2, calcium-mediated light emission, (a) and (c) 1·9 nlm/cm, (b) 3·8 nlm/cm; trace 3, isometric tension response, (a) 5 g/cm, (b) and (c) 2·5 g/cm; trace 4, 1 V calibration pulse and stimulus mark. Intensity, (a) 3·5 V, (b) 4·0 V, (c) 3·0 V. Nominal duration (msec), (a) 200, (b) 350, (c) 40. Horizontal calibration, 100 msec/cm. Resting light emission, (a) 0·64 nlm, (b) 1·1 nlm, (c) 0·27 nlm; temperature, 11–12°C. Resting potentials −50 to −56 mV. Calibration bar, 1 cm (from Ashley and Ridgway, 1970b).

A typical trace presented in figure 1(a) illustrates the fibre responses to a single depolarizing pulse after injection of aequorin. Trace 1 illustrates the membrane response to an imposed depolarizing pulse and represents the electrical event on the fibre membrane. Trace 2 illustrates the changes in the light emission and hence internal calcium as a result of the applied stimulus pulse. Trace 3 presents the isometric tension, which is the mechanical response of the fibre. Trace 4 has a calibration pulse followed by a stimulus mark which represents the precise onset and cessation of the stimulus pulse.

The rapid change in intracellular calcium (the *calcium transient*) is an intermediate chemical event in the sequence of reactions which constitute excitation–contraction (E–C) coupling. The calcium transient at these stimulus durations has two main phases, an S-shaped rising phase mediated by the membrane response and a rapid exponential falling phase initiated during membrane repolarization and having a time constant of about

80 milliseconds at 11–12°C. The rising and falling phases of the calcium transient represent respectively a rise and fall in the sarcoplasmic calcium concentration. The peak of the calcium transient represents a steady state where the rates of release and removal of calcium from the sarcoplasm are equal.

Perhaps the most interesting observation about the traces in figure 1(a) concern the time relations between the calcium transient and the rising phase of the tension response. The peak of the calcium transient occurs at the maximum rate of rise of the tension response, whilst the calcium transient has returned to virtually the resting value at peak tension. Relaxation thus occurs at *resting* calcium concentrations. The calcium transient resembles the first derivative of the rising phase of the tension response, so that tension should be the sum of the calcium transient and this implies that the contractile machinery is integrating the available calcium. Calcium therefore controls the *rate* of development of tension in these fibres. Similar results have been obtained from the large fibres from the crab *Maia* (Ashley, 1969 and unpublished experiments).

At longer stimulus durations the calcium transient is no longer biphasic, but has an additional intermediate falling phase (figure 1(b)). The rapid rising and slowly falling phases are both mediated by the membrane response, whilst the exponential falling phase is initiated during membrane repolarization. The tension response continues to rise, but at a reduced rate during the slow falling phase. In the case where a brief tetanic burst is applied to the fibre (figure 1(c)), all the fibre responses are partially fused. However, the calcium transient responses are not all the same size; the third and fourth transients are smaller than the second at this stimulus intensity. Moreover, an envelope bounding the peak values of the four transients resembles the shape of the longer duration transient presented in figure 1(b). Finally, each calcium transient in the tetanus produces a distinct change in the rate of development of tension. The decline in the intracellular calcium concentration appears dependent on the membrane depolarization, for it is not observed at low stimulus intensities, and may be brought about by a change in the properties of a coupling membrane which links depolarization of the outer membrane to the release of calcium from the SR. The mechanism may involve the delayed rectifying current observed in these fibres by voltage clamp methods (Hagiwara, Hayashi and Takahashi, 1969).

The optical method using the photoprotein aequorin can give information about internal calcium movements both in the normal muscle and also in fibres where the normal processes of E–C coupling have been disturbed. In particular, when the fibre has been treated for fairly short periods of time with hypertonic salines, there is a rapid loss of tension responses. Under these circumstances, there is a concomitant decline in the calcium transient response suggesting that the interruption in coupling occurs at the mechanism linking the membrane depolarization to the release of calcium by the SR (Ashley and Ridgway, 1970a and b).

Estimates from the aequorin experiments (Ashley, 1970) suggest that the resting calcium concentration in barnacle fibres is $0{\cdot}8\text{–}2{\cdot}0 \times 10^{-7}$M, whilst for detectable tension responses the calcium concentration rises some 2–3 times to $1{\cdot}6\text{–}6{\cdot}0 \times 10^{-7}$M. Maximum activation of the fibres by immersion in raised potassium salines results in a calcium concentration that is $0{\cdot}8\text{–}2 \times 10^{-6}$M. These experiments also suggest that two calcium ions are required per tension site for activation.

In vertebrate muscle, Jöbsis and O'Connor (1966) indicated, in a preliminary series of experiments using the calcium-indicator murexide, that there was an early release of calcium before the onset of detectable tension. Unlike the results on *Balanus*, however, this was an averaged result from some twenty contractions on whole toad sartorius muscle which had been loaded with the indicator by intraperitoneal injection of the intact animal.

B. Role of Extracellular Calcium

1. *Vertebrates*

Hill, as has already been mentioned, seriously questioned the part played by an influx of extracellular calcium in causing direct mechanical activation in frog muscle. However, a flux of calcium does occur associated with the electrical and chemical events of the coupling process. In particular, the calcium influx observed by Bianchi and Shanes (1959) during a single twitch was about $0{\cdot}2$ pmole/cm^2 and is presumably a result of calcium entry following the depolarization of the cell membrane by the action potential. A figure of 44 pmole/cm^2 was also reported as the amount of calcium entering during a potassium contraction. Frank (1958) suggested that perhaps the calcium influx played some intermediate role in the initiation of a mechanical response in muscle. He observed a rapid abolition of the contraction initiated by 25 mM potassium if the muscle had been pretreated with calcium-free saline. The response was almost completely eliminated within 1 minute (Frank, 1964), yet the frog toe muscle was still contractile as judged by its response to caffeine saline. Similar results have been reported by other workers (Jenden and Reger, 1963; Luttgau, 1963) who have suggested that perhaps the main reason for the loss of tension responses in calcium-free salines was a drop in the resting potential which might block the process coupling membrane depolarization to calcium release by the SR. Additional experiments by Frank (1964) indicated that the loss of a contractile response preceded the drop in membrane potential and also confirmed his earlier findings that many multivalent cations could replace calcium and restore the muscle response to raised potassium (Frank, 1964). These findings were supported by the experiments of Curtis (1963), again on frog toe muscle, who made the interesting observation that if the calcium was reduced to about 100 μM, there was no significant decrease in the resting potential, even over long periods of time, yet the fibre response to raised potassium saline was abolished.

It seems that at least three processes are initiated when external calcium is removed. The first involves the loss within 3–4 minutes of a superficial coupling-calcium essential to the function of E–C coupling in the normal muscle. The second involves the gradual change in the membrane properties in calcium-free saline. This was typified by a decrease in the membrane potential and membrane resistance which seemed to occur more rapidly in the single frog semitendinosus fibres used by Luttgau (1963), than in Frank's experiments on toe muscle. Thirdly, a much slower phenomenon which appeared to be due to a depletion of the calcium stores within the longitudinal elements of the SR. This was observed as a gradual unresponsiveness of toe muscle to repeated applications of calcium-free caffeine saline or a gradual reduction of the response of the muscle to contractions supported by other divalent ions in the absence of calcium. Both of these effects could be readily reversed by exposures to normal calcium saline.

In summary, it appears that an influx of external calcium does play a part in the normal coupling processes leading to contraction and this is independent of its effect on the resting potential. It is interesting to speculate that the release of perhaps a membrane-bound calcium fraction is the physiological triggering process for calcium release by the main calcium stores of the cell (see also Frank, 1964; Lorkovic, 1967). Certainly the calcium influx experiments of Bianchi and Shanes suggest that the calcium entering as a result of the action potential, if evenly distributed, would be too small to cause even a threshold tension response (see also Sandow, 1965). One final point of interest is that the internal calcium stores can be reduced by as much as 50 per cent by soaking for up to 5 hours in a calcium-free saline (Gilbert and Fenn, 1957). It is difficult to see why the membrane calcium required for the coupling process is not replenished by the steady calcium efflux from the fibre under these conditions. Here the findings of Van der Kloot (1968) seem relevant since disruption of the T-tubules by osmotic shock made little difference to the rate of calcium efflux in frog. The area of the T-tubules in frog has been estimated as between 5–7 times the area of the surface membrane (Peachey, 1965), so that any fall in the efflux by osmotic disruption would be readily detected. It seems therefore that the major calcium efflux does not occur across the membranes of the T-system. If the sites at which the triggering or coupling calcium is lost lie deeper in the fibre, perhaps at the level of the triad, then they would not be replenished by the calcium efflux occurring across the surface membrane.

The points raised in the above section concern the participation of external or membrane-bound calcium as a possible trigger for the main process of calcium release. The experiments of Brecht and Gebert (1966) suggest that it is possible to obtain sufficient entry of calcium from salines containing a lower than normal calcium content to produce a detectable tension response, independent of direct membrane excitation. It is not certain whether these findings are related to the general increase in membrane permeability observed when the calcium concentration in the bathing saline is reduced, and which is characterized by a drop of the resting potential and decrease in membrane resistance.

2. *Invertebrates*

The situation as far as the removal of external calcium is concerned seems similar to the vertebrate situation already discussed.

The evidence is less well documented but OCa salines applied for 5–10 minutes almost completely abolished the contraction initiated by raised potassium saline in fibres from the barnacle *Balanus* (Edwards and Lorkovic, 1967; Ashley, unpublished observations) and from the crab *Paralithoides* (Ashley, unpublished observations). Probably a short term exposure to calcium-free salines results in the depletion of the same fraction of calcium as in the experiments on frog toe muscle. In certain invertebrate fibres, however, the inward current phase of the active membrane response is not mediated by sodium but by a divalent cation. Evidence for the participation of divalent ions in the inward current phase of the action potential was suggested by Fatt and Ginsburg (1958) and by Werman, McCann and Grundfest (1961) and Werman and Grundfest (1961) under certain conditions. More recently, Hagiwara and Naka (1964) and Hagiwara, Chichibu and Naka (1964) obtained more convincing evidence that in isolated *Balanus* fibers, calcium ions constitute the main inward current phase when the inside of the fibre is treated with a calcium-binding agent such as EGTA. In addition ^{45}Ca tracer experiments (Hagiwara and Naka, 1964) suggest that sufficient calcium may enter the muscle fibre across the surface and cleft system membranes during the passage of an action potential to raise the internal calcium concentration to about 1×10^{-6}M (Edwards and Lorkovic, 1967). This concentration is sufficient to produce a contractile response by the fibre (Hagiwara and Nakajima, 1966; Ashley, 1967). In intact fibres, however, the membrane responses are usually only graded (Hoyle and Smyth, 1963) and it is difficult to determine how far calcium entry during neural stimulation of the intact fibre will be of direct importance in initiating a a contractile response.

C. Action of Caffeine and Related Compounds on Calcium Movements

As has already been mentioned, caffeine is able to cause a contractile response in skeletal muscle cells, and this occurs without any major change in the resting potential (Taylor, 1953; Axelsson and Thesleff, 1958; Conway and Sakai, 1960). The contraction can occur in completely depolarized muscle or even when the compound is introduced directly inside a living fibre by micro-injection (Caldwell and Walster, 1963). Such experiments have led to the suggestion that caffeine may act by releasing 'bound calcium' from sites within the muscle (Frank, 1962, 1964); the most likely position would be from the longitudinal elements of the SR. To test this suggestion caffeine mixed with various proportions of a calcium-binding agent (either EGTA or EDTA) were injected into single *Maia* muscle fibres. The results obtained from such experiments with EGTA are illustrated in figure 2. Each injection represents the first injection into the fibre. Injections of 100 mM caffeine or 100 mM $CaCl_2$ produce about the same extent of shortening

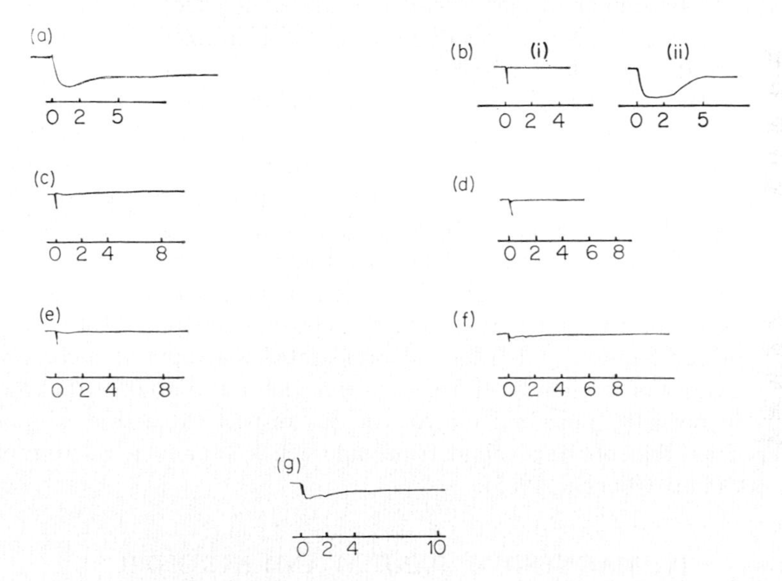

Figure 2. Illustrates the shortening responses of single crab muscle fibres following the injection of caffeine–EGTA mixtures. The shortening is produced as a result of the injection of the mixture over a 10 mm length of the fibre. The small vertical line which often precedes the shortening response represents an injection artifact. Injections: (a) 100 mM caffeine alone, (b)(i) 110 mM caffeine–100mM EGTA, (ii) 100 mM $CaCl_2$ ten minutes later, (c) 110 mM caffeine–100 mM EGTA, (d) 120 mM caffeine–100 mM EGTA, (e) 130 mM caffeine–100 mM EGTA, (f) 140 mM caffeine–100 mM EGTA and (g) 150 mM caffeine–100 mM EGTA. Time (min); temperature, 20–22°C. Traces (a) to (g) represent injections into separate fibres.

(figure 2 (a) and (b)(ii)), whilst injection of about equimolecular amounts of caffeine: EGTA results in the suppression of the fibre contractile response (for example figure 2(c) and (d)). If the amount of caffeine in the mixture is gradually increased the fibre contractile responses are at least partially restored (figure 2 (e), (f) and (g)) (Ashley and Caldwell, 1964). These experiments strongly support the idea that caffeine acts by releasing calcium from intracellular binding sites and a simple interpretation implies that the release ratio is close to 1:1 (Ashley, 1965). Moreover, the work of Herz and Weber (1965) indicated that caffeine causes release from and inhibits uptake of calcium by isolated SR vesicles from frog.

Injections using the compound EDTA instead of EGTA were complicated by the interaction of the chelating agent with the fibre magnesium as well as with the calcium released by the caffeine. In these experiments much lower caffeine to EDTA ratios were observed in the region of 1:3 to 1:4 for just-threshold suppression of a detectable shortening response (Ashley and Caldwell, 1964). The mechanism of action of other closely related xanthines appears to be similar to that of caffeine. The contraction produced by the compound theophylline can also be suppressed if injected with about equimolecular amounts of EGTA (Ashley, 1965).

The work of Fleckenstein, Hille and Adam (1951) and Conway and Sakai (1960) suggested that the alkaloid quinine causes contraction by a mechanism independent of membrane depolarization. The work on single *Maia* fibres (Ashley, 1965) indicated that, as with caffeine, quinine contractions can be suppressed almost completely by intracellular injections of EGTA. This, together with the fact that the quinine contraction can occur in depolarized fibres, strongly suggests that the site of action of quinine is on the longitudinal elements of the SR and that it acts by releasing calcium. More recently, Isaacson and Sandow (1967) have demonstrated that quinine increases the rate of release of calcium from frog sartorius and it also doubles the calcium influx in normally polarized muscle. In this respect the action of quinine differs from that of the alkaloid ryanodine which stimulates calcium efflux but not influx (Bianchi, 1963).

IV. MAGNESIUM, BARIUM AND STRONTIUM

Unfortunately, little is known about the movements of magnesium ions in the resting or in the active muscle. Gilbert (1960) examined the influx of magnesium ions into resting frog sartorius using the isotope ^{28}Mg. He was able to demonstrate that the uptake took place in three main stages lasting about 0·5, 30, and 300 minutes. The first rapid stage was interpreted as surface absorption, the second as an entry into an extracellular phase and the third, the entry of the tracer inside the muscle cell. The maximum exchangable intracellular magnesium was estimated from these experiments as 1·1 mM/kg, so that some 80 per cent of the total fibre magnesium appeared to be in a non-exchangable form. Recently Schied, Straub and Hermenau (1968) have confirmed the two slower phases of magnesium uptake by a different method than Gilbert's. Estimates of the magnesium activity inside single muscle fibres have not been made, but calculations by Nanninga (1961) imply that of the total fibre magnesium in frog of 11 mM/kg, only 3–3·4 mM/kg is in a free form, a value that is a little higher than the figure derived from the influx experiments of Gilbert. Magnesium ions, however, appear to play an essential role in the regulation of the biochemical events of contraction (Weber, Herz and Reiss, 1969).

Heilbrunn and Wiercinski (1947) reported that besides calcium, intracellular injections of barium ions caused a contractile response in frog. More recently, Caldwell and Walster (1963) found that both barium and strontium injected intracellularly into isolated *Maia* fibres also produced a contractile response. Additionally, Van der Kloot (1965) has reported that isolated vesicles of the SR from the lobster (*Homarus*) were able to accumulate strontium, but he was unable to demonstrate any increase in the influx of strontium during stimulation of crayfish thoracic muscle (Van der Kloot, 1966).

Intracellular injections of ^{89}Sr into the large fibres from the barnacle *Balanus nubilus* (Ashley, 1967) strongly suggested that strontium was bound and mobilized in a manner similar to that of calcium ions. In figure 3(a), the time course is illustrated of the loss of ^{89}Sr from a single fibre. The rate constant for the efflux rose rapidly to a peak value soon after injection and this was associated with a small contraction. The rate constant fell rapidly over a period of an hour to a low value, characteristic of the resting fibre. Application of 200 mM potassium saline produced a contraction and initiated a rapid rise and fall of the rate constant, although the fibre remained depolarized. It seemed possible that the increase in the loss of the isotope in the high potassium saline could be attributed to a physical squeezing-out from the cleft system of tracer that had accumulated during the early phases of the experiment. However, application of subcontraction levels of potassium (figure 3(b)) also produced a marked rise in the rate constant for efflux, the rate constant only started to fall when the membrane was repolarized. Additionally, either caffeine or quinine salines increased the efflux of strontium from the fibres. The rapid fall in the rate constant after the initial injection implies a rapid removal of free strontium ions from the fibre sarcoplasm. It seems likely that the ions are accumulated into the SR and can be released together with calcium from this binding site by agents that normally cause the release of calcium. Similar observations have been reported by Lowe and Caldwell (1964) studying the efflux of ^{45}Ca from

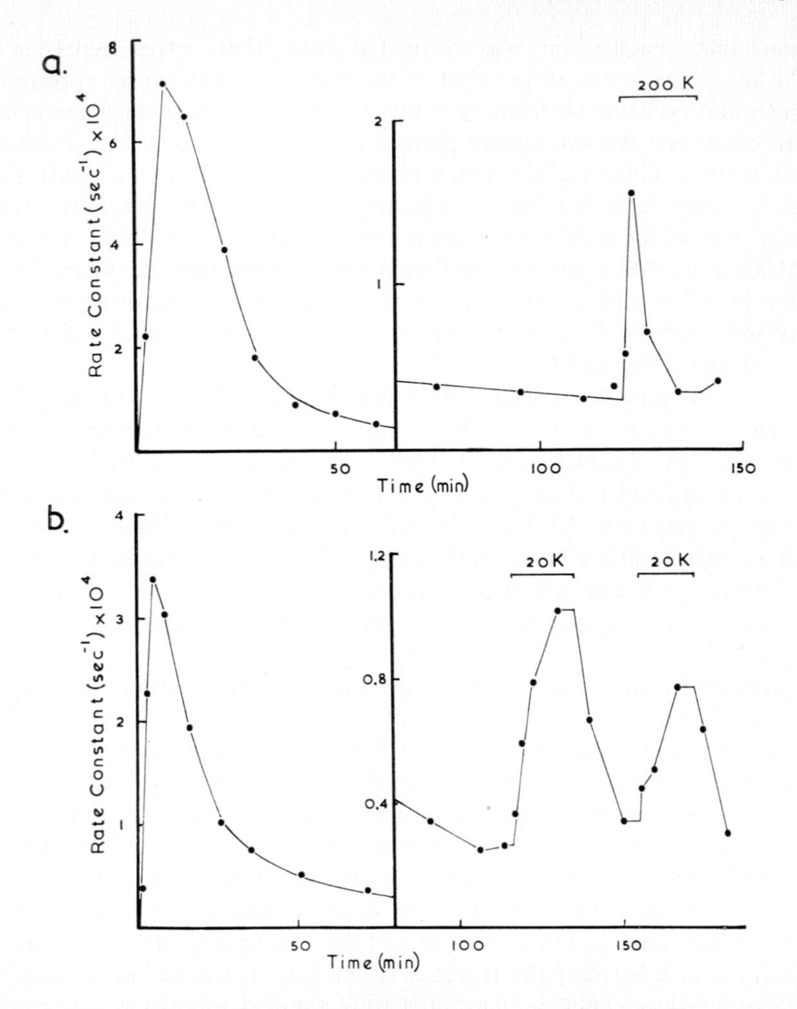

Figure 3. Illustrates the efflux of ^{89}Sr from single barnacle muscle fibres. The isotope was injected initially at a concentration of <10 mM and caused a slight contraction. (a) response of the resting efflux to the application of 200 mM potassium saline applied for the period indicated; mean resting potential, —47 mV, temp. 23–24°C. (b) response of the resting efflux to two applications of 20 mM potassium saline applied for the periods indicated, mean resting potential, —52 mV, temperature 23–24°C (from Ashley, 1967).

single *Maia* fibres when subjected to caffeine-containing salines. It is of interest that the rate constant for the efflux in high potassium saline (figure 3(a)) rose by some 5–6 times and fell rapidly before the fibre membrane was repolarized by the re-application of normal saline. The rapid rise and fall in the rate constant implies a rapid release followed by the removal of strontium from the fibre sarcoplasm. This rapid removal is characteristic of the inactivation of the E–C coupling mechanism of phasic muscle fibres when challenged with high potassium salines (see Hodgkin and Horowicz, 1960b). In contrast, the response of the efflux to subcontraction levels of potassium suggested that the efflux began to fall only when the membrane was repolarized. Perhaps the membrane had not been sufficiently depolarized to initiate the inactivation process. Alternatively, the concentration of calcium and strontium ions in the sarcoplasm may not have reached some critical value which triggers the inactivation process.

There is evidence, that has already been discussed in section IIIB, concerning the ability of strontium and other divalent ions to support E–C coupling in calcium free salines. Edwards, Lorkovic and Weber, (1966) have also demonstrated that strontium ions can activate the myofibrillar ATPase, although not as effectively as calcium. In addition, Hagiwara and Naka (1964) have reported that strontium ions can replace calcium ions in the inward current phase of the action potential in the muscle fibres from the barnacle, *Balanus nubilus*.

V. CHLORIDE AND PHOSPHATE

A. Chloride

In the early experiments of Boyle and Conway (1941) chloride ions were considered to be distributed in a simple Donnan fashion across the muscle membrane separating a single intracellular compartment from the bathing saline. A more complicated interpretation of the situation was suggested by Harris (1958) from experiments using labelled chloride (^{36}Cl) to follow the efflux from frog sartorius. The model proposed that two intracellular compartments existed, one of which was freely accessible to the bathing saline. This compartment contained similar concentrations of sodium, potassium and chloride as the external medium and was in Donnan equilibrium with this medium. The true intracellular compartment was considered to be in Donnan equilibrium with both the intermediate compartment and the external medium but was accessible to potassium and chloride ions only. A similar suggestion of two intracellular compartments was made by Simon, Shaw, Bennett and Muller (1957) from experiments on ionic movement in toad sartorius, although they also suggested that the true intracellular potassium might be bound. As has already been discussed, there is little experimental evidence to support the suggestion of intracellular potassium binding. The loss of tracer chloride from frog muscle can be divided into two main components, a rapid phase which can be in part attributed to exchange of chloride in the intermediate region and in part to exchange of chloride in the intracellular compartment. This is followed by a slower phase which becomes exponential after about 20 minutes and represents mostly intracellular exchange (Harris, 1958; Adrian, 1961). An apparently diphasic efflux curve was also reported in the early experiments of Levi and Ussing (1948) in frog sartorius. The two phases are more easily seen at lower temperatures, as the exchange of intracellular chloride, the slower phase, has a high temperature coefficient (Harris, 1958). Additionally, replacement of part of the chloride in the saline by nitrate also affected the slower phase (Harris, 1958), and both Adrian (1961) and Harris (1963) reported that perchlorate and thiocyanate ions behave in a similar way to nitrate. Adrian has used both nitrate and thiocyanate, completely replacing chloride in the saline, to slow the exchange of intracellular chloride. In these circumstances, extrapolation of the slow exponential phase of the efflux to zero time predicted an intracellular chloride concentration of 3·1–3·8 mM/kg. The fact that in frog, nitrate and other larger anions are able to slow the exchange of intracellular chloride has prompted Hodgkin and Horowicz (1959b) and Hutter and Padsha (1959) to suggest that, as there is no appreciable change in the membrane potential on going from chloride to nitrate saline, nitrate may simply reduce the chloride permeability to a value similar to that of the nitrate ion.

A major part of the resting conductance in frog muscle, some 200–250μ mho/cm^2 (Fatt and Katz, 1951), has been attributed to chloride ions (Hodgkin and Horowicz, 1959b). Adrian (1961) on the basis of the tracer chloride experiments has estimated that the chloride flux is 10 pmole/cm^2/sec which is equivalent to a resting conductance of only 38μ mho/cm^2 assuming that both the influx and efflux are independent of each other. This is ten times too small to account for the observed leakage current. In crustacean muscle, however, DeMello and Hutter (1966) report that replacement of chloride by nitrate in *Astacus* had little effect on the resting conductance. Similar results have been reported for single *Maia* fibres by Richards (1969) where chloride appears to account for most of the resting fibre conductance. Discrepancies exist in frog and in other tissues (Caldwell and and Keynes, 1960) between the resting and chloride conductances and it seems possible that part of the leakage current in frog may be carried by other ions than chloride.

The decrease in the resting membrane conductance in frog where chloride has been replaced by larger anions is reflected by a prolongation of the falling phase of the action potential as a result of the increase in membrane resistance (Etzensperger and Bretonneau, 1956; Harris, 1958). This prolongation of the repolarization phase results in prolongation of the *active state* in frog muscle and has been examined in detail by Sandow and his collaborators (Kahn and Sandow, 1950, 1955) as well as by Hill and Macpherson (1954), Harris (1958) and Hodgkin and Horowicz (1960c).

Katz (1948) and Hodgkin and Horowicz (1959b, 1960a) reported that the phenomenon of anomalous rectification could be observed in frog muscle, where the part of the membrane permeable to potassium ions showed a high permeability to an inward current but a low permeability to an outward current. Adrian and Freygang (1962a and b), and Freygang, Goldstein and Hellam (1964) suggested that the main chloride permeability was across the surface membrane of the fibre which was relatively impermeable to potassium and sodium ions. The main potassium permeability was across the walls of the intermediate space (most probably the wall of the T-system) and this was the site of anomalous rectification. Adrian and his collaborators also suggested that separating the lumen of the intermediate space from the outside saline there was a membrane permeable to sodium and potassium ions. As has been mentioned, there is no evidence for a membrane at the mouth of the T-system, although the lumina of the tubules appear to be filled with an electron dense material which might be relatively immobile, charged protein molecules. More recently, Girardier, Reuben, Brandt and Grundfest (1963) and Reuben, Brandt, Garcia and Grundfest (1967) have suggested that in crayfish the main potassium permeability is across the outer membrane, whilst the walls of the T-system, especially at the level of the excitatory diads, are permeable to chloride ions. It is suggested that this chloride permeability may well be important in the triggering of calcium release by the longitudinal elements of the SR.

B. Phosphate

The influx of orthophosphate into frog muscle has been examined by Causey and Harris (1951) who reported evidence for an intermediate fraction which equilibrated rapidly with the orthophosphate in the saline and was attributed to surface bound phosphate. They also observed that this fraction, representing only 1 per cent of the total muscle phosphate, was probably involved in the initial rapid efflux of phosphate from sartorii previously soaked in labelled orthophosphate salines. Incorporation of some of the labelled phosphate into an acid-soluble organic phosphate fraction was also observed. More recently, Lowe (1964) and Lowe and Caldwell (personal communication) have indicated that in *Maia* fibres, there is an intermediate phosphate fraction which equilibrates with the external saline. This fraction, which contains little or no high energy phosphate, represents some 1 per cent of the total muscle phosphate and can tentatively be assigned to the space occupied by the cleft and T-system. The internal orthophosphate appeared to exchange rather slowly with the external saline but once inside was apparently rapidly incorporated into ATP and arginine phosphate. Janke, Fleckenstein, Marmier and Koenig (1966) found that the incorporation of labelled orthophosphate from the saline into ATP was retarded by dinitrophenol. Fleckenstein, Janke and Davies (1956) have also observed that if the internal concentration of orthophosphate is increased by contraction, here is a slowing of the incorporation of the ^{32}P-orthophosphate from the bathing saline into ATP. Fleckenstein and his collaborators suggested that this slower incorporation was brought about partly by an increase in the gradient against phosphate influx, requiring the expenditure of additional free energy for the active transport of the phosphate inwards. Also partly by a decrease in the available free energy, brought about by a decrease in the phosphagen level inside the muscle cell during the contraction. It is also of interest, in view of the work of Kernan on active sodium extrusion, that insulin has been reported to increase the rate of phosphate entry into rat diaphragm (Sachs and Sinex, 1953).

An increase in the efflux of orthophosphate from contracted muscles was observed by Fleckenstein, Janke and Davies (1956) and this has been confirmed by the work of Caldwell and Lowe (personal communication) in single *Maia* muscle fibres. They also observed that after injection the rate constant for the efflux of labelled orthophosphate fell over a period of an hour to a value of $1\text{–}10 \times 10^{-5}$/min. This seems to correspond with the decrease in the specific activity of the orthophosphate and the increase in the activity of the ATP and arginine phosphate fractions. The rate constants for the efflux, after the initial hour are equivalent to an efflux of about 0·7 pmole/cm^2/sec. Although this value is much lower than the sodium efflux in *Maia*, Caldwell (1968) has pointed out that it is similar in value to the resting calcium efflux from these fibres (about 1·8 pmole/cm^2/sec.). It seems possible that calcium moves out of the muscle cell in conjunction with orthophosphate as an accompanying anion.

As has already been mentioned, orthophosphate rapidly equilibrates with ATP and phosphagen both in *Maia* fibres and in squid axon. (Caldwell, Hodgkin, Keynes and Shaw (1964); Caldwell and Walster, 1962). Similar results have also been obtained by Fleckenstein and his collaborators (Fleckenstein, Janke and Davies (1956)) using ^{18}O-labelled water and by following its incorporation first into orthophosphate and then into the oxygens of the gamma phosphate of ATP. Caldwell and his associates have been able to calculate the rate constants for the exchange of phosphorus between a three compartment system consisting of orthophosphate, ATP and the phosphagen. The rate constants for the exchange of phosphorus between orthophosphate and ATP is somewhat slower than the exchange between ATP and arginine phosphate in both *Maia* muscle and squid nerve.

In order to demonstrate a net breakdown of ATP during a single contraction in frog, Cain, Infante and Davies (1962) inhibited the rapid ATP–phosphagen exchange reaction with a specific inhibitor of the enzyme, creatine phosphoryl transferase. More recently, Mommaerts and Wallner

(1967) have confirmed the findings of Davies and his collaborators, but report that all the ATP breakdown occurs during the contraction phase and they were not able to detect ATP breakdown during relaxation. In addition, Cain, Infante and Davies (1962) and Kushmerick, Larson and Davies (1969) report that very little of the shortening and activation heat in frog sartorius muscle can be attributed to the breakdown of ATP or phosphagen.

VI. HYDROGEN AND BICARBONATE

The intracellular activity of hydrogen ions in large nerve and muscle cells has been measured by Caldwell (1954, 1958) using ion-selective glass electrodes. In the muscles fibres from the crabs *Maia* and *Carcinus* the intracellular pH was close to 7, implying that the concentration of both hydrogen and bicarbonate ions is not that expected from a simple Donnan distribution. It is difficult to alter the intracellular pH except by the application of CO_2 in the bathing saline. CO_2 at one atmosphere was found to rapidly reduce the pH of *Maia* fibres to values close to 6. In addition, Caldwell (1958) was not able to detect any major change in pH during potassium-induced contraction in these fibres. As he pointed out, however, changes of less than 0·1 of a pH unit might have occurred. Caldwell's results on squid axon and in invertebrate muscle fibres have been confirmed by Kostyuk and Sorokina (1960) in frog, in that the intracellular pH is also close to 7. Butler, Waddell and Poole (1967) using the weakly dissociated acid DMO have also come to similar conclusions, although Carter, Rector, Campion and Seldin (1967) using pH sensitive electrodes have suggested that the intracellular pH of rat muscle may be close to 6.

Although Caldwell could not detect a pH change during contraction, Jöbsis (1968) using an intracellular indicator bromcresol purple, and Distèche (1960a and b) using thin walled pH electrodes applied to the surface of the cell, have reported changes associated with contraction in vertebrate muscle. Both methods are so far uncalibrated and the absolute change in pH is unknown.

VII. CONCLUSIONS

In summary, it appears that the properties of certain ions within muscle cells resemble closely their behaviour in free solution, that is, the surface membrane of the cell acts as the main barrier to their diffusion. In particular, the efflux of the isotopes of sodium, potassium and chloride when introduced directly inside the cell, follows a simple exponential time course and the kinetics are not apparently influenced to an appreciable extent by intracellular binding. The flux experiments in general can be complicated, in certain cases, by the presence of an intermediate compartment between the intracellular and extracellular phases. In vertebrate muscle this appears to be part of the sarcoplasmic reticulum, whilst in invertebrates there is in addition a large compartment formed by the invaginations of the outer membrane.

The kinetics of the efflux of calcium and strontium ions, however, are influenced by intracellular binding and this is characterized in large muscle cells by a fairly rapid, non-exponential fall in the efflux soon after the initial injection. This binding process appears to be related to the mechanism that maintains the calcium concentration in the sarcoplasm of normal fibres at a very low value, and is associated with the longitudinal elements of the sarcoplasmic reticulum.

Acknowledgements

This article was written during tenure of a grant from the Medical Research Council. The author thanks the editors of the *Journal of Physiology* and the *American Zoologist* for permission to publish Figures 1 and 3.

REFERENCES

Adrian, R. H. (1961) *J. Physiol.*, **156**, 623
Adrian, R. H. and W. H. Freygang (1962a) *J. Physiol.*, **163**, 61
Adrian, R. H. and W. H. Freygang (1962b) *J. Physiol.*, **163**, 104
Adrian, R. H. and C. L. Slayman (1964) *J. Physiol.*, **175**, 49P
Ashley, C. C. (1965) Dissertation for Ph.D., University of Bristol
Ashley, C. C. (1967) *Amer. Zool.*, **7**, 647
Ashley, C. C. (1969) *J. Physiol.*, **203**, 32P
Ashley, C. C. (1970) (in the press)
Ashley C. C. and P. C. Caldwell (1964) Cited by P. C. Caldwell *Proc. Roy. Soc. Ser B*, **160**, 512
Ashley, C. C., P. C. Caldwell, A. G. Lowe, C. D. Richards and H. Schirmer (1965) *J. Physiol.*, **170**, 32P
Ashley, C. C. and E. B. Ridgway (1968) *Nature*, **219**, 1168
Ashley, C. C. and E. B. Ridgway (1969) *J. Physiol.*, **200**, 74P
Ashley, C. C. and E. B. Ridgway (1970a) In A. Cuthbert (Ed.) *Calcium and Cellular Function*, Macmillan, London, p. 42
Ashley, C. C. and E. B. Ridgway (1970b) *J. Physiol.*, **209** (in the press)
Axelsson, J. and S. Thesleff (1958) *Acta Physiol. Scand.* **44**, 55
Baker, P. F., A. L. Hodgkin and T. I. Shaw (1961) *Nature*, **190**, 885
Baker, P. F., A. L. Hodgkin and T. I. Shaw (1962) *J. Physiol.*, **164**, 355
Beaugé, L. A. and R. A. Sjodin (1967) *Nature*, **215**, 1307
Bendall, J. R. (1951) *J. Physiol.*, **114**, 71
Bennett, H. S. and K. R. Porter (1953) *Amer. J. Anat.*, **93**, 61
Bianchi, C. P. (1963) *J. Cell. Comp. Physiol.*, **61**, 255
Bianchi, C. P. and A. M. Shanes (1959) *J. Gen. Physiol.*, **42**, 803
Bittar, E. E. (1966) *J. Physiol*, **187**, 81
Bittar, E. E., P. C. Caldwell and A. G. Lowe (1967) *J. Mar. Biol. Ass. U.K.*, **47**, 709
Boyle, P. J. and E. J. Conway (1941) *J. Physiol.*, **100**, 1
Brading, A. F. and A. W. Jones (1969) *J. Physiol.*, **200**, 387
Brecht, K. and G. Gebert (1966) *Experientia*, **22**, 713
Brinley. F. J. (1968) *J. Gen Physiol.*, **51**, 445

Butler, T. C., W. J. Waddell and D. T. Poole (1967) *Fedn. Proc. Fedn. Am. Socs Exp. Biol.*, **26**, 1327
Cain, D. F., A. A. Infante and R. E. Davies (1962) *Nature*, **196**, 214
Caldwell, P. C. (1954) *J. Physiol.*, **126**, 169
Caldwell, P. C. (1958) *J. Physiol.*, **142**, 22
Caldwell, P. C. (1960) *J. Physiol.*, **152**, 545
Caldwell, P. C. (1968) *Physiol. Rev.*, **48**, 1
Caldwell, P. C. (1970) In E. E. Bittar (Ed.) *Membranes and Ion Transport*, Wiley, London, Vol. 1, p. 431
Caldwell, P. C., A. L. Hodgkin, R. D. Keynes and T. I. Shaw (1959) *J. Physiol.*, **147**, 18P
Caldwell, P. C., A. L. Hodgkin, R. D. Keynes and T. I. Shaw (1960) *J. Physiol.*, **152**, 561
Caldwell, P. C., A. L. Hodgkin, R. D. Keynes and T. I. Shaw (1964) *J. Physiol.*, **171**, 119
Caldwell, P. C. and R. D. Keynes (1959) *J. Physiol.*, **148**, 8P
Caldwell, P. C. and R. D. Keynes (1960) *J. Physiol.*, **154**, 177
Caldwell, P. C. and G. E. Walster (1962) *J. Physiol.*, **163**, 15P
Caldwell, P. C. and G. E. Walster (1963) *J. Physiol.*, **169**, 353
Carey, M. J. and E. J. Conway (1954) *J. Physiol.*, **125**, 232
Carey, M. J., E. J. Conway and R. P. Kernan (1959) *J. Physiol.*, **148**, 51
Carter, N. W., F. C. Rector, D. S. Campion and D. W. Seldin (1967) *Fedn Proc. Fedn. Am. Socs. Exp. Biol.*, **26**, 1322
Causey, G. and E. J. Harris (1951) *Biochem. J.*, **49**, 176
Conway, D. and T. Sakai (1960) *Proc. Natl. Acad Sci. U.S.*, **46**, 897
Conway, E. J. (1960) *Ciba Fdn. Study Group*, **5**, 2
Conway, E. J., R. P. Kernan and J. A. Zadunaiski (1961) *J. Physiol.*, **155**, 263
Cross, S. B., R. D. Keynes and R. Rybova (1965) *J. Physiol.*, **181**, 865
Curtis, B. A. (1963) *J. Physiol.*, **166**, 75
Davies, R. E. (1963) *Nature*, **199**, 1068
De Mello, W. C. and O. F. Hutter (1966) *J. Physiol.*, **183**, 11P
Desmedt, J. E. (1953) *J. Physiol.*, **121**, 191
Distèche, A. (1960a) *Nature*, **187**, 1119
Distèche, A. (1960b) *Mem. Acad. Roy. Belg.*, **32**, 3
Edwards, C. and H. Lorkovic (1967) *Amer. Zool.*, **7**, 615
Edwards, C., H. Lorkovic and A. Weber (1966) *J. Physiol.*, **186**, 295
Endo, M. (1964) *Nature*, **202**, 1115
Etzensperger, J. and Y. Bretonneau (1956) *C.R. Seanc. Soc. Biol.*, **150**, 1777
Fatt, P. and B. L. Ginsberg (1958) *J. Physiol.*, **142**, 516
Fatt, P. and B. Katz (1951) *J. Physiol.*, **115**, 320
Fenn, W. O. and D. M. Cobb (1936) *Amer. J. Physiol.*, **115**, 345
Fleckenstein, A., H. Hille and W. E. Adam (1951) *Pflügers Arch. Ges. Physiol.*, **253**, 264
Fleckenstein, A., J. Janke and R. E. Davies (1956) *Arch. Exp. Path. Pharmak.*, **228**, 596
Foulks, J. G., J. A. Pacey and F. A. Perry (1965) *J. Physiol.*, **180**, 96
Frank, G. B. (1958) *Nature*, **182**, 1800
Frank, G. B. (1962) *J. Physiol.*, **163**, 254
Frank, G. B. (1964) *Proc. Roy. Soc. Ser., B*, **160**, 504
Frazier, H. S. and R. D. Keynes (1959) *J. Physiol.*, **148**, 362
Freygang, W. H., D. A. Goldstein and D. C. Hellam (1964) *J. Gen. Physiol.*, **47**, 929
Gilbert, D. L. (1960) *J. Gen. Physiol.*, **43**, 1103
Gilbert, D. L. and W. O. Fenn (1957) *J. Gen. Physiol.*, **40**, 393
Girardier, L., J. P. Reuben, P. W. Brandt and H. Grundfest (1963) *J. Gen. Physiol.*, **47**, 189
Hagiwara, S., S. Chichibu and K. Naka (1964) *J. Gen. Physiol.*, **48**, 163
Hagiwara, S., H. Hayashi and K. Takahashi (1969) *J. Physiol.*, **205**, 115
Hagaiwara, S. and K. Naka (1964) *J. Gen. Physiol.*, **48**, 141
Hagiwara, S. and S. Nakajima (1966) *J. Gen. Physiol.*, **49**, 807
Harris, E. J. (1958) *J. Physiol.*, **141**, 351
Harris, E. J. (1963) *J. Physiol.*, **166**, 87
Harris, E. J. and G. P. Burn (1949) *Trans. Faraday Soc.*, **45**, 508

Hasselbach, W. and M. Makinose (1961) *Biochem. Z.*, **333**, 518
Hasselbach, W. and M. Makinose (1962) *Biochem. Biophys. Res. Commun.*, **7**, 132
Hasselbach, W. and M. Makinose (1963) *Biochem. Z.*, **339**, 679
Hasselbach, W. (1966) *Ann. N.Y. Acad. Sci.*, **137**, 1041
Heilbrunn, L. V. and F. J. Wiercinski (1947) *J. Cell. Comp. Physiol.*, **29**, 15
Herz, R. and A. Weber (1965) *Fedn Proc. Fdn. Am. Socs. Exp. Biol.*, **24**, 208
Hill, A. V. (1948) *Proc. Roy. Soc., Ser. B*, **135**, 446
Hill, A. V. (1949) *Proc. Roy. Soc., Ser. B*, **136**, 399
Hill, A. V. and L. Macpherson (1954) *Proc. Roy. Soc., Ser. B*, **143**, 81
Hinke, J. A. M. (1959) *Nature*, **184**, 1257
Hinke, J. A. M. (1969) In M. Lavallée, O. F. Schanne and N. C. Hébert (Eds.) *Glass Microelectrodes*, Wiley, New York, p. 349
Hodgkin, A. L. (1958) *Proc. Roy. Soc. Ser. B.*, **148**, 1
Hodgkin, A. L. (1964) *The conduction of the nervous impulse*, Liverpool University Press, Liverpool
Hodgkin, A. L. and P. Horowicz (1959a) *J. Physiol.*, **145**, 405
Hodgkin, A. L. and P. Horowicz (1959b) *J. Physiol.*, **148**, 127
Hodgkin, A. L. and P. Horowicz (1960a) *J. Physiol.*, **153**, 370
Hodgkin, A. L. and P. Horowicz (1960b) *J. Physiol.*, **153**, 386
Hodgkin, A. L. and P. Horowicz (1960c) *J. Physiol.*, **153**, 404
Hutter, O. F. and S. M. Padsha (1959) *J. Physiol.*, **146**, 117
Huxley, A. F. and R. E. Taylor (1958) *J. Physiol.*, **144**, 426
Huxley, H. E. (1694) *Nature*, **202**, 1067
Hoyle, G. and T. Smyth (1963) *Comp. Biochem. Physiol.*, **10**, 291
Isaacson, A. and A. Sandow (1967) *J. Gen. Physiol.*, **50**, 2109
Janke, J., A. Fleckenstein, P. Marmier and L. Koenig (1966) *Arch. Ges Physiol.*, **287**, 9
Jenden, D. J. and J. F. Reger (1963) *J. Physiol.*, **169**, 889
Jöbsis, F. F. (1968) In E. Ernst and F. B. Straub (Eds.) *Symposium Biologica Hungarica*, Akademiai Kiado, Budapest, Vol. 8, p. 151
Jöbsis, F. F. and M. J. O'Connor (1966) *Biochem. Biophys. Res. Commun.*, **25**, 246
Kahn, A. J. and A. Sandow (1950) *Science*, **112**, 647
Kahn, A. J. and A. Sandow (1955) *Ann. N.Y. Acad. Sci.*, **62**, 137
Katz, B. (1948) *Proc. Roy. Soc., Ser. B*, **135**, 506
Kernan, R. P. (1962) *J. Physiol.*, **162**, 129
Kernan, R. P. (1968) *J. Gen. Physiol.*, **51**, 204 s.
Keynes, R. D. (1951) *J. Physiol.*, **114**, 119
Keynes, R. D. (1954) *Proc. Roy. Soc., Ser. B*, **142**, 359
Keynes, R. D. and G. W. Maisel (1954) *Proc. Roy. Soc., Ser. B*, **142**, 383
Keynes, R. D. and R. C. Swan (1959) *J. Physiol.*, **147**, 591
Kostyuk, P. G. and Z. A. Sorokina (1960) In A. Kleinzeller and A. Kotyk (Eds.) *Membrane Transport and Metabolism*, Academic Press, New York, p. 193
Kostyuk, P. G., Z. A. Sorokina and Yu. D. Kholodova (1969) In M. Lavallée, O. F. Schanne and N. C. Hébert (Eds.) *Glass Microelectrodes*, Wiley, New York, p. 322
Kushmerick, M. J., R. E. Larson and R. E. Davies (1969) *Proc. Roy. Soc., Ser. B*, **174**, 293
Kushmerick, M. J. and R. J. Podolsky (1969) *Science*, **166**, 1297
Lev, A. A. (1964) *Nature*, **201**, 1132
Levi, H. and H. H. Ussing (1948) *Acta Physiol. Scand.*, **16**, 232
Ling, G. N. (1960) *J. Gen. Physiol.*, **43**, 149
Ling, G. N. (1962) *A Physical Theory of the Living State*, Blaisdell, New York
Lorkovic, H. (1967) *Comp. Biochem. Physiol.*, **22**, 799
Lowe, A. G. (1964) Dissertation for Ph.D., University of Bristol
Lowe, A. G. and P. C. Caldwell (1964) Cited by P. C. Caldwell *Proc. Roy. Soc., Ser. B*, **160**, 512
Luttgau, H. C. (1963) *J. Physiol.*, **168**, 679
McLaughlin, S. G. A. and J A. M. Hinke (1966) *Can. J. Physiol. Pharmacol.*, **44**, 837
McLaughlin, S. G. A. and J. A. M. Hinke (1968) *Can. J. Physiol. Pharmacol.*, **46**, 247

Mines, G. R. (1913) *J. Physiol.*, **46**, 188
Mommaerts, W. F. H. M. and A. Wallner (1967) *J. Physiol.*, **193**, 343
Nanninga, L. B. (1961) *Biochim. Biophys Acta*, **54,** 338
Ohnishi, T. and S. Ebashi (1964) *J. Biochem., Tokyo*, **55**, 599
Oikawa, T., C. S. Spyropoulos, I. Tasaki and T. Teorell (1961) *Acta Physiol. Scand.*, **52**, 195
Peachey, L. D. (1965) *J. Cell Biol.*, **25,** 209
Porter, K. R. and G. E. Palade (1957) *J. Biophys. Biochem. Cytol.*, **3**, 269
Portzehl, H., P. C. Caldwell and J. C. Rüegg (1964) *Biochim. Biophys. Acta*, **79,** 581
Portzehl, H., P. Zaoralek and A. Grieder (1965) *Pflügers Arch. Ges. Physiol.*, **286**, 44
Reuben, J. P., P. W. Brandt, H. Garcia and H. Grundfest (1967) *Amer. Zool.*, **7**, 623
Richards, C. D. (1969) *J. Physiol.*, **202**, 211
Ringer, S. (1883) *J. Physiol.*, **4,** 29
Ridgway, E. B. and C. C. Ashley (1967) *Biochem. Biophys. Res. Commun.*, **29**, 229
Sachs, J. and F. M. Sinex (1953) *Amer. J. Physiol.*, **175**, 353
Sandow, A. (1952) *Yale J. Biol. Med.*, **25**, 176
Sandow, A. (1965) *Pharmacol. Rev.*, **17**, 265
Scheid, P., R. W. Straub and W. Hermenau (1968) *Pflügers Arch. Ges. Physiol.*, **301**, 124
Selverston, A. (1967) *Amer. Zool.*, **7**, 515
Shanes, A. M. and C. P. Bianchi (1960) *J. Gen. Physiol.*, **43**, 481
Shaw, J. (1958) *J. Exp. Biol.*, **35**, 902
Shimomura, O. and F. H. Johnson (1969) *Biochemistry*, **8**, 3991
Shimomura, O., F. H. Johnson and Y. Saiga (1962) *J. Cell. Comp. Physiol.*, **59**, 223
Shimomura, O., F. H. Johnson and Y. Saiga (1963) *J. Cell Comp.. Physiol.*, **62**, 1
Simon, S. E., F. H. Shaw, S. Bennett and M. Muller (1957) *J. Gen. Physiol.*, **40**, 753
Tasaki, I., A. Watanabe and T. Takanaka (1962) *Proc. Natn. Acad. Sci. U.S.A.*, **48**, 1177
Taylor, R. E. (1953) *J. Cell. Comp. Physiol.*, **42**, 103
Van der Kloot, W. G. (1965) *Comp. Biochem. Physiol.*, **15**, 547
Van der Kloot, W. G. (1966) *Comp. Biochem. Physiol.*, **17**, 1019
Van der Kloot, W. G. (1968) *Comp. Biochem Physiol.*, **26**, 377
Ware, F., A. L. Bennett and A. R. McIntyre (1955) *Fedn. Proc. Fedn. Am. Socs. Exp. Biol.*, **14,** 158
Weber, A., R. Herz and I. Reiss (1964) *Proc. Roy. Soc. Ser. B*, **160**, 489
Weber, A., R. Herz and I. Reiss (1969) *Biochemistry, N.Y.* **6,** 2266
Werman, R. and H. Grundfest (1961) *J. Gen. Physiol.*, **44**, 997
Werman, R., F. McCann and H. Grundfest (1961) *J. Gen. Physiol.*, **44**, 979
Woodward, A. A. (1949) *Biol. Bull, Woods Hole*, **97**, 264
Zadunaisky, J. A. (1966) *J. Cell Biol.*, **31**, C11
Zadunaisky, J. A., J. F. Gennaro, N. Bashirelahi and M. Hilton (1968) *J. Gen Physiol.*, **51**, 290s

CHAPTER 2

Ion movements in smooth muscle

P. J. Goodford

The Biophysics and Biochemistry Department

The Wellcome Research Laboratories, Beckenham, Kent, England

I. INTRODUCTION

The sliding filaments in cardiac and striated muscles are arranged in a regular array which is responsible for the characteristic histological appearance of the tissue, and any muscle without these striations may be called 'smooth'. This classification in its broadest sense therefore includes a wide range of tissues, but 'visceral and vascular muscles are so different from other non-striated muscles that they can be considered as the only true smooth muscles' (Prosser, 1962). The present discussion will relate to these two types alone, with particular emphasis on the visceral smooth muscles since their ionic distribution has been most frequently studied.

The general characteristics of the ionic distribution in smooth muscle are similar to those in other excitable tissues. The potassium content is higher and the sodium content lower than that of blood plasma, and most evidence suggests that the fundamental processes responsible for this distribution are the same as those in nerve, cardiac or skeletal muscles. Indeed, visceral smooth muscles combine a nervous and a muscular function, since they can not only contract but can also conduct excitation from cell to cell (Gunn and Underhill, 1914; Alvarez and Mahoney, 1922; Gasser, 1926; Bozler, 1938; Ferguson, 1940; Klinge, 1951; Evans and Schild, 1953; Prosser and coworkers, 1955; Prosser and Sperelakis, 1956; Bülbring and coworkers, 1958; Davson, 1959; Goto and coworkers, 1960; Woodbury, 1960; Bortoff, 1961).

The total ionic content of a smooth muscle is the sum of the contents of each individual part, and ideally each part of the living tissue should be investigated independently. However, smooth muscle cells are so small that such an approach has not yet been possible, and the whole tissue has usually been studied. The results have been interpreted by apportioning the ions between different regions in the tissue, the simplest method being to treat the muscle as two homogeneous regions, the extracellular and intracellular spaces. However, it is now generally accepted that these regions are not homogeneous, and a more detailed description of the ionic distribution is needed. This can be obtained by studying the tissue with as many different experimental techniques as possible, so that analytical, histological and physiological observations may all be used to describe its overall behaviour.

In this chapter the structure and function of smooth muscle will first be described, and its total ionic composition will then be considered. Using this as background information, the exchange of radioactive ions will then be examined one by one, until the conclusion is reached that at least four separate types of exchange can occur in the tissue. These will then be used to provide a realistic interpretation of the ionic distribution which can be tested quantitatively, since it is compatible with well-established physical laws.

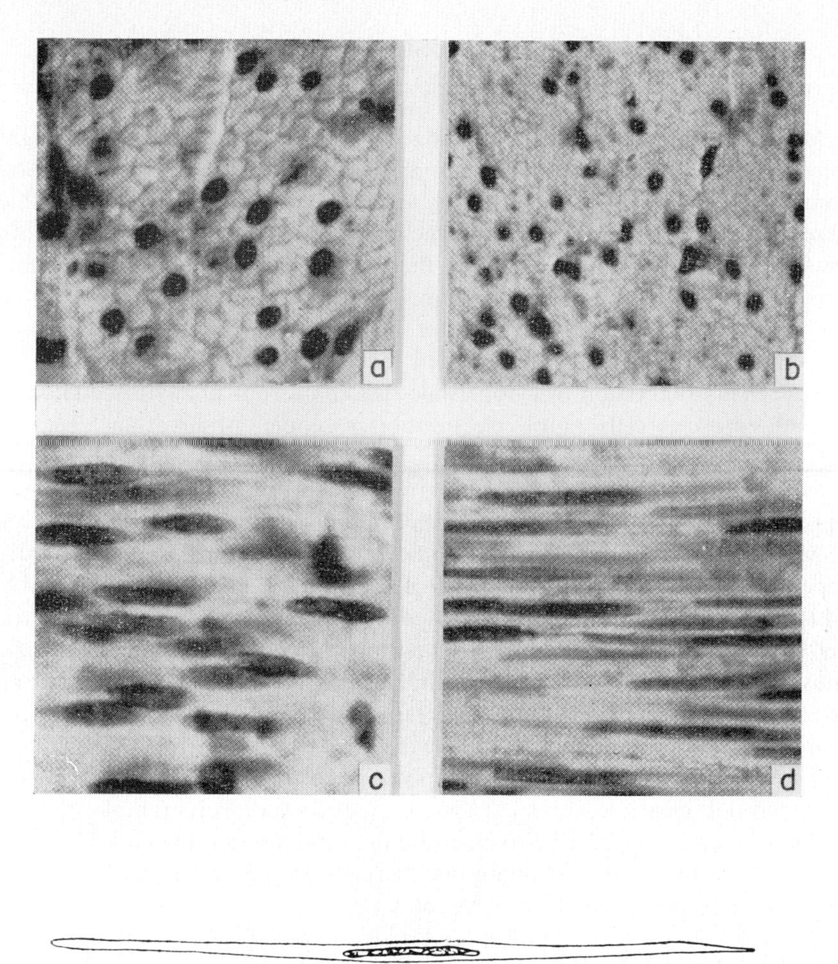

e

Figure 1. a–d are histological sections of the smooth muscle of the guinea-pig taenia coli fixed at two different lengths (Freeman-Narrod and Goodford, 1962). a and c are at 50 per cent of the length *in vivo* and b and d are at 130 per cent. a and b are cross sections and c and d are longitudinal sections. The shape of the cell and of the nucleus were dependent upon the degree of stretch. e is an idealized smooth muscle cell (From figure 4, C. McGill, *Am. J. Anat.*, **9**, 512).

II. SMOOTH MUSCLE STRUCTURE

A. Light Microscopy

McGill (1909) observed a wide range of smooth muscles under the light microscope, and her paper lists many references to the earlier literature. She commented on the large cell nucleus and other subcellular bodies, on the absence of striations, and on the syncytial arrangement of many smooth muscles. It appeared to her that there were well-defined protoplasmic anastomoses between the smooth muscle cells, so that the cytoplasm of one cell apparently ran directly into that of its neighbour with no cell membrane in between, although McGill conceded that some single cells might also be present. This suggestion of cytoplasmic fusion was not finally discounted in smooth muscle cells until electron-miscroscopic observations became possible in the last decade. It has now been shown instead that smooth muscle cells can really form junctions with each other, but that their cytoplasms are always separated by a membranous structure.

McGill described an idealized smooth muscle cell as a long cylinder tapering at both ends with a diameter of some 4 to 8 microns in the middle. This region sometimes tended to bulge slightly due to the presence of the cell nucleus, but the shape of the cell and the nucleus changed together as the muscle contracted or relaxed. Both became longer and thinner during stretching (figure 1), and the average radius r of the cell was related to the extension of the whole muscle sample (figure 2).

The information in light micrographs can be quantitated, and correlated with physiological measurements in the same tissue. For example, the cell radius (figure 2) can be used to calculate the total amount of cell surface in a given weight of muscle, assuming that the cells are thin cylinders of length l, and their volume V would then be given by

$$V = \pi r^2 l \tag{1}$$

and their surface area A by

$$A = 2\pi r l \tag{2}$$

The ratio of cell surface area to cell volume would be

$$\frac{A}{V} = \frac{2}{r} \tag{3}$$

and this depends only on the measured radius r of the cells. Since r is approximately 3 microns for a smooth muscle cell (McGill, 1909; Freeman-Narrod and Goodford, 1962), the ratio is some $0{\cdot}666\mu^{-1}$, implying that each cubic micron of cytoplasm has some 0·666 square microns of cell surface associated with it. On the simple two-compartment hypothesis this would be the surface

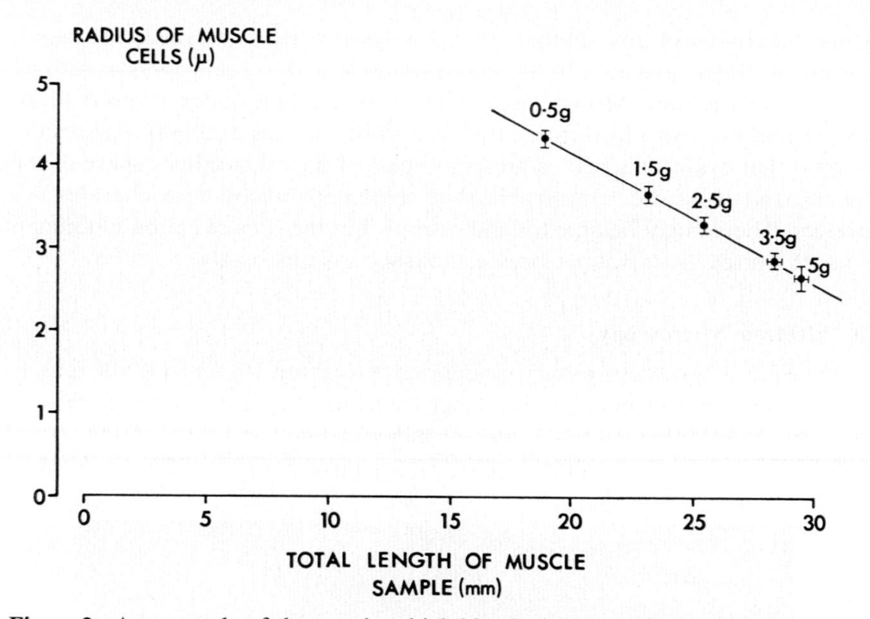

Figure 2. An example of the way in which histological observations can be quantitated. Light micrographs were prepared from pieces of taenia coli fixed under different physiological conditions (Freeman-Narrod and Goodford, 1962). The length of the muscle samples increased and the radius of the cells fell when the applied tension was raised from 0·5 to 5 grammes. Reproducible results can be obtained if sufficient sections are measured, and each point on the diagram was calculated from one hundred and fifty observations. The standard error of the mean is shown where it exceeds the size of the printed symbol.

through which ions moved between the intracellular and the extracellular regions, but such an interpretation would be a gross over-simplification in the present case, although it affords an important example of the way in which histological and ionic observations can be correlated. Interdisciplinary liaisons of this type are worthy of further development. For instance, the briefest examination of the light micrographs in figure 1 shows that the nucleus forms a substantial proportion of the total tissue volume, and one might even be able to estimate its volume from histological sections after taking appropriate precautions. It is almost inconceivable that this region is completely free of ions, and one must assume *a priori* that there may be other subcellular structures as well. Each of these may have its own particular ionic distribution.

Burnstock and coworkers (1963) made an histological examination of the circular smooth muscle layer of the stomach of the toad *Bufo marinus*. They were not able to demonstrate any gross amount of connective tissue, and could barely detect any sulphated mucopolysaccharide. On the other hand, the extracellular substance in the smooth muscle of the taenia coli was stained intensely by the Hale–Mowry method, and this stainable material was reduced by pretreatment with hyaluronidase (Goodford and Leach, 1966). The results suggest that hyaluronic acid is present in part of the extracellular space of the taenia, and it is also conceivable that sulphated mucopolysaccharides are present. These might be able to bind cations, but the sites of cation binding in smooth muscle have not yet been established unequivocally.

B. Electron Microscopy

Figure 3 is an electron-micrographic cross-section of the smooth muscle of the guinea-pig taenia coli. The heterogeneous nature of the tissue may at once be appreciated at this scale of magnification, and neither the extracellular nor the intracellular space is a uniform region. The unbroken

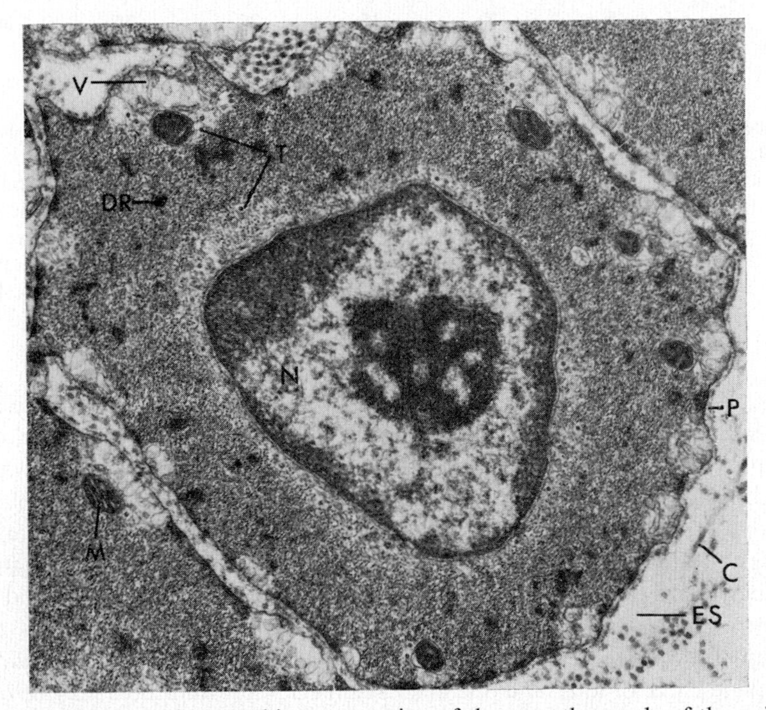

Figure 3. Electron-micrographic cross-section of the smooth muscle of the guinea-pig taenia coli (× 20,000). This shows the distribution of the nucleus (N), mitochondria (M), tubules (T), vesicles (V) and the extent of the contractile elements (the finely stippled background with dark regions (DR)). There is an extracellular space (ES) containing collagen fibres (C) and mucopolysaccharide (P). The heterogeneous nature of the tissue is clearly shown (By courtesy of G. S. Wells).

membrane surrounding the cell is clearly defined, and the extracellular space outside it is clear in some regions, and is faintly shaded elsewhere which may be indicative of mucopolysaccharide. Dark extracellular collagen fibres are present. The largest intracellular structure is the nucleus, and the nuclear membrane is not continuous but small openings between the nucleus and cytoplasm can be observed. Most of the cytoplasm is structured due to the contractile mechanism, and the occasional dark regions may indicate a close association between actin and myosin, although no actomyosin striations can be seen. There is no regular endoplasmic reticulum, but fine tubules can be seen in cross-section close to the mitochondria and the nucleus. The mitochondria are observed near the cell membrane, which itself shows a variety of structures: in some places it is smooth and lightly stained; in some smooth but more densely marked; and in others it is bunched into patches of vesicles. It has been suggested that each type of membrane may be responsible for one particular property of smooth muscle as an excitable tissue.

Attempts have been made to measure the relative sizes of the extracellular and intracellular regions by measuring smooth muscle electron micrographs, but it is an open question how much distortion may be produced by the histological pretreatment. Furthermore, there is always a risk of bias when three-dimensional volumes are estimated from two-dimensional areas, unless many serial sections are cut, but nevertheless valuable comparative observations may be made. For instance, it has been established that the conduction velocity of the action potential, the relative refractory period, and the electrical resistance of the muscle in the sucrose gap, are all correlated with the histologically measured extracellular space (Prosser and coworkers, 1960).

Electron micrographs afford the opportunity to re-assess the ratio of cell surface area to cell volume (A/V) on a more realistic basis. The cell membrane must have a larger surface area than was originally assumed because it is infolded at the vesicular regions to form pockets, and the cells are not in fact perfect cylinders. Rhodin (1962) has calculated that this may increase the surface area by twenty-five per cent, and our own observations suggest that a still larger correction approaching one hundred per cent may be applicable in visceral smooth muscle (Goodford and coworkers, 1967 and unpublished observations, Goodford, 1970). It is probable that most values of the ratio A/V recorded in the literature on the basis of light micrographs are low estimates, and one may only conclude that the true value is in the range $0{\cdot}1$–$1\,\mu^{-1}$ for mammalian smooth muscles.

C. Conclusions

Several attempts have been made to correlate the structure of smooth muscle with its function and with the distribution of ions in the tissue.

Specific experiments are now being designed in which light or electron-microscopic sections are prepared under known physiological conditions, and it is probable that more worth-while relationships will be established in this way.

III. SMOOTH MUSCLE FUNCTION

A. The Contractile Mechanism

The contractile proteins of mammalian uterine smooth muscle have been extensively studied, and in many respects they resemble the well-known actomyosin system of skeletal muscle (Csapo, 1948; Needham, 1962, 1964 and 1967; Needham and Shoenberg, 1967). They are dissociated by ATP into actin and myosin, and are present in the same proportions as in skeletal muscle (Huys, 1960). Skeletal muscle myosin will interact with uterine actin (Needham and Williams, 1963b). Trypsin will split smooth muscle myosin into light and heavy meromyosins, but another soluble protein, tropomyosin, also occurs in smooth muscle although it does not interfere with the reaction between actin and myosin. However, three important differences may be emphasized between mammalian skeletal and smooth muscles. First, there is little or no evidence to show that the myosin of relaxed mammalian smooth muscle aggregates to form thick filaments analogous to those of skeletal muscle (Shoenberg and coworkers, 1966; Elliott, 1967; Needham and Shoenberg, 1967), although many efforts have been made to detect such filaments. It is almost universally assumed that a sliding filament mechanism is responsible for mammalian smooth muscle contraction, and this must be accepted but the present lack of conclusive evidence is unsatisfactory. The second main difference is in the amount of smooth muscle actomyosin, which is only some 6–10 g/kg wet weight, whereas skeletal muscles may contain ten times this amount (Needham and Williams, 1963a). A part of the discrepancy may be due to the larger extracellular space of mammalian smooth muscles, so that a given weight of smooth muscle tissue may genuinely contain a smaller volume of contractile cells than its skeletal equivalent, but this factor is insufficient to account for the whole difference. Furthermore, it is surprising that smooth muscles can develop a maximum tension which is comparable to that of skeletal muscle. Thirdly, the ATPase activity of smooth muscle actomyosin is low (Needham and Williams, 1959 and 1963b), and the divalent cations Ca^{2+} and Mg^{2+} which greatly enhance this activity in skeletal muscle have a much smaller effect in the present case.

Figure 4 shows a train of spontaneous action potentials in the smooth muscle of the guinea-pig taenia coli, and each electrical event is followed by a slow contractile response of some seconds duration to maximum tension. It is clear that each electrical event immediately precedes the corresponding

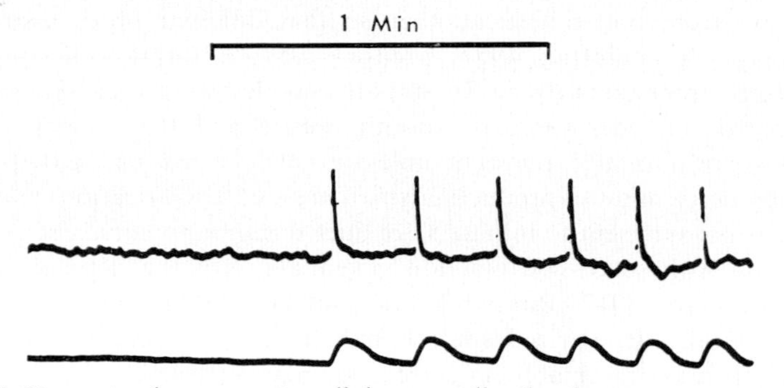

Figure 4. Upper tracing: an extracellular recording by the sucrose-gap method of electrical activity in the smooth muscle of the guinea-pig taenia coli. Lower tracing: the tension developed by the muscle (Axelsson and Bülbring, 1961). There is a quiescent period followed by a train of spontaneous action potentials, and each electrical event is followed by a slow contractile response (see text).

contraction, but it is also clear that the onset of tension development is much slower than in skeletal muscle. This could be due to the low ATPase activity which might limit the rate at which energy becomes available for the contraction of smooth muscle, or to diffusion delays for an intracellular transmitter, or to the special visco-elastic properties of the tissue (Åberg, 1967).

The contractile mechanism of mammalian smooth muscle differs significantly in these various ways from other contractile tissues, and in particular from skeletal muscle. The differences may arise because visceral smooth muscles function under conditions of approximately constant mechanical tension, although the muscle length may vary widely, whereas skeletal muscles are constrained to work at almost constant length since they are fixed to the bony structure of the body. It is necessary at the present time to assume that the fundamental mechanisms are the same, but it is also necessary to maintain an open mind and to investigate the structure and function of smooth muscle without constraining them to follow the predetermined pattern in other tissues.

B. Electrical Properties of Smooth Muscle

Many mammalian smooth muscles show a continuous spontaneous electrical discharge, and it has been suggested that this may continue throughout the life of the individual in his gastro-intestinal tract (Bozler, 1962). In such tissues the membrane potential is never quite constant, but local depolarizations or pre-potentials are observed which can lead to a pace-making activity, as in the heart. The pace-maker region may be localized as in the ureter (Bozler, 1942), or in uterine and intestinal smooth muscle it may move from one cell to another so that different fibres assume the controlling role (Bülbring, 1957; Melton, 1957; Bülbring and coworkers, 1958). Such spontaneously active smooth muscles behave as syncytia, but another type includes vascular smooth muscle and the urinary bladder, and these are normally quiescent unless excited by nervous activity. Each excitatory nerve impulse produces a small muscle depolarization, and if one of these is not sufficient to initiate a conducted action potential there may be summation with successive junction potentials until the depolarization is sufficient (Bozler, 1942; Burnstock and Holman, 1961). Finally it must be noted that fully depolarized smooth muscles can also contract and relax although there is presumably no corresponding electrical activity (Evans and coworkers, 1958; Falk and Landa, 1960), and mechanical and electrical events can therefore be separated under these conditions.

There is evidence that the basic ionic distribution is similar in both quiescent and spontaneously active mammalian smooth muscles (Casteels, 1968), and that the variations between these types may be due to the varied innervation of the tissues. It would therefore be as well at this point to consider the distribution of the ions one by one, and then to attempt a synthesis of these observations into a general interpretation which should, ideally, help to rationalize our understanding of the contractile and electrical properties of the tissue.

C. Conclusions

It is confidently assumed that the fundamental biochemical and physiological mechanisms in smooth muscle are comparable to those in other contractile tissues. This is, however, a non-proven assumption, and observations of the ionic distribution may be useful in providing further evidence on this point.

IV. TOTAL SMOOTH MUSCLE ELECTROLYTE

A. Total Tissue Electrolyte *in vivo*

Table 1 lists the total electrolyte content of a range of smooth muscles, measured either *in vivo* or immediately after killing an animal. The tissue samples were weighed to give the 'fresh' weight, and were then analysed in order to establish the weight of each ion which they contained. The original results have been recalculated, where possible, so that the results in Table 1 are in mmole/kg fresh weight, the tacit assumption being that the weight of tissue is proportional to the weight of ion which it contains. Figure 5 shows that this is a reasonable assumption, at least in so far as the magnesium content of the smooth muscle of the guinea-pig taenia coli is concerned.

Table 1. Total smooth muscle electrolyte *in vivo*. Some observations of the electrolyte content of smooth muscle samples *in vivo*, or immediately after the death of an animal. The original observations have been recalculated, where possible, to express the results in mmole/kg fresh weight of tissue, with the standard error of the mean and the number of observations in parenthesis. These are summarized results. The original papers should be consulted for further details. Unweighted column averages are calculated to indicate the approximate *in vivo* composition of a typical smooth muscle (see text).

Muscle	Species	Potassium	Sodium	Chloride	Calcium	Magnesium	References
Stomach	Frog						
	(*R. temporaria*)	66·5±0·9 (29)	40·2±1·1 (29)	33·9±0·5 (30)			Armstrong (1964)
	(*R. pipiens*)	68·0±4·9 (13)	29·8±6·4 (13)				Armstrong (1965)
	(*R. catesbiana*)	83·0	32·0	34·0	1·05		Meigs and Ryan (1912)
	Toad (*B. marinus*)	81·2±2·6 (5)	64·3±3·9 (5)	54·3±4·8 (5)			Burnstock and coworkers (1963)
	Dog	72·2±2·0 (23)	73·0±1.0 (23)				Davenport (1963)
	Hen	91·0	31·0	24·0			Constantino (1911)
	Turkey	117·0	32·0	25·0			Constantino (1911)
	Ox	94·0	39·0	30·0			Constantino (1911)
	Pig*	20·3*	121·1*	51·8*	6·1*		Saiki (1908)
Taenia coli	Guinea-pig	89·1±1·7 (10)	70·7±1·7 (10)	65·0			Daniel (1958)
		84·2±7·3	71·6±4·1		1·14±0·16		Nagasawa (1963)
					1·8±0·1 (53)		Bauer and coworkers (1965)
				56–76			Goodford (1964)
						6·13±0·08 (12)	Sparrow (1969)
	Rabbit	90·6±1·6 (5)	74·7±5·8 (5)	54·0±2·5 (5)			Daniel (1958)
Uterus	Rabbit						
	(immature)	100·7±4·0 (11)	87·4±4·0 (11)	66·6±3·0 (11)			Kao (1961)
	(+oestrogen)	108·6±3·0 (13)	78·6±3·0 (13)	66·7±2·0 (13)			Kao (1961)
	(+progesterone)	111·2±2·0 (16)	82·7±2·0 (16)	71·5±2·0 (16)			Kao (1961)
	Human	74·3		59·7			Manery and Hastings (1939)
Intestine	Cat	77·0±2·5 (27)	61·3±1·5 (27)	66·6±2·2 (13)			Barr (1959)
	Rabbit		61·1±1·45 (16)	67·7±1·05 (16)			Bitman and coworkers (1959)
Bladder	Dog	81·1±5·7 (9)	73·2±4·9 (9)		3·65±0·3 (9)	5·08±0·3 (9)	Wein and coworkers (1967)
Artery	Rat				4·0		Tobian and Chesley (1966)
Unweighted average contents		88	59	52	2·3	5·6	

*These values were not included in the averages (see text)

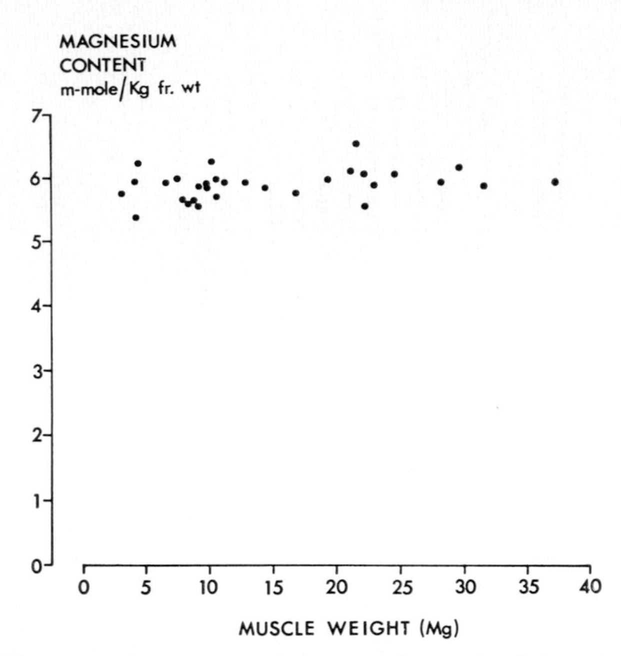

Figure 5. The magnesium content of the smooth muscle of the guinea-pig taenia coli, determined immediately after dissection from freshly killed animals. The ordinate is the magnesium content of the sample expressed as the ratio (weight of magnesium)/(weight of sample). The absolute size of the sample does not affect the final result (see text).

Smooth muscles contain, in general, more potassium than sodium and more magnesium than calcium, but there are some notable exceptions to these generalizations. Some of the more obvious anomalies should probably be rejected, and the figures for ox and pig stomach muscles make an interesting contrast (table 1). It is reasonable to suggest that the high sodium and low potassium contents reported for pig stomach muscle are due either to the lapse of some time between killing the animal and muscle dissection, or to maltreatment of the tissue. This conclusion is strengthened by the high muscle calcium which was also observed, since careful control observations have repeatedly shown that damage always causes smooth muscle sodium and calcium to rise (Kao, 1961; Goodford, 1962; Bauer and coworkers, 1965). Cooling has a similar effect and anomalies such as that described for pig stomach were frequently reported in the nineteenth century, before the importance of careful muscle treatment was fully appreciated.

There is substantial evidence that the ionic composition of vascular smooth muscle is modified by hypertension (Tobian and Fox, 1956; Tobian and Chesley, 1966), potassium falling while sodium and calcium increase.

This is the typical picture for damaged muscle, but it might be of interest to establish the exact site at which the changes take place since the observations may have some clinical importance. Parasympathomimetic agents sometimes produce similar effects (Davenport, 1963; Paton and Rothschild, 1965).

The tissue contents of each cation have been averaged at the bottom of table 1, in order to give rather arbitrary values for the *in vivo* ionic composition of a typical smooth muscle. The typical cation content is thus $88+59+2\,(2{\cdot}3+5{\cdot}6) = 162{\cdot}8$ mequiv./kg fresh wt, which exceeds the chloride content of 52 mequiv./kg fresh wt by some 111 mequiv./kg fresh wt. Some of this 'anion deficit' must be due to bicarbonate ions, but no detailed cation/anion balance sheet has yet been established for smooth muscle in the way that Boyle and Conway (1941) considered skeletal muscle. However, it is generally agreed that smooth muscles may contain bound or sequestered cations, and possibly unknown anions as well (Kao, 1961; Freeman-Narrod and Goodford, 1962; Garrahan and coworkers, 1965; Stephenson, 1967).

It would be desirable to measure the exchange of radioactive isotopes *in vivo* in order to understand the significance of the observed ionic distribution, but the whole living animal is such a complex system that the necessary techniques have not yet been mastered. A worth-while preliminary step would be to establish methods of controlling the specific activity of isotopes in the blood, but in the absence of such an approach it has been necessary to rely on tracer observations *in vitro* and to measure the total tissue electrolyte under the same conditions.

B. Total Tissue Electrolyte *in vitro*

Total tissue electrolyte determinations *in vitro* show a much wider range of variation than those made immediately after killing an animal, and some reasons for this scatter should be briefly considered. A prime cause is that the tissue is subject to more handling *in vitro*, with a consequently increased risk of tissue damage, and this may largely account for the differences between tables 1 and 2. Other factors are the wide range of *in vitro* conditions used by different experimenters, and the differing units used to express the results.

The *in vitro* content of smooth muscle has frequently been expressed relative to some parameter measured at the end of the experiment, like the wet or dry weight, but such a reference parameter may itself have been affected by the experimental conditions to which the muscle was subjected, and in this case it would not give an unbiased point of reference (Goodford, 1968; Brading and Setekleiv, 1968). The simplest method of ensuring that the *in vitro* conditions have not affected the reference point is to use a reference determined before the experiment actually begins, such as the fresh weight of the muscle measured immediately after the animal is killed.

Table 2. Total smooth muscle electrolyte *in vitro*. Some observations of the electrolyte content of smooth muscle samples *in vitro*. The original observations have been recalculated, where possible, to express the results in mmole/kg wet weight of tissue, with the standard error of the mean and the number of observations in parenthesis. These are summarized results. The original papers should be consulted for further details. Unweighted column averages are calculated to indicate the approximate *in vitro* composition of a typical smooth muscle.

Muscle	Species	Potassium	Sodium	Chloride	Calcium	Magnesium	References
Taenia coli	Guinea-pig	69·56±1·26	61·24±1·19	79·49±1·6			Brading and Setekleiv (1968)
		73·67±1·51	79·40±2·11	94·16±3·50			Brading and Setekleiv (1968)
		75·4±3·0	72·6±2·3	57·6±6·3			Barr and Malvin (1965)
		83·0±1·6 (23)	62·0±1·4 (23)	75·0±1·8 (24)			Casteels (1966)
		86·4±1·3 (13)	62·9±1·1 (13)	77·3±1·2 (13)			Casteels and Kuriyama (1966)
		89·4±0·9 (20)	64·8±1·3 (20)	81·2±0·9 (11)			Casteels and Kuriyama (1966)
		90·0±1·0 (49)	84·0±2·0 (30)	95·0±1·0 (5)			Goodford (1964)
		89·0	71·2	65·0			Daniel and Singh (1958)
		69·8±5·9	86·3±7·3		1·66±0·48		Nagasawa (1963)
		51·7 (2)	103·1 (2)				Daniel (1958)
		72·0±1·0 (40)	91·5±1·3 (31)				Freeman-Narrod and Goodford (1962)
		74·0±1·0 (45)	97·0±3·0 (45)				Goodford and Hermansen (1961)
					1·78±0·14 (8)		von Hattingberg and coworkers (1966)
					2·5±0·05 (114)		Bauer and coworkers (1965)
					2·22±0·06 (9)		Karaki and coworkers (1967)

					4·2±0·2		Schatzmann (1961)
					3·39±0·03 (9)		Urakawa and Holland (1964)
						5·11±0·05 (60)	Sparrow (1969)
	Rabbit	38·6 (3)	141·3 (3)	115·6 (3)			Daniel (1958)
		85·7	69·4	64·7			Daniel and Singh (1958)
		79·0±5·7 (6)					Matthews and Sutter (1967)
Stomach	Frog						
	(*R. temporaria*)	57·1±5·2 (6)	72·5±4·8 (6)	34·8±2·0 (9)			Armstrong (1964)
	(*R. pipiens*)	69·0±5·0 (20)	67·0±6·7 (25)	43·0±3·0 (25)			Bozler and coworkers (1958)
					1·54±0·13 (16)		Bozler (1963)
	Toad						
	(*B. marinus*)	62·9±3·5 (8)	82·6±3·0 (8)				Burnstock and coworkers (1963)
					1·05±0·01		Sparrow and Simmonds (1965)
Intestine	Dog	75·4±3·0	72·6±2·3	57·6±6·3			Barr and Malvin (1965)
	Guinea-pig	85·0±2·1 (15)	78·1±1·7 (15)		4·84±0·13 (15)		Paton and Rothschild (1965)
		79·3±1·6 (8)	77·7±3·2 (8)		5·79±0·35 (8)		Paton and Rothschild (1965)
Artery	Dog	45·8±1·2 (18)	102·0±2·1 (18)	85·8±1·1 (18)			Villamil and coworkers (1968)
Aorta	Rat	35·6±3·3 (30)	88·9±1·1 (30)				Hagemeijer and coworkers (1965)
Vein	Rabbit	29·0±3·6 (6)					Matthews and Sutter (1967)
Unweighted average contents		69	81	73	2·9	5·1	

However, it may reasonably be objected that this additional handling of the tissue can only increase trauma, and in cases where this is a serious objection it is necessary to choose the most stable reference parameter which is available. There is some evidence (Goodford and Vaughan-Williams, 1962) that the weight of a tissue sample after drying to constant weight at 105 °C is a suitable reference point, at least for heart muscle, and the 'dry' weight has often been used with success in smooth muscle measurements (Casteels and Kuriyama, 1966; Casteels, 1966). It has recently been suggested that the nucleic acid content of each sample might form the most unbiased point of reference, if this could be determined with sufficient accuracy.

There is a further possible source of bias in smooth muscle measurements because the total electrolyte content sometimes appears to depend on the absolute size of the muscle sample, irrespective of whether the fresh or wet weight of the tissue is taken as the reference point. It is therefore important to work with small muscle samples of approximately the same size *in vitro* (about 15 mg is suitable) since the bias can be serious when slightly larger specimens are used. A piece of muscle can be so thick that the central cells are inadequately perfused with bathing medium and therefore become anoxic (Creese and coworkers, 1958), and such a danger should always be considered with smooth muscle specimens weighing more than 50 mg.

Values have again been calculated in table 2 to give arbitrary values for a typical smooth muscle electrolyte content *in vitro*, and the results for each ion differ appreciably from the corresponding *in vivo* values. In fact there are no known experimental conditions which will allow the normal *in vivo* composition to be maintained, and *in vitro* measurements must therefore be regarded as a poor second-best since the fundamental object of smooth muscle research is to establish how the tissue works in the living animal. However, there is the inestimable experimental advantage *in vitro* that radioactive observations can be carried out under steady-state conditions, since the ionic composition of the smooth muscle does not change appreciably over a period of several hours (figure 6), and this simplification to some extent compensates for the inherent *in vitro* disadvantages.

C. Osmotic Effects

The osmotic behaviour of smooth muscle is complex (Bozler, 1959, 1961 and 1962), and it has been suggested that there may be strong internal linkages which can distort the normal pattern (Bozler, 1964). In hypertonic solutions, prepared by adding sucrose to normal Kreb's medium, the smooth muscle cells of the guinea-pig taenia coli shrank as perfect osmometers (Brading and Setekleiv, 1968), while in hypotonic media they swelled slightly and then apparently burst.

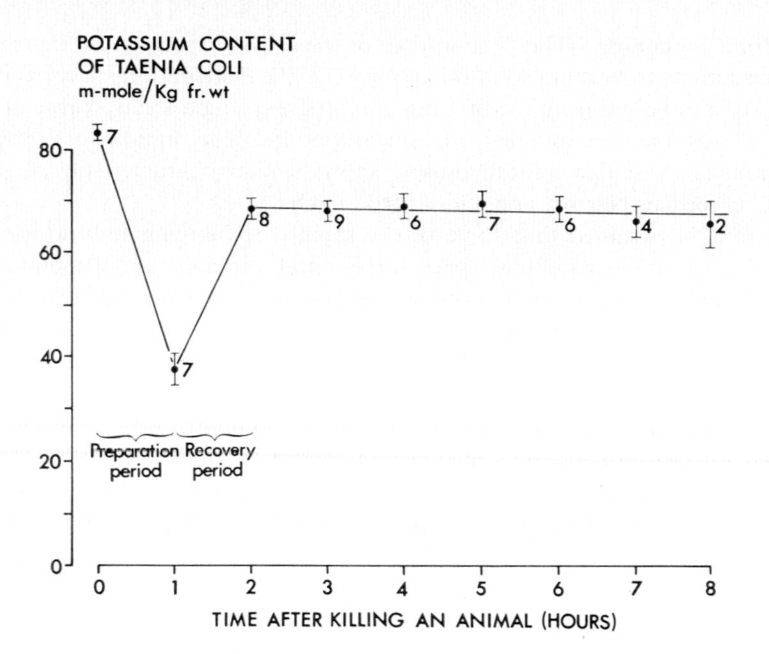

Figure 6. The potassium content of the smooth muscle of the guinea-pig taenia coli *in vitro* (Goodford and Hermansen, 1961). Muscle samples were dissected from freshly killed animals and were analysed after varying immersion times in Kreb's solution. Fresh muscles had a high potassium content which fell while the samples were being prepared. However, a steady state was reached in oxygenated Kreb's solution at 35°C, and this was maintained for the duration of the experiment.

D. Conclusions

The total electrolyte composition of smooth muscles resembles that of other excitable contractile tissues, especially when observations *in vivo* are compared, but substantial quantitative differences have been observed which may be related to the specific functional requirements of smooth muscle.

V. RADIOACTIVE SODIUM EXCHANGE

A. ^{22}Na and ^{24}Na Uptake

Figure 7 shows the uptake of ^{22}Na by the smooth muscle of the frog stomach at 20–24°C. Muscle samples weighing 18–35 mg were dissected from freshly killed *Rana pipiens* and were pre-equilibrated in normal bicarbonate Ringer's solution, after which they were immersed in a similar medium containing ^{22}Na. The uptake of tracer was expressed as the ratio of tissue activity to bathing solution activity (i.e. the distribution space) and this increased rapidly during the first few minutes reaching a steady value of 0·39 which was then maintained for several hours. The uptake of ^{24}Na by mammalian vascular smooth muscle shows similar characteristics at 4° or 20°C (Freeman-Narrod and Goodford, 1962).

It may be assumed that some of the rapidly exchanging sodium is freely dissolved in the extracellular space surrounding the cells, and attempts have therefore been made to determine the size of this space by transferring muscle samples from a normal Ringer's solution to one containing a solute which does not enter the cells. Inulin was the solute of choice in early experiments. The inulin content of the sample should then give a measure of its extracellular space, but evidence has now accumulated which suggests that inulin was an unfortunate substance to choose in the first place because it gives results which are systematically low (Barr and Malvin, 1965; Goodford and Leach, 1966; Law and Phelps, 1966; Villamil and coworkers, 1968).

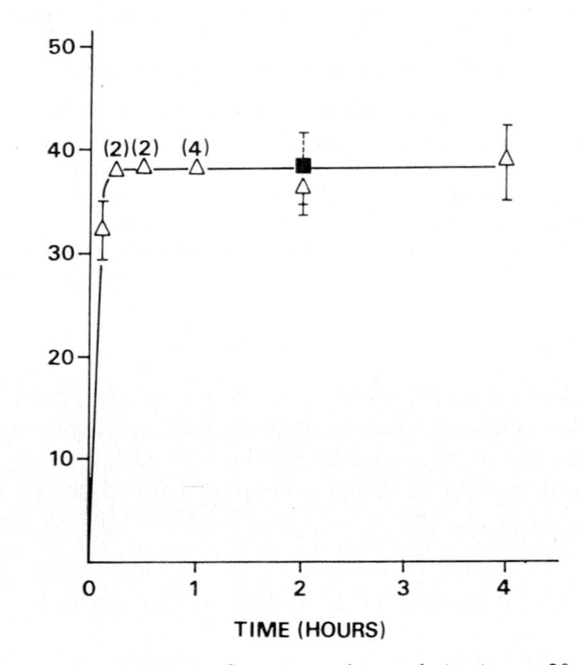

Figure 7. The uptake of ^{22}Na by frog stomach muscle *in vitro* at 20–24°C (Stephenson, 1967). The abscissa is the time in hours after transfer from non-radioactive to ^{22}Na Ringer's solution and the ordinate is 100 × the ratio of tissue activity to bathing solution activity (i.e. the distribution space. See text). The ^{22}Na was rapidly distributed through some 390 ml/kg wet weight after which there was no further significant uptake.

Inulin is a large molecule and the histological evidence has shown that parts of the extracellular space also contain large mucopolysaccharide molecules. The two large molecular species cannot occupy the same region at the same time (Ogston, 1958; Ogston and Phelps, 1961), so that inulin is partially excluded from the extracellular space although the smaller monovalent ions have ready access throughout. Thus the use of inulin introduced a bias which distorted early interpretations of the ionic distribution in smooth muscle, since it led to erroneous values of the intracellular ionic concentrations.

The use of the extracellular space may be illustrated by considering a simple system in which smooth muscle samples are equilibrated with a Ringer's solution containing a low concentration of inulin and 137 mmole/l sodium. Assuming that analysis of the tissues shows an inulin space of 330 ml/kg wet weight (Goodford, 1964), one may calculate that the weight of extracellular sodium in the tissue should be $137 \times 330/1000 = 45$ mmole sodium/kg wet weight. This calculation assumes that:

i) the average inulin concentration in the extracellular space is the same as that in the bathing medium;
ii) inulin occupies the whole extracellular space at that average concentration;
iii) inulin is not fixed extracellularly; and
iv) that inulin does not enter the cells.

If the analysis also shows that the total sodium content is 70 mmole sodium/kg wet weight, one may conventionally conclude that the weight of intracellular sodium is $70 - 45 = 25$ mmole sodium/kg wet weight. However, when smaller molecules such as [^{14}C]-labelled sucrose, sorbitol or mannitol are used instead of inulin, the extracellular space measurement is some 400 ml/kg wet weight (figure 8; Goodford and Leach, 1966), which leads by the same calculation to an intracellular content of only 15 mmole sodium/kg wet weight. This is about half the value found on the basis of the inulin space, and it may be seen that the calculated intracellular content depends critically on the extracellular space measurement. Any small systematic bias may produce substantial distortions, and it is not unknown for negative intracellular sodium contents to be calculated.

Two points should be emphasized. First, it is important to choose an extracellular marker which has molecules of a comparable size to the ions under investigation, and to establish that the marker does not enter the cells significantly and that it is not metabolized. Secondly, it is illogical to conclude that a cation must be intracellular, just because conventional extracellular space calculations do not account for it (Schoffeniels and coworkers, 1966). All the evidence for each particular ion must be considered, in order to formulate the best interpretation of its distribution in the tissue.

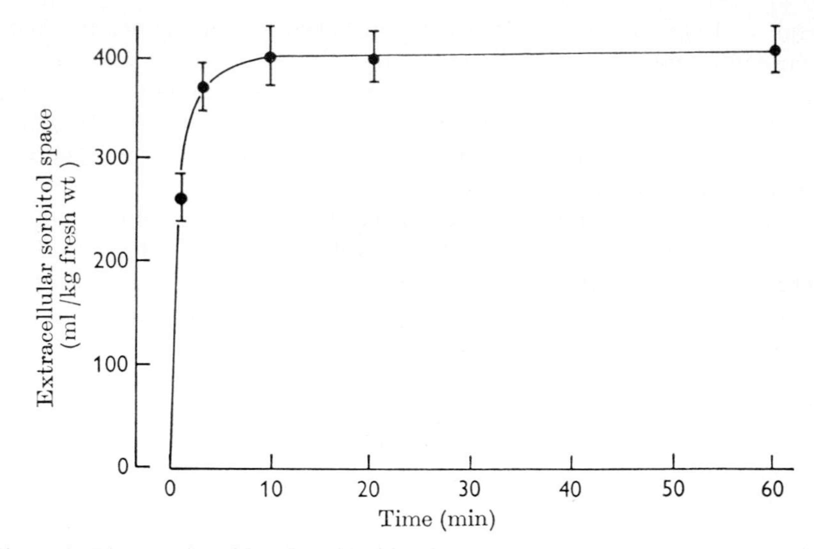

Figure 8. The uptake of [^{14}C]-sorbitol by the taenia coli *in vitro* at 35°C (Goodford and Leach, 1966). The abscissa is the time in minutes after transfer from non-radioactive to [^{14}C]-sorbitol Kreb's solution and the ordinate is the extracellular sorbitol space. Note that the uptake of sorbitol reaches a steady value within a few minutes (see text).

B. ^{22}Na and ^{24}Na Wash-out

Sodium uptake in mammalian smooth muscle at body temperature soon reaches a point at which the relative activity of the tissue no longer differs significantly from 1, and since no useful information is obtained after this time, emphasis has been placed on radio-sodium wash-out experiments. Figure 9 shows the results of an experiment in which a sample of dog carotid artery was equilibrated with Kreb's solution containing ^{22}Na for ninety minutes, and was then re-immersed in a series of non-radioactive Kreb's solutions. The amount of radioactivity left in each sample of Kreb's solution was measured after the muscle had been moved on, and the amount of radioactivity remaining in the tissue at the end of the experiment was also determined. It would then have been possible to plot the observations directly against time to give a 'differential effluent count' (Goodford, 1968), but in the present case the counts were added together in the reverse order to calculate the total amount of radioactivity in the muscle sample itself at each stage of wash-out. These 'integral counts' were plotted in figure 9 (with a logarithmic ordinate normalized to one hundred per cent at the start of the experiment when the muscle was first removed from radioactive solution), and the points fell on a steep curve which straightened out after the first sixty minutes of sodium efflux.

If ^{22}Na were being lost from a single region in the tissue at a rate proportional to the 'amount' of ^{22}Na in that region, there would be a simple proportionality:

$$\frac{da}{dt} = -\lambda a \tag{4}$$

in which a represents the amount of ^{22}Na in the region,
λ is a rate constant, and

$\frac{da}{dt}$ is the rate of loss of tracer from the region.

Integration leads to the equations:

$$\text{Log}_e a = -\lambda t + \text{Log}_e a_o \quad \text{and} \quad a = a_o e^{-\lambda t} \tag{5a; b}$$

in which a_o is a constant.

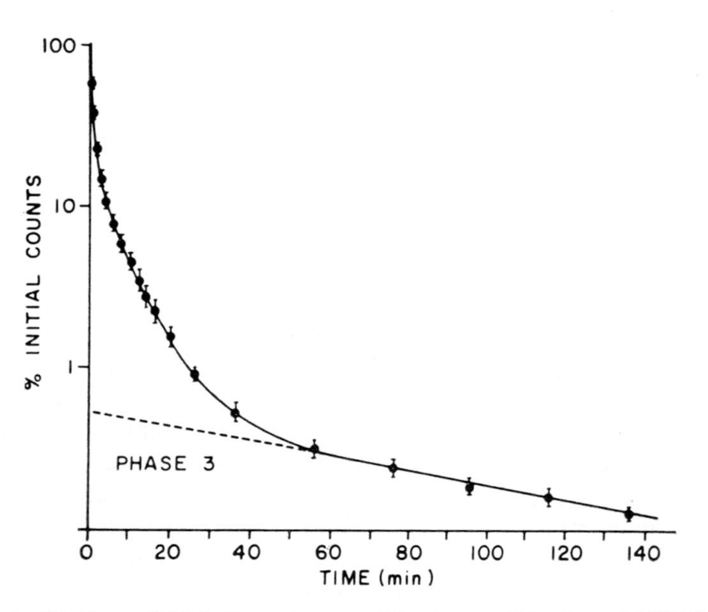

Figure 9. The loss of ^{22}Na from dog carotid artery wall *in vitro* at 37°C (Garrahan and coworkers, 1965). Pieces of tissue were pre-equilibrated with ^{22}Na Kreb's solution for ninety minutes. The abscissa is the time in minutes after transfer to non-radioactive Kreb's solution and the logarithmic ordinate is the amount of radioactivity remaining in the tissue expressed as per cent of the amount at the start of efflux. The curve can be described as a series of exponentials, resolving into a single one after the first sixty minutes (see text).

This shows that the simple proportionality hypothesis applied to a single region should lead to a logarithmic (or exponential) law and to a straight line on the semilogarithmic axes of figure 9. The results are not compatible with this interpretation.

It is of course a gross over-simplification to treat a smooth muscle as a single region. If it were treated as two independent regions there would be two proportionality equations each leading to an exponential term, and the sum of these exponentials should describe the loss of tracer. With three independent hypothetical regions there would be three exponential terms and so on, and the exchange in any system of regions can in fact be described by a sum of sufficient exponential terms. This is both the strength and the weakness of the method; a strength because any efflux curve can be described in this way; a weakness because the description would be misleading if the initial proportionality hypothesis were not justified. In the present case (figure 9) the continuous curve is the sum of more than three exponentials, but fortunately additional evidence is available since the efflux of sodium has been studied under varied physiological conditions. The interpretation does not therefore depend upon algebraic analysis alone, and it has been tentatively concluded after consideration of all the evidence that:

i) the most rapid period of exchange, within the first five minutes or less, corresponds to the diffusion of sodium through the extracellular space. This may be described by a special diffusion equation but it is worth noting that the extracellular space can be treated as a large number of regions each exchanging exponentially. The diffusion equation is, in fact, equivalent to many exponentials added together and their total sum should measure the amount of extracellular sodium.

It is a characteristic of diffusion processes that they are less sensitive to changes of temperature than many chemical reactions, but more sensitive than other physical phenomena. They have a characteristic Q_{10} of about 1·5, meaning that the rate of diffusion would increase by fifty per cent if the temperature were raised 10 °C, and the first phase of sodium exchange shows just this effect. Furthermore, it is not sensitive to drugs such as ouabain, it is unaffected by altering the ionic composition of the medium, and other ions such as lithium show a similar kinetic picture. There is therefore substantial evidence that a first phase really corresponds to the exchange of extracellular sodium, but it is possible that it is not all freely dissolved in the extracellular space. Some may be present as a counter-cation at fixed anionic groups (Headings and coworkers, 1960; Goodford, 1966), and it is an open question whether this should or should not be regarded as a separate phase of its own.

ii) The results after the first minute or two can be described as the sum of only two exponentials. The first of these in figure 9 accounts for some sixteen per cent of the total tissue sodium exchanging with an exponential half-life of some five minutes at body temperature. Its rate of exchange decreases dramatically on cooling, and the size of the phase increases both absolutely and as a percentage of the total tissue sodium at low temperatures. Its properties depend on the ionic composition of the medium, and it is inhibited in potassium-free solutions. It is inhibited by ouabain and 2,4-dinitrophenol, and apparently corresponds in all respects to the exchange of sodium through the smooth muscle cell membrane (Hagemeijer and coworkers 1965; Garrahan and coworkers, 1965).

It is not easy to measure the size of this phase accurately since it merges imperceptibly into the extracellular diffusion exchange, and simple analysis of the exponential curves may give biased values (Huxley, 1960; Weatherall, 1962; Goodford, 1966). Inclusion of a little extracellular material would distort the calculations very significantly, and transmembrane sodium fluxes exceeding 200 pmole/cm^2 sec were worked out when the inulin space was still being used (Goodford and Hermansen, 1961). The true flux is at least an order of magnitude less than this (Buck and Goodford, 1966; Casteels, 1968), but it is too early to be dogmatic about its exact value, and fluxes as low as 0·18 pmole/cm^2 sec have been calculated for the flux in arterial smooth muscle (Keatinge, 1968b).

iii) By sixty minutes all the rapid exponential processes (i and ii) are virtually complete, and the final phase of exchange only accounts for some one per cent of the tissue sodium exchanging with a half time of seventy minutes. On the semilogarithmic coordinates of figure 9, the corresponding points fall on a single straight line, and since the phase is both small and slowly-exchanging it may be treated to a first approximation as unchanging sequestered sodium within the tissue.

It has been shown that the observations in figure 9 can be explained on the basis of three regions in the tissue, and similar results have been obtained from other smooth muscles (Burnstock and coworkers, 1963; Buck and Goodford, 1966). Nevertheless, it is important to appreciate that the detailed interpretation is still questionable, and that further evidence may lead to a re-interpretation of the different exponential phases. Indeed, the very nature of the exponential analysis might require modification if the simple proportionality hypothesis were not valid, and the real justification of the present treatment is that it brings together a range of experimental observations into a single unified scheme.

C. Conclusions

The exchange of sodium in smooth muscle can be described on the basis of at least three phases: the extracellular phase, the transmembrane phase and the phase of the sequestered sodium. Furthermore it is an open question whether there are other components to the extracellular phase since it is not easy to measure the most rapid exchange with sufficiently great accuracy. The possibility must be held in mind that some sodium might be loosely bound extracellularly, in excess of the free extracellular cation.

VI. RADIOACTIVE CALCIUM EXCHANGE

A. ^{45}Ca Uptake

When samples of intestinal smooth muscle are transferred from a normal Ringer's solution to one containing radioactive calcium, there is a rapid uptake of tracer reaching an equilibrium value within five or ten minutes (Bauer and coworkers, 1965; von Hattingberg and coworkers, 1966). With healthy samples of taenia coli this corresponds to complete exchange of tracer (figure 10) and to a relative activity of virtually 1, but in cold or damaged samples the exchange is less complete. The uptake in figure 10 may

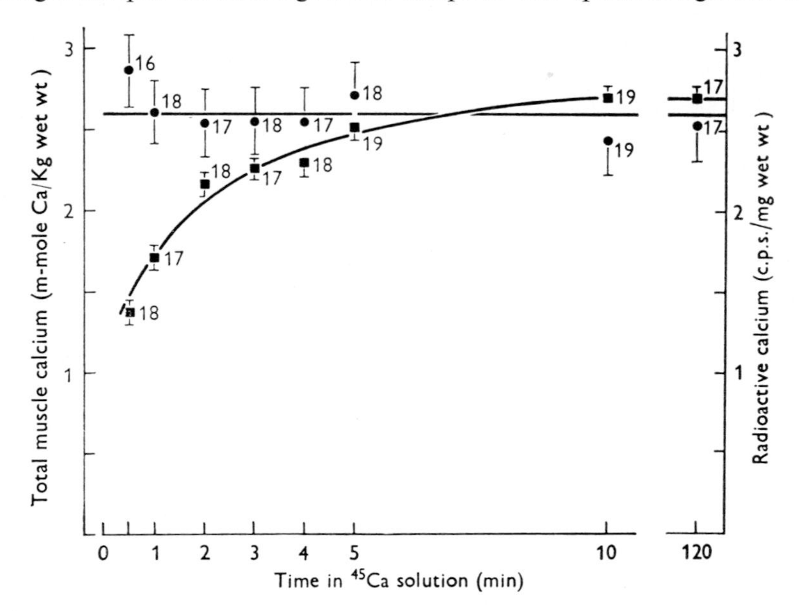

Figure 10. The total tissue calcium and uptake of ^{45}Ca by the taenia coli *in vitro* at 35°C (Bauer and coworkers, 1965). The abscissa is the time in minutes after transfer of the muscles from non-radioactive to ^{45}Ca Kreb's solution; the left ordinate is the calcium content of the tissue and the right ordinate is the ^{45}Ca content on a scale selected so that the points representing total and radioactive calcium would coincide when there had been complete exchange. The number of observations and standard error of the mean are shown where possible. Note the very rapid initial extracellular uptake followed by a somewhat slower phase. Exchange was complete within ten minutes.

be described by a rapid diffusion process followed by a single approximately exponential phase with a half time of about two minutes, and it was suggested by Schatzmann (1961) that some of this calcium may be associated with the cell membrane. It may be in equilibrium with the ionized extracellular calcium (Bauer and coworkers, 1965), and the view is now widely but by no means generally accepted that this membrane calcium is available to the contractile mechanism for the initiation of tension.

On such an hypothesis it is conveniently taken for granted that tension development involves a calcium-activation step in smooth muscle, as it apparently does in other contractile tissues. The evidence is by no means so good in the present case (Lüllmann, 1969) and all that can be said with confidence is that both actin and myosin are present. If, however, the stoichiometric relation between the weight of calcium needed to activate a given weight of actomyosin is the same in smooth and skeletal muscles, then there is sufficient calcium in the postulated membrane phase to fully activate the contractile mechanism several times over (Goodford, 1967).

It is well established that electrical action potentials can be recorded from smooth muscles which have been immersed for long periods in sodium-deficient or sodium-free solutions, as long as calcium is present (Holman, 1958; Axelsson, 1961; Bülbring and Kuriyama, 1963). More recently, action potentials have been seen in a calcium-deficient solution containing sodium (Keatinge, 1968a; Golenhofen and Petranyi, 1969), and such results have led to a controversy as to whether the ions carrying charge across the smooth muscle cell membrane are sodium (as they are in many other excitable tissues during the rising phase of an action potential) or calcium. It would now appear that this question has no definite answer because it presupposes that one ion is right and the other wrong, whereas the inward current may, in all probability, be carried by either ion. Indeed, if ions such as calcium are actually present in or near the smooth muscle cell membrane, it is reasonable to consider whether they may be displaced by the arrival of other ions during an action potential, so that an initial flux of sodium might be converted into a joint sodium–calcium flux, or *vice versa*. There is evidence that this type of interaction between mono- and divalent ions does take place in smooth muscle (Goodford, 1965b, 1968), but it is not the simple sodium:calcium competition which in heart muscle leads to the $[Ca]/[Na]^2$ law (Nash and coworkers, 1966). Thus the size of the second phase of radioactive *calcium* uptake in the taenia coli is reduced when the *potassium* concentration in the bathing medium is raised (figure 11), which would be expected if K^+ ions were displacing Ca^{2+}, and in general any cation may take part to a greater or lesser extent. The occurrence of such ionic interactions in smooth muscle should therefore be noted, although their detailed mechanism and location in the tissue has not yet been established.

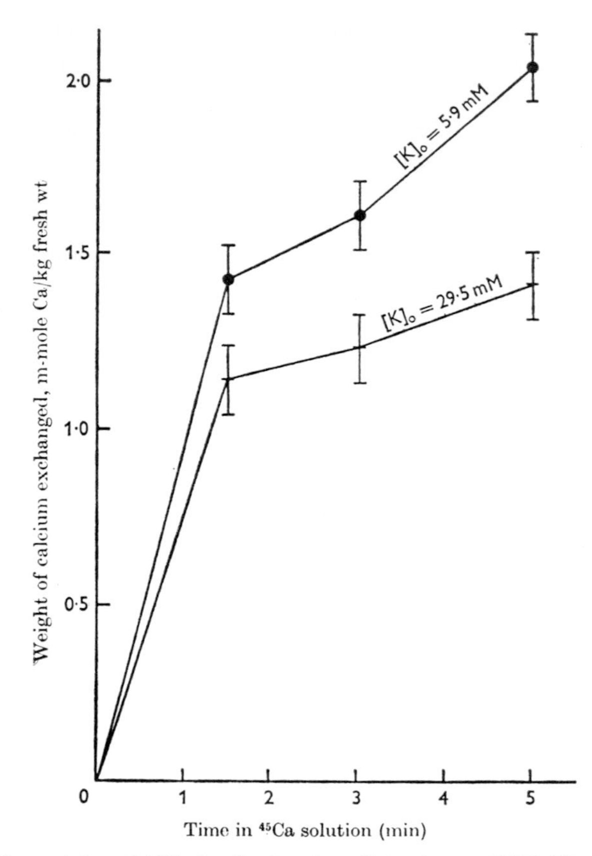

Figure 11. The uptake of ^{45}Ca by the taenia coli *in vitro* at 35°C (Goodford, 1965). The abscissa is the time in minutes after transfer from non-radioactive to ^{45}Ca solution and the ordinate is the uptake of ^{45}Ca. Observations were made in normal Kreb's solution ($[K]_0 = 5{\cdot}9$ mM) and in high-potassium Kreb's. In high potassium the second phase of uptake was inhibited, but the first extracellular phase (which can be estimated by retropolating the points back to zero time) was not significantly affected (see text).

There is some evidence that pilocarpine (Chujyo and Holland, 1962), barium (Karaki and coworkers, 1967), 2,4-dinitrophenol and papaverine (von Hattingberg and coworkers, 1966) may increase the rate of uptake of calcium, but it is not easy to be sure that changes of extracellular space have not influenced these effects.

B. ^{45}Ca Efflux

The uptake of tracer calcium by healthy mammalian smooth muscle resembles sodium in so far as the relative activity soon approaches 1 in two rapid phases, but a small amount of slowly exchanging calcium can be detected as a third phase in efflux experiments. In order to do this it is necessary to load the tissue with radioactive calcium for long periods because it is impossible to measure efflux from a region until the tracer has entered in the first place, and if it is slow to leave it will be slow to enter. This has been elegantly demonstrated by Schatzmann (1964) who has also shown that the amount of slowly exchanging calcium is greatly increased if phosphate ions are incorporated in the bathing medium (figure 12). The effect is so marked that it has been observed in uptake experiments when phosphate ions are present (Schatzmann, 1961; Bauer and coworkers, 1965), and it

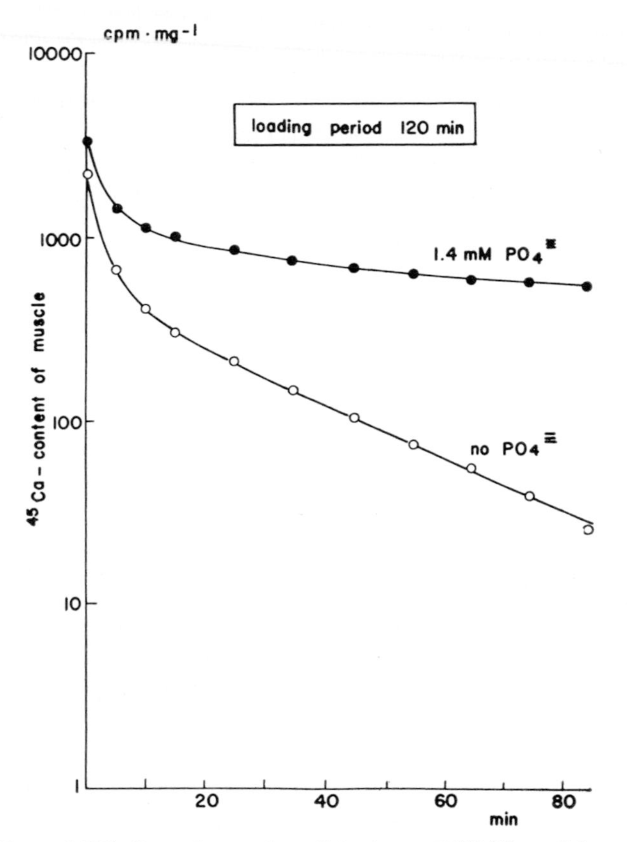

Figure 12. Loss of ^{45}Ca from the taenia coli *in vitro* at 37°C (From Schatzmann, 1964, with permission from the Pergamon Press). Pieces of tissue were pre-equilibrated with ^{45}Ca solution. The abscissa is the time in minutes after transfer from ^{45}Ca solution and the logarithmic ordinate is the radioactive calcium content of muscle. Calcium exchange was compared in the presence and absence of phosphate, and the tissue always contained more tracer calcium when phosphate was present.

appears as if some very slowly exchanging calcium is added to the tissue. The presence or absence of phosphate does not affect contractility (Karaki and coworkers, 1966) and in all probability microcrystallites of calcium phosphate may precipitate while the normal exchange of calcium is unaffected and goes on in parallel (Goodford, 1964). Thus it has been shown that the calcium content increases rapidly when the pH rises above 7·0 in the presence but not in the absence of phosphate, and this may be associated with the transition:

$$Ca(H_2PO_4)_2 + 2OH^- \rightarrow HPO_4^{2-} + 2H_2O + CaHPO_4 \downarrow$$

which occurs just above this pH value (Goodford, 1967).

Warming (Goodford, 1966), or the application of ethylene diamine tetra-acetate (EDTA) which chelates Ca^{2+} ions, can increase the rate of loss of ^{45}Ca from smooth muscle, and high potassium concentrations, histamine, adrenaline and 5-hydroxytryptamine (5-HT) have similar effects on the early stages of tracer loss (Nagasawa, 1965), but it is not easy to eliminate the effects of a changing extracellular space upon the diffusion pattern at this time.

C. Conclusions

There have been substantial difficulties in the measurement of calcium and its tracers which have led to an appreciable experimental variation, and this variability has been increased because the distribution of calcium in smooth muscle depends upon the viability of the tissue. In unfavourable conditions, substantial quantities of very slowly exchanging calcium may accumulate, and the behaviour of these deposits may be related to the physicochemical rather than the physiological properties of the calcium ion. Physiological stimuli only produce small changes on the slow phase even under normal conditions (Schatzmann, 1964; Goodford, 1964, 1965a) and the effect of altering the temperature is similar both upon the slow calcium efflux and upon the rate of solution of calcium phosphate (Goodford, 1964). Furthermore, the first phase of calcium exchange is apparently extracellular, and is therefore of little physiological significance.

Sparrow and Simmonds (1965) demonstrated a substantial loss of total tissue calcium when toad stomach muscle was immersed in calcium-free solution. The contractile response to acetylcholine was simultaneously lost, but could be restored when only nine per cent of the tissue calcium was replaced. Thus the acetylcholine effect depended on a small proportion of the total tissue calcium which could be restored in a relatively short time after depletion. It is important to establish the location and exchange kinetics of this physiologically active fraction.

VII. RADIOACTIVE MAGNESIUM EXCHANGE

A. ^{28}Mg Uptake

The radioisotope ^{28}Mg has a half-life of twenty-one hours, and is only available at present with specific activities suitable for uptake but not for efflux experiments. One may see from figure 13 that magnesium exchange in normal conditions is described by a single exponential function after the first few minutes. It has a surprisingly long half-time of four hundred minutes so that the process is not complete even after long experimental periods. If it

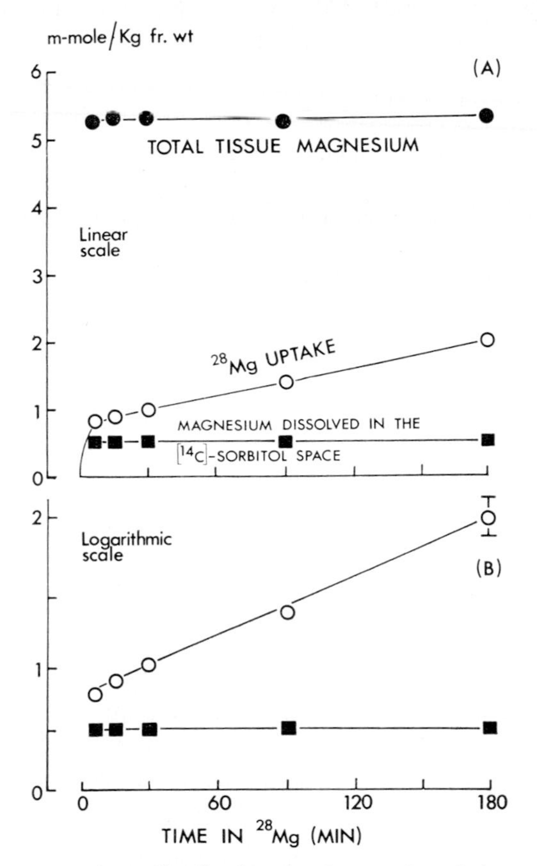

Figure 13. The magnesium distribution in the taenia coli immersed *in vitro* in normal Kreb's solution at 37°C (Sparrow, 1969). The abscissa is the time in minutes after transfer to radioactive ^{28}Mg solution and the ordinate is linear in upper figure A and logarithmic in lower figure B. The last fifteen minutes of immersion were always in a solution containing radioactive [^{14}C]-sorbitol. There is a slow phase of magnesium exchange, and a rapid phase which exceeds the amount of magnesium freely dissolved in the [^{14}C]-sorbitol space (see text).

corresponds to the exchange of cellular magnesium across the cell membrane, one may then calculate that the flux is approximately 0·02 pmole/cm^2.sec; that the average intracellular concentration is 7·3 mmole/kg fresh weight; and that the magnesium equilibrium potential is -23 mV when activity coefficients are neglected (Sparrow, 1969).

The slowly exchanging magnesium corresponds to a substantial proportion of the total tissue content, but when the straight line in figure 13B is extrapolated back to zero time it still gives a surprisingly high intercept. This shows that the amount of rapidly exchanging magnesium in the tissue is more than can occur in free solution in the extracellular space, which is itself calculated from the [^{14}C]-sorbitol observations shown on the same figure. It may therefore be tentatively concluded that there are at least three phases of magnesium exchange:

i) the rapidly exchanging magnesium freely dissolved in the [^{14}C]-sorbitol space;
ii) the 'excess rapidly exchanging magnesium' not accounted for by the [^{14}C]-sorbitol space;
iii) the slowly exchanging magnesium.

Figure 14 shows the *in vitro* uptake of ^{28}Mg into pieces of taenia coli which had been pre-equilibrated at 37 °C with an isotonic sucrose solution containing 2 mM $MgCl_2$. Under these conditions it is not easy to describe the uptake as a single slow exponential process (Sparrow, 1969), and the three separate phases can be clearly seen. Sparrow has shown, furthermore, that the size of the first phase is proportional to the concentration of Mg^{2+} ions in the bathing solution, as would be expected if it represented magnesium ions freely dissolved in the extracellular space, but the size of the intermediate phase is not so simply described. It depends on the concentrations of the other ions which are present and apparently corresponds to a tissue region in which there is competition between Mg^{2+} and Ca^{2+} and K^+ ions for fixed anionic sites.

The size of the slowest and largest phase was approximately constant in all Sparrow's experiments.

B. Conclusions

The exchange of magnesium resembles that of sodium and calcium, in that three separate phases can be described, but the last of these is many times larger for magnesium. Furthermore, calcium and magnesium share the common property that one phase of their exchange is an ionic interaction with other cations. However, unexchangeable magnesium has not yet been detected in smooth muscle.

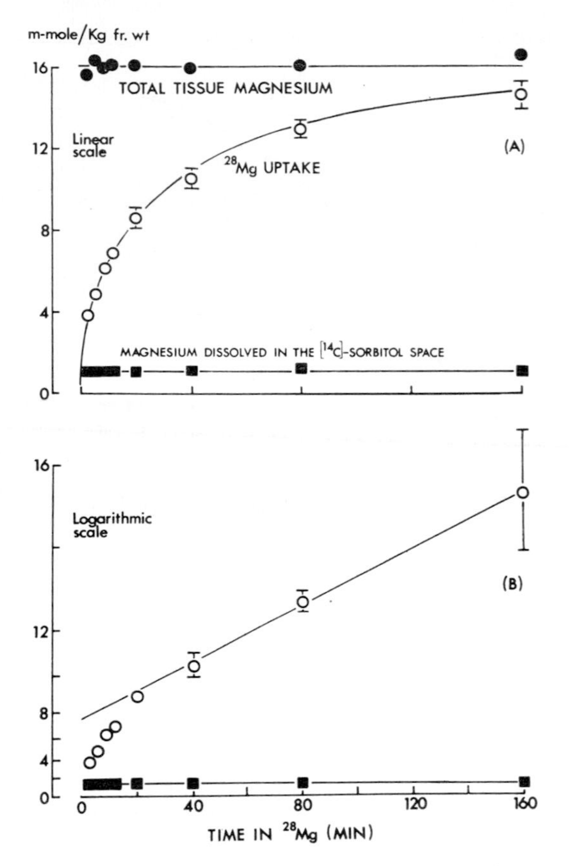

Figure 14. The magnesium distribution in the taenia coli immersed *in vitro* in isotonic sucrose solution containing 2 mM $MgCl_2$ at 37°C (Sparrow, 1969). The abscissa, ordinates and symbols are the same as figure 13. In the absence of other cations the total tissue magnesium has increased from some 5 to 16 mmole Mg/kg fresh weight. The ^{28}Mg observations could not be fitted to a single exponential function, and there is clear evidence of a three-compartment exchange (see text).

VIII. RADIOACTIVE POTASSIUM EXCHANGE

A. ^{42}K Uptake

Figure 15 shows the uptake of radioactive ^{42}K by the smooth muscle of the guinea-pig taenia coli. As with magnesium, it is a slow process extending over several hours (Born and Bülbring, 1956), and the calculated trans-membrane flux of potassium is between 2 and 4 pmole/cm^2.sec (Goodford and Hermansen, 1961). There is evidence that some of the potassium in the tissue is sequestered and exchanges still more slowly than the rest (Buck and

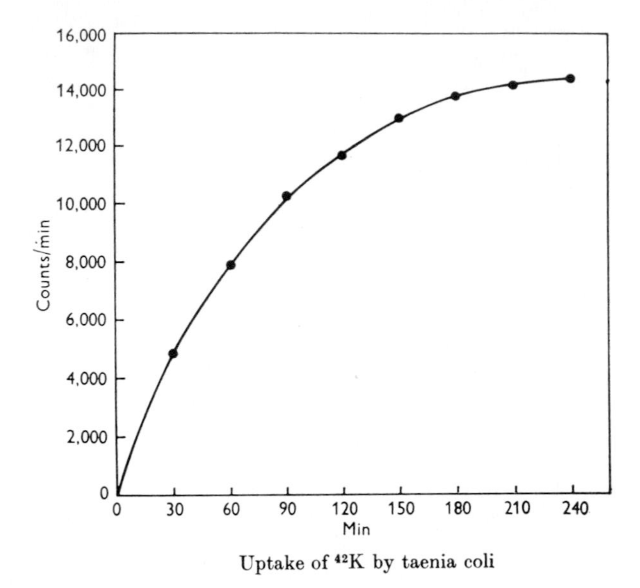

Figure 15. The uptake of ^{45}K by the taenia coli *in vitro* in Kreb's solution (Born and Bülbring, 1956). The abscissa is the time in minutes after transfer from non-radioactive to ^{42}K solution and the linear ordinate is ^{42}K uptake. The rapidly exchanging tracer was washed out of the tissue before each measurement, and the results fall on a single uptake curve with an exponential half time of fifty-five minutes.

Goodford, 1966), and some potassium is of course dissolved in the extracellular space. Moreover, in solutions of reduced ionic strength, another component of rapidly exchanging potassium has been detected, and the size of this component is determined by competition with other cations. Thus the exchange of smooth muscle potassium can be described as a four component system.

B. ^{42}K Efflux

The initial loss of radioactive ^{42}K from smooth muscle shows a rapid phase which may correspond partly to the dissolved potassium in the extracellular space, and partly to potassium on anionic sites. Within a few minutes, however, a long slow exponential efflux can be observed which continues for several hours under steady-state conditions. This has been extensively studied because the observations are highly reproducible, and it is therefore possible to detect any small changes which may be produced by physiological stimuli. The dashed line in figure 16 shows control observations of the potassium efflux at five minute intervals during this slow phase, and the action of acetylcholine ($1{\cdot}4 \times 10^{-5}$M) may be clearly seen. A similar effect is observed when carbachol or catecholamines are applied to depolarized intestinal smooth muscle (Durbin and Jenkinson, 1961; Jenkinson and Morton, 1967) but under normal conditions the catecholamine effect is not so obvious and may even be reversed (Bülbring and coworkers, 1966).

The simplest interpretation of a change of flux is that it is caused by either a change in membrane permeability, or a change in transmembrane potential, or some special but undefined mechanism such as the stimulation of a potassium extrusion pump. One may thus conclude that the action of acetylcholine and carbachol do not depend solely on changes of the transmembrane potential, because their effects are observed in depolarized tissues. It seems improbable *a priori* that potassium is actually pumped out of the tissue, from a region of high concentration to one of low, and it is therefore assumed that the parasympathomimetic effects are due to changes in the membrane permeability. In fact it appears that carbachol increases the permeability of the smooth muscle membrane to all cations (Durbin and Jenkinson, 1961). Catecholamines may act similarly in cold depolarized

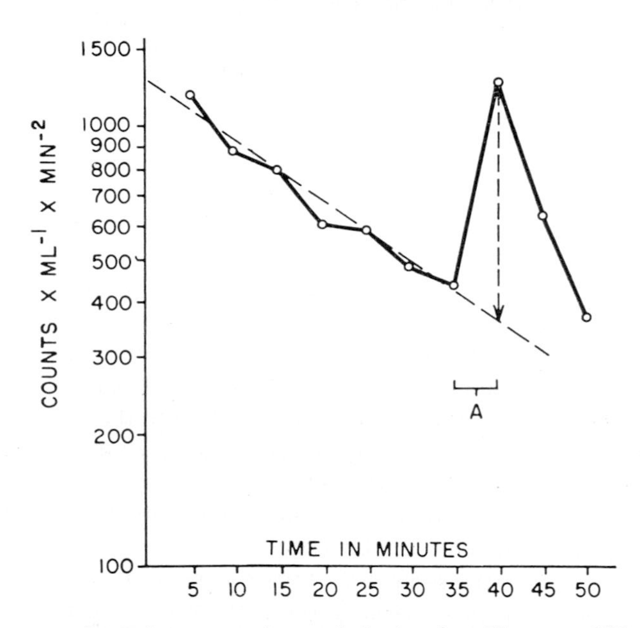

Figure 16. Loss of ^{42}K from smooth muscle *in vitro* (von Hagen and Hurwitz, 1967). Pieces of tissue were pre-equilibrated with ^{42}K solution. The abscissa is the time in minutes after transfer from ^{42}K solution and the logarithmic ordinate is the rate of loss of tracer potassium from the muscle. For the first thirty-five minutes the rate of loss of tracer was approximately exponential, but the loss increased dramatically when acetylcholine (A) was applied (see text).

tissues, but it is not easy to differentiate between the three possible modes of action at body temperature, since sympathetic stimulation can change both the membrane potential and the metabolic activity of the tissue. It is therefore reasonable to suppose that the catecholamines may influence all three mechanisms.

The presence of loosely bound cations at anionic sites in smooth muscle has already been mentioned, and it has been pointed out that ionic interactions can take place at such sites (Goodford, 1966). Such a cationic interaction has recently been demonstrated by von Hagen and Hurwitz (1967), who find that the rate constant for ^{42}K efflux from the smooth muscle of the guinea pig ileum is reduced when the sodium or calcium concentrations in the bathing solutions are raised. Their observations suggest that ionic interactions may influence the behaviour of the cell membrane itself, assuming that membrane passage is the rate-limiting step for potassium efflux, and any model of the ionic distribution must take their findings into account. It should also explain the significant finding that hyperpolarization tends to reduce the potassium efflux, and that depolarization due to ouabain or potassium causes it to rise (Setekleiv, 1967).

IX. A MODEL SYSTEM

A. Concordance

The ionic distribution in smooth muscle has been observed with increasing frequency in recent years, and it would now be appropriate to make special mention of two of the findings. They are selected because they demonstrate certain characteristic properties of the tissues, and not because they are any more important than the other experimental results which have been described, and which often supply valuable corroborative evidence.

Barr and Malvin (1965) studied the diffusion of urea, arabinose, mannitol, sucrose, raffinose, inulin and serum albumin into intestinal smooth muscle. Each substance approached a plateau concentration in the tissue after three or four hours, but the height of every plateau was different and fell into one of four general classes. While urea seemed to come to a concentration equilibrium in all the tissue water, the serum albumin might not even be able to penetrate the whole extracellular space, and while the arabinose and mannitol might penetrate some way into the cells, it was the sucrose, raffinose and inulin which provided good methods of estimating the extracellular space. Since inulin is a substance of uncertain composition (Nilwarangkur and Berlyne, 1965), the best estimates come from unmetabolized molecules about as large as sucrose, but the presence of the other plateaux show that *there are at least four water compartments in intestinal smooth muscle.*

The observations of Garrahan and coworkers (1965) demonstrate that the loss of radioactive tracer sodium consists of not less than three exponential phases (figure 9), the slowest of which may be called the sequestered cation while the intermediate phase apparently corresponds to transmembrane exchange. However, the most rapidly exchanging material may be subdivided into two separate components, one representing the monatomic cations freely dissolved in the extracellular space and the other cations attached to fixed negative sites in the tissue. There is competition for these sites between cations, and this property can be used to differentiate them clearly from those in free solution although their exchange kinetics may be similar (Goodford, 1966; Sparrow, 1969). Thus, in general, *the ion exchange properties of smooth muscle are those of a four component system* consisting of:

i) the rapidly exchanging ions freely dissolved in the extracellular space;

ii) the 'excess rapidly exchanging ions' which are competing for superficial anionic sites;

iii) the transmembrane exchange, and

iv) the sequestered material.

It should perhaps be emphasized that the size and rate of each component will be different for every ion, even in the same tissue.

B. The Compartmental Concept

In order to construct a model for the ionic distribution in smooth muscle it is assumed that the ions are present in a number of uniform compartments in the tissue. The concept of uniformity is important because it implies that the composition of each compartment is the same right up to its limiting boundary, and that there is then an infinitely sharp spatial interface with the adjoining uniform region. This is of course completely unrealistic, but the concept of compartmentalization is well established among ion-exchange physiologists. It offers the advantage that the various phases of tracer exchange can be assigned to the various compartments in the tissue, and in the present case it is tempting to speculate that the four water compartments of Barr and Malvin (1965) correspond to the four phases of ionic exchange. However, it would be rash to accept this simple interpretation entirely on the basis of the present evidence.

At some time a final break will have to be made with compartmental theory. This time will be reached when it no longer describes the observations adequately, and when it can no longer be used in order to construct new hypotheses for experimental testing. The ultimate justification for any model is its usefulness in these ways, and the time is not yet ripe to reject the concept of compartmentalization out of hand.

C. An Intercompartmental Transport Hypothesis

The traditional hypothesis which has been used in order to interpret movements between compartments is that '*the rate at which material tends to leave a region is proportional to the amount in the region.*' Leaving aside for a moment the difficulties in defining the term 'amount', this enables one factor to be isolated from the multitude of influences which control a biological system. Furthermore, it enables a quantitative approach to be taken because a proportionality constant λ can be assigned to each flow of material. Consider two adjacent uniform regions 1 and 2, from each of which the rates of flow to the other are

$$\frac{\mathrm{d}}{\mathrm{d}t}a_1 = -\lambda_1 a_1 \text{ and } \frac{\mathrm{d}}{\mathrm{d}t}a_2 = -\lambda_2 a_2 \qquad (4')$$

At equilibrium these will be equal, so that

$$a_1\lambda_1 = a_2\lambda_2 \qquad (7)$$

and there will be an equilibrium constant

$$K = \lambda_1/\lambda_2 \qquad (8)$$

The physicochemical interpretation of this behaviour would be that an energy barrier exists between the regions, and that material in transit has to pass this barrier. It can only do so if sufficient thermal energy is available, the amount required depending on the height to be overcome which may not be the same in one direction as in the other. A simple analogy would be jumping over a hedge from one field to another at a lower level since more energy would be needed for the higher return jump, and in the case of two physicochemical regions with an energy difference L separated by a barrier of height H above the higher region (figure 17), it can be shown that the probability of a molecule having the requisite energy is

$$\mathrm{e}^{-H/RT} \text{ and } \mathrm{e}^{-(H+L)/RT} \qquad (9)$$

in each direction respectively, where

R = the gas constant = 8·314 joule/degree mole
T = the absolute temperature (311 Å at body temperature), and
e = the base of natural logarithms (2·71828).

These probabilities are proportional to the rate constants λ_1 and λ_2 so that equations 4′, 7 and 8 may be rewritten

$$\frac{\mathrm{d}}{\mathrm{d}t}a_1 = -a_1\mathrm{e}^{-H/RT} \text{ and } \frac{\mathrm{d}}{\mathrm{d}t}a_2 = -a_2\mathrm{e}^{-(H+L)/RT} \qquad (4'a)$$

$$a_1\mathrm{e}^{-H/RT} = a_2\mathrm{e}^{-(H+L)/RT} \qquad (7a)$$

and

$$\frac{a_2}{a_1} = K = \mathrm{e}^{L/RT} \qquad (8a)$$

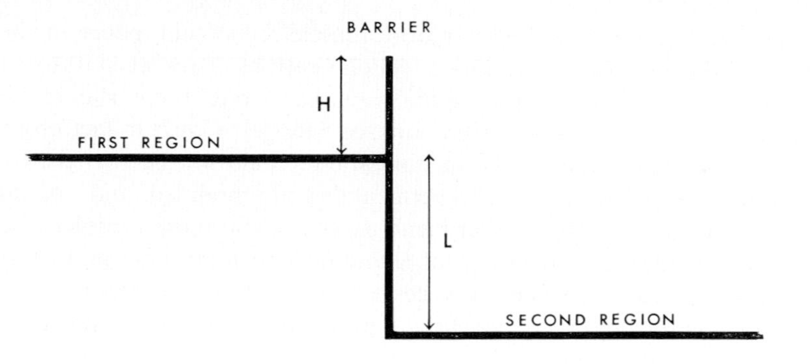

Figure 17. Potential energy levels at the sharp boundary separating two uniform regions. Movement from the first to the second can be achieved by molecules having an energy greater than H, but a still larger energy $(H+L)$ is needed for movement in the reverse direction. There will therefore be an overall tendency for molecules to accumulate in the second region (see text).

It is clear from these equations that an increase in the height H to $H+h$ will reduce both unidirectional fluxes by the same factor $e^{-h/RT}$, and that this *increase in barrier height is analogous to a reduced permeability* in the traditional sense. On the other hand, *the equilibrium constant is dependent upon the difference of levels L* but not on H, and a *modified hypothesis* may be proposed for the movements between compartments. It is based on the assumption that '*the rate at which material tends to leave a region is proportional to the amount that can cross the energy barriers surrounding the region,*' and this enables the rate constant λ to be considered quantitatively, as well as the total amount of material a in the region.

The effect of an electrical potential difference E between regions may be interpreted on the modified hypothesis because it gives an energy difference:

$$L = nFE \tag{10}$$

for an ion of valency n, where F is Faraday's constant (96,500 Coulomb/equiv), and so the equilibrium thermodynamic approach is compatible with the modified hypothesis which leads straight from equations 8a and 10 to the Nernst equation

$$E = -\frac{RT}{nF} \log_e \frac{a_1}{a_2} \tag{11}$$

The effects of electrical potential changes upon the unidirectional fluxes can be predicted on the modified hypothesis. For instance, a depolarization of the membrane surrounding a smooth muscle cell should reduce the height L which influences the rate of loss of cellular cations. Thus, even if the height H and the membrane permeability were unaltered, there should be an increased passive loss of potassium ions, as Setekleiv (1967) in fact observed. On the other hand, the uptake of potassium should not be changed by the depolarization, unless H and the permeability are modified, and this can be tested in potassium uptake experiments so that a still more complete picture can be obtained. If considerations of H and L alone are insufficient to account for the observations, it may be necessary to think about interactions with other regions in the tissue, but *ad hoc* processes need only be postulated if this extension also fails.

D. A Smooth Muscle Model

Experimental methods are now so refined that at least four separate regions have been detected in smooth muscle, and the full value of the observations will not be realized if they are forced to fit an over-simplified model. In fact the most straightforward yet plausible system consists of:

i) The extracellular space, containing solutes in free diffusive equilibrium with the outside surroundings;
ii) An extracellular array of fixed ions, with associated counter-ions in solution;
iii) An intracellular space with ions in free solution, and
iv) An intracellular sequestering region.

It must be emphasized that even this seemingly complicated system is trivial in comparison with the complexities of the living tissue, and it is tempting to propose at least a fifth region containing an intracellular array of fixed ions and counter-ions as well. However, the four-compartment system affords a reasonably simple model.

X. CONCLUSIONS

Observations on the ionic distribution in smooth muscle lead almost irresistibly to one of two conclusions. Either the attempt to apply compartmental concepts to these results should be abandoned, or the complexities of a multi-compartment system should be accepted. The second alternative is proposed for reasons already discussed, and the strength of this approach is three-fold:

i) the movements of materials between compartments can be quantitatively defined by algebraic equations and established physical laws;

ii) the numerical values of the constants in the equations can be measured experimentally, at any rate in principle and usually in practice; and

iii) a real physical meaning can be assigned to each of the constants.

The efflux of a single substance from a single compartment can, with simplifying assumptions, be described by a single constant λ as in equation 4, but at least one more constant is needed if the reverse flow is to be considered. When a general four-compartment system is to be studied it is necessary to use over a dozen constants for each material, and since many different interacting substances may be present in a biological tissue a great many constants may be needed. It is therefore possible to make the model fit almost any experimental observations by choosing suitable values for the constants, and it may be desirable to reduce this generality by restricting the system. Restrictions which may be suggested for smooth muscle include:

i) the use of only four compartments, and the abandonment of the *ad hoc* proposal of a fifth;

ii) the assignment of these compartments in series, one behind the other;

iii) an extracellular space in equilibrium with the outside surroundings, so that its composition is exactly specified; and

iv) the use only of precise quantitative physical laws to define kinetic and distribution properties.

The number of disposable constants is drastically reduced by these not-unreasonable restrictions, but it may be desirable to limit the system still further by more empirical postulates. Some which may be considered are:

v) that the intracellular sequestered material is present as counter-ions; and

vi) that these counter-ions are associated with an intracellular array of fixed ions, which is identical to the corresponding extracellular array.

The system is by now tightly restrained, and Goodford (1969) has shown that the four-compartment distribution of K^+, Na^+, Ca^{2+}, Mg^{2+} and Cl^- ions is then defined by:

i) the composition of the surrounding medium and the cytoplasm;

ii) the concept of a regional electro-neutrality;

iii) the Nernst equation; and

iv) some half a dozen parameters, of which three are already measured.

Thus the number of disposable constants is reduced from many dozens to very few and when these have been determined it will be possible to test the model against a wide range of quantitative observations. There will be no disposable constants available to secure a better fit when discrepancies are observed, but the model may itself be modified. New experiments may be designed to test it once again, and the overall approach will lead to the accumulation of relevant observations and therefore to a better understanding of the muscle.

It is, finally, relevant to consider an important proviso. Any 'passive' system of this type will eventually run down unless there is an input of free energy. The conventional method of introducing this is to postulate sodium, potassium, chloride, calcium or hydrogen ion-pumps, but the quantitative performance characteristics of these are rarely defined. Furthermore, there is a tendency to postulate a new pump *a priori*, whenever the distribution of a new ion is investigated in detail. The present model might be increased in elegance if fewer separate energy inputs were required and an ideal system might require only one such source of free energy, while all other ion flows occurred passively between compartments. (Goodford and coworkers, 1967).

Smooth muscle physiologists may now be faced with a choice: either the complexities of a multicompartment system must be accepted, with the associated advantage that its kinetics can be quantitatively defined by known physical laws; or the simplicity of a two-compartment system may be retained at the expense of postulating an increasing number of ill-defined *ad hoc* pump-type mechanisms.

REFERENCES

Åberg, A. K. G. (1967) *Acta Physiol. Scand.*, **69**, 348
Alvarez, W. C. and L. J. Mahoney (1922) *Amer. J. Physiol.*, **59**, 421
Armstrong, W. McD. (1964) *Amer. J. Physiol.*, **206**, 469,
Armstrong, W. McD. (1965) *Amer. J. Physiol.*, **208**, 61
Axelsson, J. (1961) *J. Physiol. (London)*, **158**, 381
Axelsson, J. and E. Bülbring (1961) *J. Physiol. (London)*, **156**, 344
Barr, L. M. (1959) *Proc. Soc. Exptl. Biol.*, **101**, 283
Barr, L. and R. L. Malvin (1965) *Amer. J. Physiol.*, **208**, 1042
Bauer, H., P. J. Goodford and J. Hüter (1965) *J. Physiol. (London)*, **176**, 163
Bitman, J., H. C. Cecil, H. W. Hawk and J. F. Sykes (1959) *Amer. J. Physiol.*, **197**, 93
Born, G. V. R. and E. Bülbring (1956) *J. Physiol. (London)* **131**, 690
Bortoff, A. (1961) *Amer. J. Physiol.*, **201**, 203
Boyle, P. J. and E. J. Conway (1941) *J. Physiol. (London)*, **100**, 1
Bozler, E. (1938) *Amer. J. Physiol.*, **122**, 614
Bozler, E. (1942) *Amer. J. Physiol.*, **136**, 553
Bozler, E. (1959) *Amer. J. Physiol.*, **197**, 505
Bozler, E. (1961) *Amer. J. Physiol.*, **200**, 656
Bozler, E. (1962) *Physiol. Rev.*, **42**, Suppl. 5, 179
Bozler, E. (1963) *Amer. J. Physiol.*, **205**, 686
Bozler, E. (1964) *Amer. J. Physiol.*, **207**, 701
Bozler, E., M. E. Calvin and D. W. Watson (1958) *Amer. J. Physiol.*, **195**, 38
Brading, A. F. and J. Setekleiv (1968) *J. Physiol. (London)*, **195**, 107
Buck, B. and P. J. Goodford (1966) *J. Physiol. (London)*, **183**, 551
Bülbring, E. (1957) *J. Physiol. (London)*, **135**, 412
Bülbring, E., G. Burnstock and M. E. Holman (1958) *J. Physiol. (London)*, **142**, 420
Bülbring, E., P. J. Goodford and J. Setekleiv (1966) *Brit. J. Pharmacol. Chemother.*, **28**, 296
Bülbring, E, and H. Kuriyama (1963) *J. Physiol. (London)*, **166**, 29

Burnstock, G., D. J. Dewhurst and S. E. Simon (1963) *J. Physiol. (London)*, **167**, 210
Burnstock, G. and M. E. Holman (1961) *J. Physiol. (London)*, **155**, 115
Casteels, R. (1966) *J. Physiol. (London)*, **184**, 131
Casteels, R. (1968) *Proceedings of the International Union of Physiological Sciences*, **6**, 144
Casteels, R. and H. Kuriyama (1966) *J. Physiol. (London)*, **184**, 120
Chujyo, N. and W. C. Holland (1962) *Amer. J. Physiol.*, **202**, 909
Constantino, A. (1911) *Biochem. Z.*, **37**, 52
Creese, R., N. W. Scholes and W. J. Whalen (1958) *J. Physiol. (London)*, **140**, 301
Csapo, A. (1948) *Nature*, **162**, 218
Daniel, E. E. (1958) *Canad. J. Biochem. Physiol.*, **36**, 805
Daniel, E. E. and H. Singh (1958) *Canad. J. Biochem. Physiol.*, **36**, 959
Davenport, H. W. (1963) *Amer. J. Physiol.*, **205**, 413
Davson, H. (1959) *A Textbook of General Physiology*, J. & A. Churchill, London, p. 631
Durbin, R. P. and D. H. Jenkinson (1961) *J. Physiol. (London)*, **157**, 74
Elliott, G. F. (1967) *The Contractile Process*, J. and A. Churchill Ltd., London. p. 171
Evans, D. H. L. and H. O. Schild (1953) *J. Physiol. (London)*, **119**, 376
Evans, D. H. L., H. O. Schild and S. Thesleff (1958) *J. Physiol. (London)*, **143**, 474
Falk, G. and J. F. Landa (1960) *The Pharmacologist*, **2**, 69
Ferguson, J. (1940) *Amer. J. Physiol.*, **131**, 524
Freeman-Narrod, M. and P. J. Goodford (1962) *J. Physiol. (London)*, **163**, 399
Garrahan, P., M. F. Villamil and J. A. Zadunaisky (1965) *Amer. J. Physiol.* **209** 955
Gasser, H. S. (1926) *J. Pharm. Exptl. Ther.*, **27**, 395
Golenhofen, K. and P. Petranyi (1969) *J. Physiol. (London)*, **200**, 136P
Goodford, P. J. (1962) *J. Physiol. (London)*, **163**, 411
Goodford, P. J. (1964) *J. Physiol. (London)*, **170**, 227
Goodford, P. J. (1965a) *J. Physiol. (London)*, **176**, 180
Goodford, P. J. (1965b) *J. Physiol. (London)*, **180**, 19P
Goodford, P. J. (1966) *J. Physiol. (London)*, **186**, 11
Goodford, P. J. (1967) *J. Physiol. (London)*, **192**, 145
Goodford, P. J. (1968) *Handbook of Physiology Alimentary Canal* Vol. IV, p. 1743
Goodford, P. J. (1970) In E. Bülbring (Ed.), *Smooth Muscle*, Arnolds Ltd., London. p. 100
Goodford, P. J and K. Hermansen (1961) *J. Physiol. (London)*, **158**, 426
Goodford, P. J., F. R. Johnson, Z. Krasucki and V. Daniel (1967) *J. Physiol. (London)*, **194**, 77P
Goodford, P. J. and E. H. Leach (1966) *J. Physiol. (London)*, **186**, 1
Goodford, P. J. and E. M. Vaughan-Williams (1962) *J. Physiol. (London)*, **160**, 483
Goto, M., H. Kuriyama and Y. Abe (1960) *Proc. Japan Acad.*, **36**, 509
Gunn, J. A. and S. W. F. Underhill (1914) *Quart. J. Exptl. Physiol.*, **8**, 275
Hagemeijer, F., G. Rorive and E. Schoffeniels (1965) *Life Sciences*, **4**, 2141
Hagen, S. von and L. Hurwitz (1967) *Amer J. Physiol.*, **213**, 579
Hattingberg, M. von, G. Kuschinsky and K. M. Rahn (1966) *Arch. Exptl. Path. Pharmak.*, **253**, 438
Headings, V. E., P. A. Rondell and D. F. Bohr (1960) *Amer. J. Physiol.*, **199**, 783
Holman, M. E. (1958) *J. Physiol. (London)*, **141**, 464
Huxley, A. F. (1960) In C. L. Comar and F. Bronner (Eds.), *Mineral Metabolism*, Vol. 1, p. 163. Academic Press, New York.
Huys, J. (1960) *Arch. Intern. Physiol. Biochem.*, **68**, 445
Jenkinson, D. H. and I. K. M. Morton (1967) *J. Physiol. (London)*, **188**, 373
Kao, C. Y. (1961) *Amer. J. Physiol.*, **201**, 717
Karaki, H., M. Ikeda and N. Urakawa (1967) *Jap. J. Pharmacol.*, **17**, 496
Karaki, H., N, Urakawa and M. Ikeda (1966) *Jap. J. Pharmacol.*, **16**, 423
Keatinge, W. R. (1968a) *J. Physiol. (London)*, **194**, 169
Keatinge, W. R. (1968b) *J. Physiol. (London)*, **194**, 183
Klinge, F. W. (1951) *Amer. J. Physiol.*, **164**, 284
Law, R. O. and C. F. Phelps (1966) *J. Physiol. (London)*, **186**, 547

Lüllmann, H. (1970) In E. Bülbring (Ed.), *Smooth Muscle*, Arnolds Ltd., London. p. 151
Manery, J. F. and A. B. Hastings (1939) *J. Biol. Chem.*, **127**, 657
Matthews, E. K. and M. C. Sutter (1967) *Canad. J. Physiol. Pharmacol.*, **45**, 509
Meigs, E. B. and L. A. Ryan (1912) *J. Biol Chem.*, **11**, 401
McGill, C. (1909) *Amer. J. Anatomy*, **9**, 493
Melton, C. E. (1956) *Endocrinology*, **58**, 139
Nagasawa, J. (1963) *Tohoku J. Exp. Med.*, **81**, 222
Nagasawa, J. (1965) *Tohoku J. Exp. Med.*, **85**, 72
Nash, C. W., E. V. Luchka and K. H. Jhamandas (1966) *Canad. J. Physiol. Pharm.*, **44**, 147
Needham, D. M. (1962) *Physiol. Rev.*, **42**, Suppl. 5, 88
Needham, D. M. (1964) In E. Bülbring (Ed.), *Pharmacology of Smooth Muscle*, Vol. 6, Pergamon Press, Oxford, p. 87
Needham, D. M and J. M. Williams (1959) *Biochem. J.*, **73**, 171
Needham, D. M. and J. M. Williams (1963a) *Biochem. J.*, **89**, 534
Needham, D. M. and J. M. Williams (1963b) *Biochem. J.*, **89**, 552
Needham, D. M. and C. F. Shoenberg (1967) In R. M. Wynn (Ed.), *Cellular Biology of the Uterus*, Meredith Publishing Company, New York. p. 291
Nilwarangkur, S. and G. M. Berlyne (1965) *Nature*, **208**, 77
Ogston, A. G. (1958) *Trans. Faraday Society*, **54**, 1754
Ogston, A. G. and C. F. Phelps (1961) *Biochem. J.*, **78**, 827
Paton, W. M. D. and A. M. Rothschild (1965) *Brit. J. Pharmacol.*, **24**, 437
Prosser, C. L. (1962) *Physiol. Rev.*, **42**, Suppl. 5, 193
Prosser, C. L., G. Burnstock and J. Kahn (1960) *Amer. J. Physiol.*, **199**, 545
Prosser, C. L., C. E. Smith and C. E. Melton (1955) *Amer. J. Physiol.*, **181**, 651
Prosser, C. L. and N. Sperelakis (1956) *Amer. J. Physiol.*, **187**, 536
Rhodin J. A. G. (1962) *Physiol. Rev.*, **42**, Suppl. 5, 48
Saiki, T. (1908) *J. Biol. Chem.*, **4**, 483
Schatzmann, H. J. (1961) *Pflügers Archiv.*, **274**, 295
Schatzmann, H. J. (1964) In E. Bülbring (Ed.), *Pharmacology of Smooth Muscle*, Vol. 6, Pergamon Press, Oxford. p. 57
Schoffeniels, E., F. Hagemeijer and G. Rorive (1966) *Arch. Int. Physiol. Bio.*, **74**, 845
Setekleiv, J. (1967) *J. Physiol. (London)*, **188**, 39P
Shoenberg, C. F., J. C. Rüegg, D. M. Needham, R. H. Schirmer and H. Nemetchek-Gansler (1966) *Biochem. Z.*, **345**, 255
Sparrow, M. P. (1969) *J. Physiol. (London)*, **200**, 71P
Sparrow, M. P. (1969) *J. Physiol. (London)*, **205**, 19
Sparrow, M. P. and W. J. Simmonds (1965) *Biochim. Biophys. Acta*, **109**, 503
Stephenson, E. W. (1967) *J. Gen. Physiol.*, **50**, 1517
Tobian, L. and G. Chesley (1966) *Proc. Soc. Exptl. Biol. Med.*, **121**, 340
Tobian, L. and A. Fox (1956) *J. Clin. Invest.*, **35**, 297
Urakawa, N. and W. C. Holland (1964) *Amer. J. Physiol.*, **207**, 873
Villamil, M. F., V. Rettori, L. Barajas and C. R. Kleeman (1968) *Amer. J. Physiol.*, **214**, 1104
Weatherall, M. (1962) *Proc. Roy. Soc., London, Ser. B*, **156**, 57
Wein, A. J., J. H. Perkins, T. Rohner and H. W. Schoenberg (1967) *Investigative Urology*, **4**, 608
Woodbury, J. W. (1960) In T. Ruch and J. F. Fulton (Eds.), *Medical Physiology and Biophysics*, Saunders, Philadelphia, p. 119

CHAPTER 3

Ion movements in heart muscle

Winifred G. Nayler

Baker Medical Research Institute, Melbourne, Australia

'*Thus, from the fibres proceeds all movements of the heart, occurring as a phenomenon of its own*'. (Stensen, 1664).

I. INTRODUCTION

Although the events that link membrane depolarization, the onset of the 'active state' (Hill, 1964) and contraction are as yet incompletely understood (Sandow, 1965; Nayler, 1967a; Langer, 1968), information is rapidly accumulating to support the hypothesis that in cardiac muscle this link between depolarization and contraction reflects an increase in the intracellular concentration of ionized calcium (Ca^{2+}) and the accumulation of these ions at specific sites within each muscle cell. The intracellular concentration of ionized calcium theoretically can be increased in several ways; first, by the movement of Ca^{2+} from the extracellular phase across the cell membrane and hence into the intracellular milieu; secondly, by the release of Ca^{2+} from intracellular binding sites into the myoplasm. Alternatively, an increase in the *effective* intracellular concentration of Ca^{2+} can be achieved by expelling from the cell other ions which compete with Ca^{2+} for receptor sites.

The mechanisms which, according to current concepts, are involved in regulating the intracellular concentration of ionized calcium in cardiac muscle can be more easily and more usefully analysed if they are related to the muscle's ultrastructure. In recent years the application of electron microscopy to the detailed study of muscle structure has, in addition to confirming some of the equivocal observations made by the earlier light microscopists (Retzius, 1890), extended our appreciation of muscle structure, thereby providing a basis on which a correlation between the various aspects of muscle structure and function can be attempted. The following general conclusions relating to the structure of cardiac muscle have been agreed upon by most microscopists.

i) Structurally the myofibrils of cardiac muscle resemble those of skeletal muscle.

ii) The sarcolemma is a complex multilayered structure which forms specialized invaginations into the cell.

iii) A fine membrane-lined series of interconnecting tubules and vesicles are distributed longitudinally throughout the myoplasm and come into close contact with the myofilaments and with the sarcolemmal invaginations.

iv) Cardiac muscle is a cellular aggregate which, because of low resistance pathways between adjacent cells, functions as a syncitium.

v) The intercalated discs represent specialized lateral cell boundaries.

Each cardiac muscle fibre contains many thin, longitudinally orientated myofibrils which, in turn, are composed of even finer myofilaments. The myofilaments are of two types, *thick* (myosin) filaments which are about 100 Å in diameter and 1·5 μ long and *thin* (actin) filaments which are 50–70 Å in diameter and extend for approximately 1 μ on either side of the Z band, as shown in figure 1. Actin and myosin filaments overlap. Six actin filaments are arranged about each myosin filament so that each actin filament is equidistant from a myosin filament. Myosin filaments are separated from one another by a distance of approximately 450 Å. Cross bridges, which are approximately 20–30 Å thick, connect actin and myosin filaments in such a way that a myosin filament is joined to each of its surrounding actin filaments every 400 Å of its length. This means that these cross bridges, which provide the link between actin and myosin, project from the myosin filaments at intervals of 60–70 Å.

The basic unit of muscle contraction is the sarcomere and this is limited on either side by the Z band (figure 1). Near the Z bands the arrangement between the actin filaments alters; they lose their hexagonal array and assume a square-like distribution so that adjacent fibres are only 220 Å apart.

Biochemical studies have shown that myosin is a globulin protein with a molecular weight of approximately 530,000 and molecular dimensions of 1,500 Å×20 Å. Myosin exhibits ATPase activity which is dependent upon

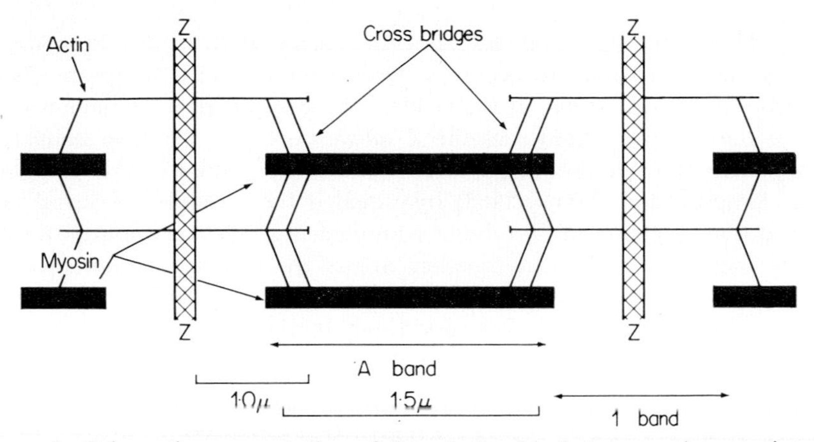

Figure 1. Schematic representation of the relationship between actin, myosin and the Z bands in cardiac muscle.

the presence of —SH groups, activated by Ca^{2+}, inhibited by Mg^{2+} and displays a high degree of specificity for the terminal phosphate group of adenosine triphosphate. Actin can exist in two forms—either as a monomer (G actin), which is globular and has a molecular weight of 62,000, or alternatively, in the presence of ATP and certain ions, as a polymer (F actin), which has a molecular weight of approximately 1·5 million. The myofilaments lie in a substance which is continuous with the sarcolemma, a multilayered structure the thickness of which varies between 300 and 425 Å. It consists of a basement membrane which is 75–200 Å thick and which is separated from a relatively thin plasma membrane (75 Å thick) by a space of approximately 150 Å. Sixty-five per cent of the cell membrane is protein, fifteen per cent lipid and the remaining twenty per cent phospholipid. The relatively high lipid content means that ions which are transported across the membrane must encounter a lipid–aqueous interface and hence, at some stage, must become lipid soluble.

At the level of the Z bands the sarcolemma invaginates and the tubules so-formed (the T tubules) extend deep into the myoplasm. It is now well established for both skeletal muscle (Huxley, 1964; Page, 1964; Franzini-Armstrong and Porter, 1965) and cardiac muscle (Simpson and Oertelis, 1962; Simpson, 1965; Simpson and Rayns, 1968) that the lumen of the T tubules communicate directly with the extracellular space and that the tubules lie in close apposition to the Z discs. If, as seems probable, these T tubules act as a conducting mechanism for the spread of electrical activity from the surface into the interior of the cell, then their close proximity to the myofibrils and to other subcellular structures, including the sarcoplasmic reticulum, is of considerable importance.

In addition to the T tubules another system of fine, membrane-lined tubules has been found to occur in cardiac muscle cells (Simpson, 1965; Simpson and Rayns, 1968) at the level of the Z band. In mammalian heart muscle these tubules, known as the *Z tubules*, are 200–500 Å in diameter; they encircle the myofibril at the level of the Z band and are closely applied to it. Simpson and Rayns (1968) investigated these tubules in detail and showed that neighbouring Z tubules communicate with one another in such a way that they are linked together across the cell by 'communicating tubules, so that they can be considered to represent a system of transversely orientated tubules at the Z level—additional to and more extensive than the T system'. Hence it must be concluded that in cardiac muscle there are at least two closely defined systems of tubules at the level of the Z bands. Theoretically both systems may participate in the rapid spread of an excitatory stimulus throughout individual cells. Since the lateral cell boundaries, represented by the intercalated discs, are known to be pathways of low resistance, it follows that an excitatory stimulus can be spread quickly throughout the muscle.

In addition to those tubules which are located at the Z band (the T and Z tubules) cardiac muscle contains a longitudinally orientated system of fine filamentous tubules (the sarcotubular system). Essentially this consists of a three-dimensional network of fine tubules which extend throughout the cell ramifying and invaginating between the myofibrils and mitochondria, forming a lace-like network around the myofibrils and extending into specialized sac-like protrusions whenever they come into close proximity with invaginations of the plasma membrane. The specialized function of these flattened sarcotubular sacs and their role in the events associated with excitation–contraction coupling and relaxation in cardiac muscle will be discussed in detail in a later section of this chapter.

II. THE TRANSMEMBRANE RESTING POTENTIAL

Quiescent cardiac muscle cells resemble those of skeletal and smooth muscle in that there is a potential difference across the cell membrane. In cardiac muscle this transmembrane potential difference is approximately 90 mV, with the inside negative with respect to the outside, and it is due largely to the relatively high intracellular concentration of K^+ ($[K^+]_i$) (Draper and Weidmann, 1951), compared with that found in the extracellular fluid.

The intracellular concentration of Na^+ ($[Na^+]_i$) in cardiac muscle is only 27–30 mM/kg cell water, and it follows that the ratio between the extra and intracellular Na^+ concentration must be of the order of 5:1. The maintenance of this gradient across the cell membrane requires either that the membrane is impermeable to Na^+ or that sodium ions are extruded from the cell at a relatively rapid rate. Conn and Wood (1959) detected a Na^+ flux of 4 $\mu\mu$ mole/cm^2 sec in *in situ* dog heart perfused at 38 °C, and Langer (1967) found that Na^+ flux in isolated dog papillary muscle perfused with Tyrode solution at 24 °C was 5·9 $\mu\mu$ mole/cm^2 sec. If fluxes of this order are involved in maintaining the resting potential and hence the ratio between intra- and extracellular concentrations of Na^+ and K^+ then it follows that considerable energy is being used to extrude Na^+ from the intracellular milieu into the extracellular space, for the process must proceed against an electrochemical gradient. The evidence available at present for cardiac muscle cells indicates that the energy used in this process is generated, in part at least, via the activity of the Na^+–K^+ activated Mg^{2+}-dependent ATPase enzyme described by Skou (1965). This enzyme, the activity of which is inhibited by Ca^{2+} and by the cardiac glycosides (Page 1962), displays a peculiar asymmetry with respect to its Na^+–K^+ activation, (Glynn, 1962; Whittam, 1962) such that K^+ stimulates its ATPase activity only at the outer surface of the cell membrane whilst Na^+ is effective in stimulating its activity only at the inner surface.

Although it is now firmly established (Page, 1965) that the flux of Na^+ and K^+ in cardiac muscle cells is linked in such a way as to maintain the ratio $\frac{[Na^+]_i \times [K^+]_i}{[Na^+]_e \times [K^+]_e}$ constant, comparatively little is known about the systems which operate to regulate this ratio, the maintenance of which appears to be fundamental to the maintenance of excitability in cardiac muscle cells. We do know, however, that two systems are involved and that only one of these systems uses energy which is derived from the membrane-located ATPase enzyme described by Skou. The other system apparently revolves around an energy-independent exchange diffusion system. Keynes and Swan (1959) found that as much as fifty per cent of Na^+ efflux in frog skeletal muscle was accountable for in terms of this energy-independent exchange diffusion, and Haas and coworkers (1964) reported that approximately one-third of measured Na^+ efflux from isolated frog atria could be accounted for this way. These two systems are shown schematically in figure 2.

The existence in cardiac muscle of this energy-independent exchange diffusion for Na^+ and K^+ makes the use of radioactive tracers to study the ATP-dependent component of ionic exchange of doubtful value in studies of *in vitro* preparations.

III. MEMBRANE DEPOLARIZATION AND THE ACTION POTENTIAL

It is generally agreed that the maintenance of a relatively stable trans-membrane potential difference during the diastolic period of non-pacemaker

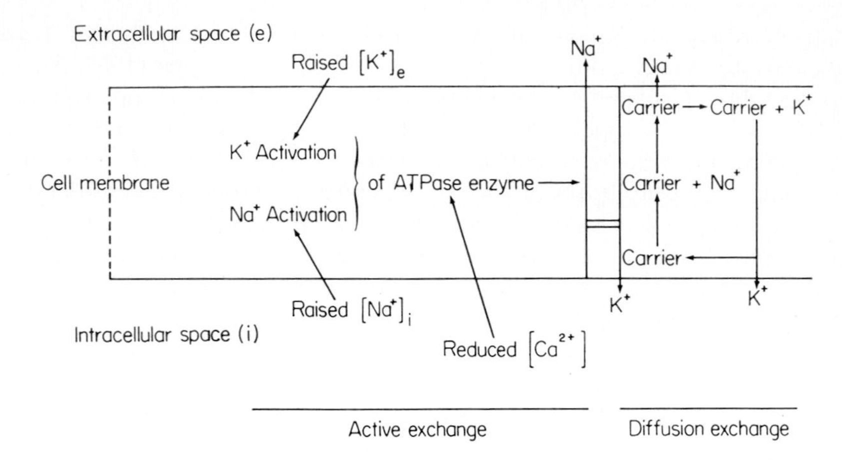

Figure 2. Schematic representation of the active and diffusion exchange of Na^+ and K^+ across cardiac cell membranes involved in maintaining the transmembrane resting potential (After Page, 1965 and Skou, 1965).

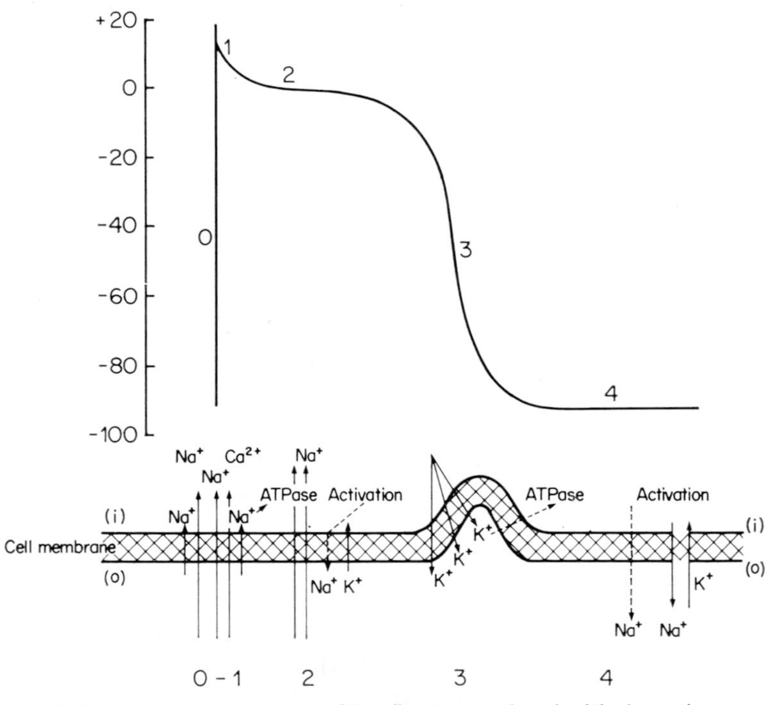

Figure 3. Schematic representation of ion fluxes associated with the action potential in cardiac muscle.

cardiac muscle cells reflects the relatively stable permeability of these cells to K^+ and their ability to expel Na^+ against an electrochemical gradient (Langer, 1968). In cardiac muscle excitation precedes contraction and occurs as is shown in figure 3, when the transmembrane potential difference is either reversed or abolished (Hoffman and Cranefield, 1960).

This change in the transmembrane potential is brought about by a sudden increase in Na^+ permeability so that Na^+ flows down the electrochemical gradient into the muscle cells, an event which is reflected in the spike of the action potential. Many electrophysiological studies (Brady and Woodbury, 1960; Noble, 1960, 1962; Deck and Trautwein, 1964; Langer, 1967) have shown that although the permeability of heart muscle cells to Na^+ is markedly increased during the spike of the action potential, this increased permeability declines during the plateau of the action potential. Noble (1962) calculated that the permeability of Purkinje fibres to Na^+ during the action potential is approximately eight times greater than that which occurs during diastole and that during each action potential the Na^+ influx is increased by as much as 77 $\mu\mu$ mole/cm^2 above the influx which occurs during the resting state. As mentioned above, the increase in Na^+ influx associated with cardiac action potentials can be divided into an initial large and rapid phase, shown schematically in figure 3, and a second phase of increased Na^+ influx which is associated with the plateau of the action potential. The increased intracellular concentration of Na^+ which must result from the rapid increase in Na^+ influx may activate the Na^+–K^+ activated ATPase enzyme and thereby help to facilitate or promote the extrusion of Na^+ from the cell into the extracellular space. There is now general agreement that the phase of rapid repolarization (stage 3 in figure 3) is associated with a marked increase in K^+ efflux. Electrophysiological studies have supported the theory that there is a region at the cell surface in which the diffusion of K^+ is restricted (Freygang and coworkers, 1964) and it seems probable the potassium ions which leave cardiac muscle cells during this phase of the action potential accumulate in the tubules of the T system, or perhaps even within the Z tubules. Such an accumulation of K^+ would theoretically further increase the activity of the membrane ATPase enzyme and hence facilitate the transport of Na^+ from the cell into the extracellular space.

Recently Niedergerke and Orkand (1966) presented evidence to show that in frog heart muscle Ca^{2+}, as well as Na^+ and K^+, contribute to the pattern of ionic fluxes associated with the spike of the action potential. If Ca^{2+} influx is accelerated during the spike of the action potential, then it may well be that these calcium ions serve as a trigger for the release of Ca^{2+} from intracellular sites, including the myofibrils and the sarcoplasmic reticulum, to activate the link between excitation and contraction. Alternatively, it may be those calcium ions which enter the muscle cell during the spike of the action potential which establish the link between excitation and contraction. Weidmann (1955) has shown that the speed of the ascending phase of the action potential is increased if the extracellular concentration of Ca^{2+} is raised. Since the ascending phase of the action potential reflects the rapid translocation of Na^{+} into the cell, it follows that the concentration of Ca^{2+} in the extracellular fluid, and hence the concentration of Ca^{2+} at the cell membrane must, in some way, influence the rate at which this translocation of Na^{+} proceeds.

In general the available data support the view that all three ions, Na^{+} K^{+} and Ca^{2+}, are involved in the genesis of the action potential and hence all contribute to the chain of events that link excitation and the development of the mechanical response in cardiac muscle. There can be little doubt (Langer, 1968) that the transport of Na^{+} and K^{+} across the cell membrane depends to a very large degree on the energy generated by the membrane-located ATPase enzyme. Since the activity of this enzyme is profoundly influenced by Na^{+}, K^{+} and Ca^{2+}, it is relatively easy to visualize a control or feedback system through which the activity of this enzyme and hence the availability of energy for the transport of these ions can be regulated. Although there are many theories to explain the interrelationship between Na^{+} and K^{+}, with respect to the ATP-generating ATPase system (Langer, 1968), the precise mechanisms involved remain obscure. Likewise the mechanisms for transporting Na^{+}, K^{+} and Ca^{2+} through the lipid-containing cell membranes have not yet been defined at a molecular level with any degree of certainty.

Termination of the plateau (phase 3) of the cardiac action potential coincides with an abrupt fall in Na^{+} permeability, so that Na^{+} exchangeability at the cell membrane returns to its resting level. Possibly, as Niedergerke and Orkand (1966b) have suggested, this decrease in the permeability of cardiac cell membranes to Na^{+} may coincide with the displacement of Ca^{2+} from a site within the actin–myosin complex and their subsequent release into either the myoplasm or the extracellular space.

In any case, the length of time for which the plateau of the cardiac action potential is maintained, and hence the duration of the time interval for which the cell membrane is maintained in a partially depolarized state reflects the relative balance which exists at that particular time between the displacement of Na^{+} and K^{+} across the cell membrane.

IV. EXCITATION–CONTRACTION COUPLING

It is already well documented that the link between membrane depolarization and contraction fails in cardiac muscle in the absence of extracellular Ca^{2+}. As early as 1907 Locke and Rosenheim reported that 'if Ca^{2+} are removed from solutions bathing isolated cardiac muscle then the spontaneous action currents remain long after the mechanical beat has become minimal'. Despite this link between the availability of Ca^{2+} and the formation of the link between excitation and contraction (excitation–contraction coupling), quiescent cardiac muscle does accumulate Ca^{2+}, with the rate of accumulation proportional to the extracellular concentration of Ca^{2+} and inversely proportional to the concentration of Na^{+}. Grossman and Furchgott (1964) found that quiescent guinea pig atria immersed in Kreb's solution containing 2·32 mM Ca^{2+} accumulated as much as 0·50±0·04 mM ^{45}Ca/kg wet weight; when the Ca^{2+} concentration in the Kreb's solution was reduced to 1·09 mM then the atria accumulated only 0·22±0·04 mM Ca^{2+}/kg wet weight during the same time interval. Similarly, Niedergerke (1963) found that frog ventricular muscle immersed in Ringer's solution containing 1 mM Ca^{2+} and 110 mM Na^{+} accumulated 0·009 $\mu\mu$ mole $^{45}Ca/cm^2$sec, compared with an uptake of 0·11 $\mu\mu$ mole/cm^2sec if half of the Na^{+} was replaced by sucrose.

The experiments of Heilbrunn and Wiercinski (1947), Niedergerke (1955) and Podolsky and Constantin (1964) have shown that increasing the intracellular concentration of Ca^{2+} in muscle cells above a critical level results in the onset of contraction. It can be argued, therefore, that the rate at which quiescent cardiac muscle cells accumulate Ca^{2+} is too slow to allow the intracellular concentration of Ca^{2+} to reach that level which is required if the process of contraction is to occur. Using a suitably buffered calcium solution Portzehl and coworkers (1964) were able to demonstrate that activation of contraction in intact skeletal muscle fibres requires an intracellular concentration of 0·3–1·4 μM Ca^{2+}. Weber and Herz (1963) and Seidel and Gergely (1963) investigated the dependence of the interaction between freshly extracted actomyosin and ATP on the availability of ionized calcium and found that contraction failed and the myofibrillar ATPase activity was markedly reduced if the concentration of Ca^{2+} fell below 0·1–1·0 μM. It seems reasonable to assume, therefore, that an intracellular concentration of Ca^{2+} of the order of 0·3–1·4 μM should facilitate contraction under *in vivo* conditions. Comparable data is not yet available for cardiac muscle but it is difficult to imagine why the amount of Ca^{2+} required to activate the process of contraction in cardiac muscle should differ markedly from that which is required for skeletal muscle.

If we accept the hypothesis (Weber and coworkers, 1964) that 'actomyosin that contains bound calcium contracts with ATP, whereas it relaxes if part of this calcium has been removed', then it seems probable that, in the presence of ATP, the ability of heart muscle to contract in response to stimulation depends upon the availability of Ca^{2+}. The question arises then, as to whether it is those Ca^{2+} which enter the muscle cell during the spike of the action potential which activate the reaction between actomyosin and ATP, to promote contraction; alternatively those Ca^{2+} which enter the muscle cell during the spike of the action potential may cause other calcium ions to be released from intracellular storage sites and it may be these intracellularly released Ca^{2+} which activate contraction. Evidence has already been presented to show that an active accumulation of Ca^{2+} does occur during the spike (phase 0–1) of the action potential. What, then, is the evidence to support the concept that Ca^{2+} are released from an intracellular store? There is abundant evidence to show that the sarcoplasmic reticulum can and does accumulate Ca^{2+} (Hasselbach, 1964; Lee, 1965a; Langer, 1968) and the mechanisms involved in this process of accumulation will be discussed later. It is more pertinent to establish now whether or not any of the processes that are normally associated with the onset of excitation and of excitation–contraction coupling could cause the release of Ca^{2+} from storage sites within the lateral sacs of the sarcoplasmic reticulum. Two phenomena which normally accompany excitation, that is, membrane depolarization and the influx of certain ions, apparently can displace Ca^{2+} from bound sites.

A. Membrane Depolarization

In 1958 Huxley and Taylor demonstrated for the first time that the local electrical depolarization of frog skeletal muscle in the region of the Z band produces a local contractile response. Since the T tubules represent an extension of the sarcolemma into the myoplasm at the level of the Z band, and in the vicinity of the Z tubules, it follows that the electrical depolarization of the sarcolemma in the vicinity of the Z band could result in the depolarizing current being spread along the T tubules and hence throughout the cell, particularly in the vicinity of the sarcoplasmic reticulum. Studies on isolated fragments of the sarcoplasmic reticulum have shown (Lee, 1965b; Scales and McIntosh, 1968a) that the passage of square wave pulses causes a significant release of Ca^{2+} from a fraction of Ca^{2+} which previously had been bound to the sarcoplasmic reticulum. If, as seems probable, a potential change across the sarcolemma can be transmitted into the vicinity of the sarcoplasmic reticulum, and if that reticulum contains Ca^{2+}, then it is at least possible that a change in potential across the sarcolemma could result directly in the release of Ca^{2+} from the reticulum.

B. Effect of Other Ions

A cation-exchange type of binding has been postulated for fractions of sarcoplasmic reticulum isolated from skeletal muscle. Sampson and Karler (1963) compared the affinity of skeletal muscle sarcoplasmic reticulum for Na^+ and Ca^{2+} and found that if the ratio of Na^+:Ca^{2+} was 49:1, then approximately half of the Ca^{2+} which had been bound to the reticulum was displaced, indicating that Na^+ can compete with Ca^{2+} for 'receptor' sites within the reticulum. Palmer and Posey (1967) used sarcoplasmic reticulum they had isolated from rabbit heart muscle and found that Na^+, and to a lesser extent Li^+, reduced the amount of Ca^{2+} which was bound by the reticulum. Their experiments showed that the reduction in the amount of Ca^{2+} bound to the reticulum under these circumstances was not due to an inhibition of uptake but, instead, to the rapid displacement of a previously bound fraction of Ca^{2+}. According to Palmer and Posey's results 100 mM Na^+ caused the displacement of 2 μ moles Ca^{2+}/g of reticulum protein; 1·0 mM Na^+ caused as much as 1 μ mole Ca^{2+}/g reticulum protein to be displaced. Langer (1968) has extrapolated these results to show that an increase of 1·0 mM Na^+ in the myoplasm in the vicinity of the sarcoplasmic reticulum would be capable of displacing 0·02 μ mole Ca^{2+}/g wet weight of heart muscle. The two essential components of this system, an increase in the intracellular concentration of Na^+ and the close proximity of the membranes through which these Na^+ pass and the sarcoplasmic reticulum from which the Ca^{2+} is released, are common features of cardiac muscle.

Studies to compare the ability of certain cations to displace Ca^{2+} from fractions of sarcoplasmic reticulum isolated from skeletal and cardiac muscle have shown that cardiac muscle reticulum differs from that of skeletal muscle as Zn^{2+} fails to enhance the release of Ca^{2+} from sarcoplasmic reticulum prepared from cardiac muscle (Nayler and Chipperfield, 1969), although some of the heavy divalent cations, including Zn^{2+}, displace Ca^{2+} from the skeletal muscle reticulum (Carvalho, 1968). This difference in the effect of Zn^{2+} on the Ca^{2+} storing capacity of cardiac and skeletal sarcoplasmic reticulum may explain why added Zn^{2+} enhances the contractions of skeletal (Sandow and Isaacson, 1963) but not of cardiac (Ciofalo and Thomas, 1965; Nayler and Anderson, 1965) muscle, because the displacement of Ca^{2+} from the sarcoplasmic reticulum in skeletal muscle would result in more Ca^{2+} being available to activate myofibrillar ATPase activity and contraction.

In summary, therefore, if we accept the proposition that the sarcoplasmic reticulum in intact muscle cells does accumulate Ca^{2+}, the following sequence of events may occur during the process of excitation–contraction coupling in heart muscle: the depolarization of the cell membrane is transmitted along the T tubular system and hence into the vicinity of the sarcoplasmic reticulum. This depolarization, which is accompanied by an influx of Na^+ and Ca^{2+}, spreads throughout the sarcoplasmic reticulum where it, together with the raised intracellular concentration of Na^+ and Ca^{2+}, results in the displacement of Ca^{2+} from the reticulum into the myoplasm in the vicinity of the myofibrils. In the presence of ATP and Ca^{2+} the myofibrillar ATPase enzyme is activated and contraction ensues.

Various experiments in which ^{45}Ca has been used to study the distribution of ionized calcium in cardiac muscle cells (Niedergerke, 1957; Winegrad and Shanes, 1962; Langer and Brady, 1963; Langer, 1964; Shelburne and coworkers, 1967) have indicated the existence of several discrete fractions of exchangeable Ca^{2+}; one of these fractions has a half-life which is too long to be explained in terms of the extracellular space and yet is too fast to be accounted for in terms of a truly cellular compartment. It can, however, be accounted for in terms of an exchangeable fraction of Ca^{2+} localized in the sarcoplasmic reticulum. Langer (1964) defined this phase of Ca^{2+} exchange as phase 2; it has a rate constant of 0·116/min and a half-time of six minutes. The following arguments have been put forward in support of the theory that those calcium ions which affect the link between excitation and contraction are involved in this particular phase of calcium exchange.

i) The rate of decline of the tension developed during individual contractions when Ca^{2+}-free perfusion fluids are used is equal to the wash-out time for phase 2 (Langer, 1965).

ii) The increase in tissue Ca^{2+}, which accompanies an increase in the frequency with which contraction is initiated, is localized in this kinetically defined phase 2 (Langer, 1965).

iii) The increase in tissue Ca^{2+}, which occurs when heart muscle is perfused with Na^{+}-depleted solutions, is localized in phase 2 (Langer, 1964).

iv) Perfusion of dog heart muscle with Ca^{2+}-rich, Na^{+}-poor solutions favours the uptake of Ca^{2+} and results in marked dilation of parts of the sarcoplasmic reticulum, particularly in the vicinity of the lateral sacs (Legato and coworkers, 1968). This is thought to reflect over-saturation of the Ca^{2+}-storing capacity of the reticulum which, in turn, causes this part of the reticulum to imbibe water, and swell.

Despite these arguments in favour of the hypothesis that it is those calcium ions which are released from the sarcoplasmic reticulum which activate the myofibrillar ATPase enzyme and hence facilitate the interaction between actin and myosin, it is possible that calcium ions which have been displaced from cellular sites other than the sarcoplasmic reticulum may also play an important role in activating the link between excitation and contraction. This possibility is particularly applicable to heart muscle in which the sarcoplasmic reticulum is only poorly developed as, for example, in amphibian heart muscle (Fawcett and Selby, 1958; Bergman, 1960; Nayler and Merrillees, 1964; Page, 1964). Under these conditions it can be argued that during excitation calcium ions which have accumulated at superficial sites in each muscle cell, including the sarcolemma, are displaced inwards into the myoplasm in the vicinity of the myofibrils where they activate the link between actin and myosin. Hence a possible sequence of events may be:

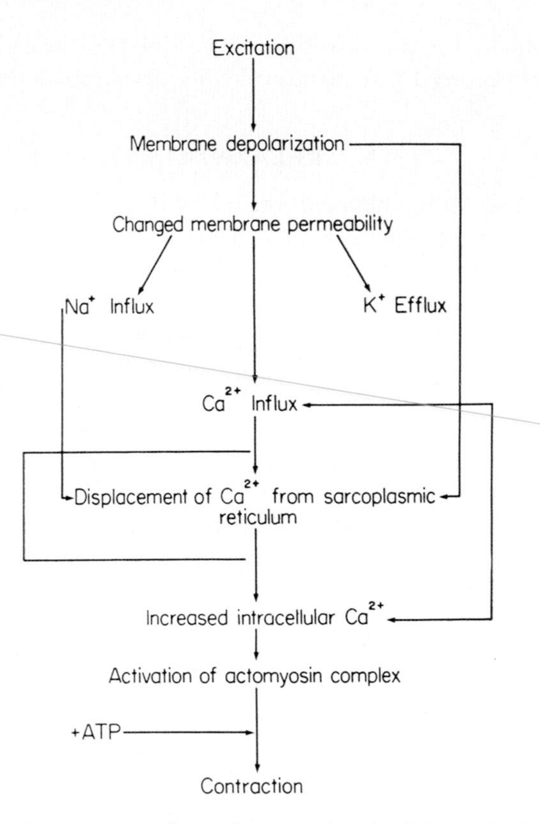

Figure 4. Schematic representation of events involved in excitation–contraction coupling in cardiac muscle.

excitation→membrane depolarization→displacement of Ca^{2+} from superficial binding sites→increased influx of Ca^{2+}→increased intracellular concentration of Ca^{2+}→activation of contraction. The alternative hypothesis, in which Ca^{2+} are released from the sarcoplasmic reticulum to activate contraction is shown schematically in figure 4. In either case competition between Na^+ and Ca^{2+} for receptor sites may involve competition at superficial sites in addition to that occurring in the sarcoplasmic reticulum.

In considering these possibilities the question naturally arises whether there is any correlation between the site from which those Ca^{2+} activating contraction are displaced, (i.e. whether it is a superficially located site, for example at the sarcolemma or an intracellular site, for example at the sarcoplasmic reticulum) and the frequency with which the heart must contract to maintain the circulation. Thus, is the displacement of Ca^{2+} from the sarcolemma and their subsequent diffusion into the vicinity of the myofibrils adequate for the slowly beating amphibian heart but inadequate for the rapidly beating heart of more highly developed animals?

V. RELAXATION

According to the sliding-filament theory of muscle contraction (Huxley, 1957) the development of tension, and hence of shortening, results from the periodic making and breaking of linkages between actin and myosin filaments. These interactions, which are thought to involve the lateral projections of the myosin filaments, require Ca^{2+} and are associated with an increased myofibrillar ATPase activity. Myosin filaments have one bound ATP molecule per lateral projection (Davies, 1963). Sommer and Spach (1963) demonstrated the existence of an ATPase enzyme in these lateral projections. Hence the enzyme, its substrate and its activator (Ca^{2+}) are all in very close proximity to one another and it seems probable that the whole reaction is initiated, or triggered, by a sudden increase in the concentration of Ca^{2+} in the immediate vicinity of the myofibrils. This increase in the intracellular concentration of Ca^{2+} may result either from the displacement of Ca^{2+} from superficially located binding sites at the cell membrane or by their release from the sarcoplasmic reticulum. In either case the amount of Ca^{2+} available for release must be directly proportional to the concentration of Ca^{2+} in the extracellular fluid and inversely proportional to the concentration of Na^{+} (Ringer, 1883; Niedergerke, 1957). The possibility that the reverse process, *relaxation*, reflects the removal of Ca^{2+} from the myofibrils and the accumulation of these Ca^{2+} at other intracellular sites or their translocation to the extracellular fluid has received widespread attention. If, as Weber and Herz (1963) have shown, 1 μmole Ca^{2+} must be removed from each gram of myofibrillar protein to achieve relaxation and if (Langer, 1968) heart muscle contains 50 mg actomyosin/g wet weight, then 0·05 μmole Ca^{2+}/g heart muscle must be removed from the vicinity of the contractile proteins and either bound to other intracellular components or expelled from the cell. Using sucrose gradients to purify fractions of crude cardiac microsomes obtained by ultracentrifugation, Katz and Repke (1967) obtained a microsomal fraction which was capable of accumulating as much as 0·05 μmole Ca^{2+}/g heart muscle. In the presence of oxalate Katz and Repke found that the Ca^{2+}-accumulating activity of these microsomal fractions was increased to 3·6 μmole/g wet weight. More recently, Weber and coworkers (1967) reported that, in the presence of ATP and absence of oxalate, sarcoplasmic reticulum which had been prepared from heart muscle could accumulate as much as 0·25 μmole Ca^{2+}/g heart muscle. The results of these and other studies (Ebashi and Lipmann, 1962; Hasselbach and Makinose, 1963; Carsten, 1964; Fanburg and Gergely, 1965; Lee, 1965) all support the hypothesis that the sarcoplasmic reticulum can accumulate sufficient Ca^{2+} to account for relaxation. Langer (1968) has emphasized that the handling of Ca^{2+} by the sarcotubules of the sarcoplasmic reticulum is not a simple phenomenon. Thus at least three distinct processes appear to be involved—*physical binding*, which proceeds without the addition of either ATP or a precipitating agent, *active transport*, which requires ATP, and *storage*, which requires the presence of a precipitating agent. Which one of these three mechanisms is operating at a particular time is not known but in any case some facility must exist whereby Ca^{2+} can be transported through lipid-containing membranes. Possibly other intracellular structures, including the mitochondria (Patriarca and Carafoli, 1968) may provide other subcellular sites at which Ca^{2+} accumulation can proceed and which therefore contribute to the scheme of events whereby the intracellular concentration is reduced to facilitate relaxation and diastole.

Consideration of the events which, according to our present concepts, are involved in the excitation–contraction–relaxation cycle, shown schematically in figure 5, coupled with the knowledge that in cardiac muscle the tension developed during contraction reflects the availability of Ca^{2+}, leads to the

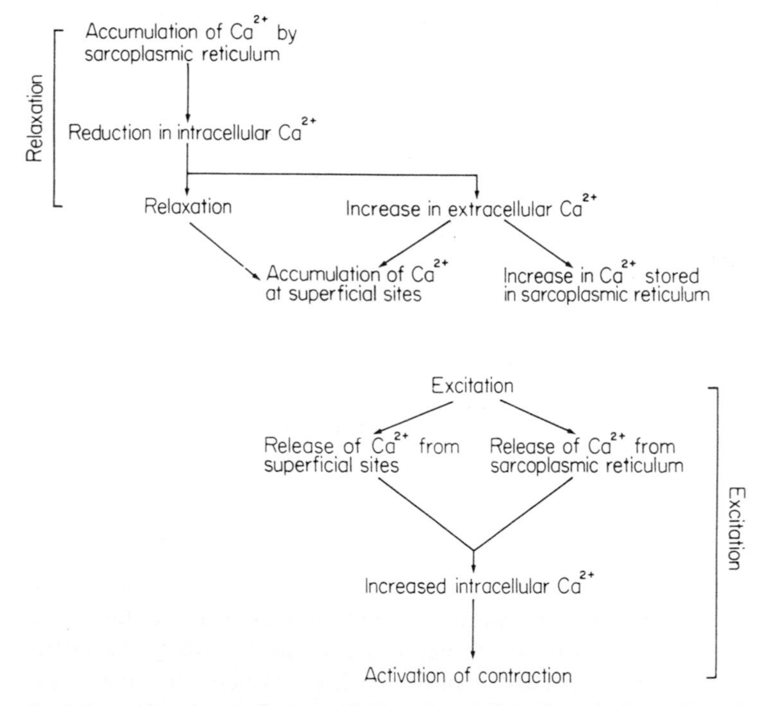

Figure 5. Schematic representation of Ca^{2+} – translocation during relaxation and excitation–contraction coupling.

conclusion that the rate at which tension is developed during an individual contraction is proportioned to the rate at which Ca^{2+} are made available in the vicinity of the myofibrils. The time interval during which contraction is maintained reflects the rapidity with which Ca^{2+} are accumulated from the myofibrils. Although this overall picture of muscle contraction in cardiac muscle is fairly well established, details of the events involved have, as yet, to be elucidated at the molecular level. For example, we need to know what happens to the calcium ions which are accumulated from the myofibrils during the process of relaxation. Are they stored in the sarcoplasmic reticulum so that they can be released into the myoplasm during subsequent excitations; alternatively, are these calcium ions taken up again at sites on superficially located membranes to be released during the next excitatory stimulus; does an equilibrium exist between the amount of Ca^{2+} stored at superficially located sites at the sarcolemma and that stored in the reticulum; and why does the release of Ca^{2+} from its storage sites within the sarcoplasmic reticulum depend upon the intracellular concentration of Ca^{2+} being in excess of a critical level?

VI. REGULATION OF MYOCARDIAL CONTRACTILITY

In the previous sections of this chapter evidence has been presented to support the theory that ionized calcium plays a unique and dominant role in maintaining the ability of cardiac muscle to contract, and that relaxation occurs if the concentration of ionized calcium in the immediate vicinity of the myofibrils falls below a critical level. Depolarization of heart muscle cells was found to be accompanied by an influx of Na^{+}. Since Na^{+} and Ca^{2+} compete for superficially located receptor sites in cardiac muscle cells, it follows that a raised extracellular concentration of Na^{+} will result in these receptor sites being occupied by Na^{+} rather than Ca^{2+}. Hence when depolarization occurs fewer Ca^{2+} will be located at the cell membrane, or in the sarcoplasmic reticulum, so that fewer Ca^{2+} will be displaced into the myoplasm and hence into the vicinity of the myofibrils. The resultant contraction, therefore, will be weak compared with that which would have occurred if more receptor sites had been occupied by Ca^{2+} when depolarization occurred. Similarly the presence of excess Ca^{2+} in the fluid bathing cardiac muscle cells will ensure that more receptors are occupied by Ca^{2+} so that when depolarization occurs an increased amount of Ca^{2+} will be available for displacement into the vicinity of the myofibrils. The resultant mechanical response therefore will be strong. Lüttgau and Niedergerke (1958) have already documented the antagonism which exists between the effect of Ca^{2+} and Na^{+} on myocardial contractility, and the effect of an increased concentration of Ca^{2+} on cardiac contractions has been recognized for more than seventy years. However, the mechanism by which the superficially located sites in cardiac muscle can discriminate between Ca^{2+} and Na^{+} has not yet been defined on a molecular basis.

The level of Na^{+} transport in cardiac muscle cells depends largely on the activity of Skou's membrane-located ATPase enzyme. Since the activity of this enzyme is inhibited by the cardiac glycosides it is not surprising to find that these drugs alter the distribution of ionized calcium in heart muscle cells. Lee and Choi (1966) and Carsten (1967) have reported that these drugs inhibit the ability of the sarcoplasmic reticulum to accumulate Ca^{+2}. If those calcium ions which activate contraction were derived from the cell membrane, rather than from the sarcoplasmic reticulum, the inhibition of uptake of Ca^{2+} by the sarcoplasmic reticulum would result in an increase in the amount of Ca^{2+} available to the myofibrils and hence result in augmented contractions. Gertz and coworkers (1967) reported, however, that ouabain increased the uptake of Ca^{2+} by the sarcoplasmic reticulum and argued that this would result in more Ca^{2+} being available for release from the reticulum during the next excitatory phase. The increased K^{+} efflux associated with the inotropic action of the glycosides must represent a secondary stage and presumably is not associated with the basic mode of action of these drugs. Other drugs, including caffeine (Nayler, 1963a; Nayler and Hasker, 1966) and nicotine (Nayler, 1963b) which have a positive inotropic effect on cardiac muscle act on the Ca^{2+}-exchange system in such a way that the overall exchangeability of Ca^{2+} is increased, so that more Ca^{2+} are available to mediate in the interaction between actin and myosin. Conversely, some drugs with negative inotropic activity, including some with β-adrenergic blocking properties (Nayler, 1966a,b, 1967b) interact with the lipids in cardiac cell membranes (including the sarcoplasmic reticulum) in such a way that the lipid solubility of Ca^{2+} is reduced. Since the uptake of Ca^{2+} into the myoplasm depends upon their transport across a lipid–aqueous interface, it follows that drugs which display this action should, as indeed they do, exert a negative inotropic effect on heart muscle (Nayler, 1967a). Other drugs such as the barbiturates (Briggs and coworkers, 1966) interact with the microsomes to interfere with their ability to transport Ca^{2+}.

The fact that the tension developed by heart muscle is proportional to the frequency with which contraction occurs is well documented (Ringer, 1883; Nayler, 1967a; Langer, 1968) and is reflected in the classical 'staircase phenomenon'. Recent investigations (Winegrad and Shanes, 1962; Niedergerke, 1963) have shown that an increased rate of Ca^{2+} flux (0·5–2·7 μmole/kg beat) is associated with an increased rate of contraction and probably this provides the basis for the augmented contractions. Presumably the increased K^{+} efflux which accompanies the 'staircase' (Hajdu, 1953) reflects a secondary change, possibly in response to the augmented intracellular concentration of

Ca^{2+} which must result from the increased Ca^{2+} flux. Langer (1968) has postulated that the loss of cellular K^+ during the 'staircase' indicates that the Na^+ pump, though increasing in activity, responds too slowly to prevent the occurrence of a small gain in intracellular Na^+. This intracellular Na^+ shift may, he argues, result in more receptors on the membrane being available for Ca^{2+} and the increased Ca^{2+} transport is mediated via these additional receptors.

VII. CONCLUSION

Evidence is now available to support the hypothesis that maintenance of the heart's ability to contract depends upon the regulated influx and efflux of Na^+, K^+ and Ca^{2+}. At the subcellular level, Ca^{2+} and Na^+ compete for receptor sites and the tension developed during each contraction, the rate at which that tension is developed and the length of time for which it is maintained reflects the amount of ionized calcium displaced into the myoplasm in the immediate vicinity of the myofibrils during the process of excitation–contraction coupling, the rate at which this displacement of Ca^{2+} proceeds and the duration of time for which the raised Ca^{2+} concentration is maintained in the vicinity of the myofibrils. The increase in contractility caused by an increase in the frequency with which excitation occurs, the negative inotropic effect of a raised extracellular concentration of Na^+ and the positive and negative inotropic action of a variety of drugs can all be explained in terms of the relevant changes in the intracellular concentration of Ca^{2+}.

Since the influx and efflux of these ions demands their translocation across lipid-containing membranes it follows that these ions must be rendered lipid-soluble. As yet little is known about the process by which this is accomplished. Now that the significance of the role played by Ca^{2+}, Na^+ and K^+ in maintaining and determining excitability and contractility in cardiac muscle is firmly established, additional data is needed to establish how these systems operate at a molecular level and to elucidate the control mechanisms involved.

Acknowledgement

The work referred to in this chapter was supported by grants-in-aid from the Life Insurance Medical Research Fund of Australia and New Zealand and the National Heart Foundation of Australia.

I am deeply indebted to the Director of our Institute, Dr. T. E. Lowe, for his continued encouragement and guidance.

REFERENCES

Bergman, R. A. (1960) *Bull. Johns Hopkins Hosp.*, **106**, 46
Brady, A. J. and J. W. Woodbury (1960) *J. Physiol. (London)*, **154**, 385
Briggs, F. N., E. W. Gertz and M. L. Hess (1966) *Biochem. Z.*, **345**, 122
Carsten, M. E. (1964) *Proc. Natl. Acad. Sci. U.S.*, **52**, 1456
Carsten, M. E. (1967) *Circulation Res.*, **20**, 599
Carvalho, A. P. (1968) *J. Gen. Physiol.*, **51**, 427
Ciofalo, F. R. and L. J. Thomas (1965) *J. Gen. Physiol.*, **48**, 825
Conn, H. L. Jr., and J. C. Wood (1959) *Amer. J. Physiol.*, **197**, 631
Davies, R. E. (1963) *Nature*, **199**, 1068
Deck, K. A. and W. Trautwein (1964) *Arch. Ges. Physiol.*, **280**, 631
Draper, M. H. and S. Weidmann (1951) *J. Physiol. (London)*, **155**, 74
Ebashi, S. and F. Lipmann (1962) *J. Cell. Biol.*, **14**, 389
Fanburg, B. and J. Gergely (1965) *J. Biol. Chem.*, **240**, 2721
Fawcett, D. W. and C. C. Selby (1958) *J. Biophys. Biochem. Cytol.*, **4**, 63
Franzini-Armstrong, C. and K. R. Porter (1964) *Nature*, **202**, 355
Freygang W. H. Jr., D. A. Goldstein, D. C. Hellam and L. D. Peachey (1964) *J. Gen. Physiol.*, **48**, 235
Gertz, E. W., M. L. Hess, R. F. Lain and F. N. Briggs (1967) *Circulation Res.*, **20**, 477
Glynn, I. M. (1962) *J. Physiol. (London)*, **160**, 18
Grossman, A. and R. F. Furchgott (1964) *J. Pharmacol. Exp. Therap.*, **143**, 120
Haas, H. G., H. G. Glitsch and R. Kern (1964) *Arch. Ges. Physiol.*, **281**, 282
Hajdu, S. (1953) *Amer. J. Physiol.*, **174**, 371
Hasselbach, W. (1964) *Federation Proc.*, **24**, 909
Hasselbach, W. and M. Makinose (1963) *Biochem. Z.*, **339**, 94
Heilbrunn, L. V. and F. J. Wiercinski (1947) *J. Cellular Comp. Physiol.*, **29**, 15
Hill, A. V. (1964) *Proc. Roy. Soc. London, Ser. B.*, **159**, 596
Hoffman, B. F. and P. F. Cranefield (1960) *Electrophysiology of the heart*. McGraw-Hill, New York
Huxley, A. F. (1957) *Progr. Biophys. Biophys. Chem.*, **73**, 255
Huxley, A. F. and R. E. Taylor (1958) *J. Physiol. (London)*, **44**, 425
Huxley, H. E. (1964) *Nature*, **202**, 1067
Katz, A. M. and D. I. Repke (1967) *Circulation Res.*, **21**, 53
Keynes, R. D. and R. C. Swan (1959) *J. Physiol. (London)*, **149**, 591
Langer, G. A. (1964) *Circulation Res.*, **15**, 393
Langer, G. A. (1965) *Circulation Res.*, **17**, 78
Langer, G. A. (1967) *J. Gen. Physiol.*, **50**, 1221
Langer, G. A. (1968) *Physiol. Rev.*, **48**, 708
Langer, G. A. and A. J. Brady (1963) *J. Gen. Physiol.*, **46**, 703
Lee, K. S. (1965a) *Federation Proc.*, **24**, 1432
Lee, K. S. (1965b) *Nature*, **207**, 85
Lee, K.S. and E. J. Choi (1966) *J. Pharmacol. Exp. Therap.*, **153**, 114
Legato, M., J. D. Spiro and G. A. Langer (1968) *J. Cell. Biol.*, **37**, 1
Locke, F. S. and O. Rosenheim (1907) *J. Physiol. (London)*, **36**, 205
Luttgau, H. C. and R. Niedergerke (1958) *J. Physiol. (London)*, **143**, 486
Nayler, W. G. (1963a) *Amer. J. Physiol.*, **204**, 969
Nayler, W. G. (1963b) *Amer. J. Physiol.*, **205**, 890
Nayler, W. G. (1966a) *J. Pharmacol. Exp. Therap.*, **153**, 479
Nayler, W. G. (1966b) *Amer. Heart J.*, **71**, 363
Nayler, W. G. (1967a) *Amer. Heart J.*, **73**, 379
Nayler, W. G. (1967b) *Circulation Res.*, **21** (Suppl. 111), 213
Nayler, W. G. and N. C. R. Merrilees (1964) *J. Cell. Biol.*, **22**, 533
Nayler, W. G. and J. Anderson (1965) *Amer. J. Physiol.*, **209**, 17
Nayler, W. G. and J. R. Hasker (1966) *Amer. J. Physiol.*, **211**, 950
Nayler, W. G. and D. Chipperfield (1969) *Amer. J. Physiol.*, **217**, 609

Niedergerke, R. (1955) *J. Physiol.* (*London*), **128**, 12
Niedergerke, R. (1957) *J. Physiol.* (*London*), **138**, 506
Niedergerke, R. (1963) *J. Physiol.* (*London*), **167**, 551
Niedergerke, R. and R. K. Orkand (1966a) *J. Physiol.* (*London*), **184**, 312
Niedergerke, R. and R. K. Orkand (1966b) *J. Physiol.* (*London*), **184**, 291
Noble, D. (1960) *Nature*, **188**, 495
Noble D. (1962) *J. Physiol.* (*London*), **160**, 317
Page, E. (1962) *J. Gen. Physiol.*, **46**, 189
Page, E. (1965) *J. Gen. Physiol.*, **48**, 949
Page, S. (1964) *J. Physiol.* (*London*), **175**, 10
Palmer, R. F. and V. A. Posey (1967) *J. Gen. Physiol.*, **50**, 2085
Patriarca, P. and E. Carafoli (1968) *J. Cellular Comp. Physiol.*, **72**, 29
Podolsky, R. J. and L. L. Constantin (1964) *Federation Proc.*, **23**, 933
Portzehl, H., P. C. Caldwell and J. C. Ruegg (1964) *Biochim. Biophys. Acta.*, **79**, 581
Retzius, G. (1890) *Biol. Untersuch.*, **1**, 51
Ringer, S. (1883) *J. Physiol.* (*London*), **4**, 29
Sampson, S. R. and R. Karler (1963) *J. Cellular Comp. Physiol.*, **62**, 303
Sandow, A. (1965) *Pharmacol. Rev.*, **17**, 265
Sandow, A. and A. Isaacson (1963) *J. Gen. Physiol.*, **49**, 937
Scales, B. and D. A. D. McIntosh (1968a) *J. Pharmacol. Exp. Therap.*, **160**, 249
Scales, B. and D. A. D. McIntosh (1968b) *J. Pharmacol. Exp. Therap.*, **160**, 261
Seidel, J. C. and J. Gergely (1963) *J. Biol. Chem.*, **239**, 3331
Shelburne, J. C., S. D. Serena and G. A. Langer (1967) *Amer. J. Physiol.*, **213**, 1115
Simpson, O. (1965) *Amer. J. Anat.*, **117**, 1
Simpson, F. O. and S. J. Oertelis (1962) *J. Cell. Biol.*, **12**, 91
Simpson, F. O. and O. Rayns (1968) *Amer. J. Anat.*, **122**, 193
Skou, J. C. (1965) *Physiol. Rev.*, **45**, 596
Sommer, J. R. and M. Spach (1963) *Federation Proc.*, **22**, 195
Stensen, N. (1664) *De musculis et glandulis observationum specimen cum epistolis duabus anatomicis*. Amstelodanum, O.P. Le Grand, p. 90 (Translated by Maurice N. Walsh, Mayo Clinic)
Weber, A. and R. Herz (1963) *J. Biol. Chem.*, **238**, 599
Weber, A., R. Herz and I. Reiss (1964) *Federation Proc.*, **23**, 896
Weber, A., R. Herz and I. Reiss (1967) *Biochim. Biophys. Acta.*, **131**, 188
Weidmann, S. (1955) *J. Physiol.* (*London*), **127**, 213
Whittam, R. (1962) *Biochem. J.*, **84**, 110
Winegrad, S. and A. M. Shanes (1962) *J. Gen. Physiol.*, **45**, 371

CHAPTER 4

Ion movements in nerve

William P. Hurlbut
The Rockefeller University
New York, New York 10021, U.S.A.

I. INTRODUCTION

Most living cells maintain across their surface membrane a difference in electrical potential of 50 to 100 mV and large concentration gradients of small organic and inorganic ions. Nerve fibers characteristically contain high concentrations of potassium ions and organic acid anions and low concentrations of sodium and chloride ions, as compared with the blood. The ionic distribution found in the squid giant axon is illustrated in table 1.

Table 1. Major ionic constituents of blood and axoplasm of axons of the squid, *Loligo pealeii*

	Concentration (mM)					
Tissue	Sodium	Potassium	Chloride	Isethionate	Aspartate	Glutamate
Blood	351	17	469	0·47	0·50	0·65
Axoplasm	69	360	157	230	68	11

(From Manery, 1939; Koechlin, 1955; Deffner, 1961.)

Besides having these attributes that are common to most cells, the nerve axon is specialized so that when stimulated with an electric current of sufficient intensity, the membrane potential briefly reverses its polarity and then quickly returns to its resting value. This sequence of changes in membrane potential is completed in about one millisecond, and this transient electrical disturbance, called the action potential, propagates rapidly and without decrement along the length of the axon. The major physiological role of mature axons is to conduct these action potentials and thereby to provide an organism with an electrical network for the rapid transmission of information throughout its body. One goal of research on nerve fibers is to provide answers, in molecular terms, to the questions: 'how are these electrical potentials generated?' and, 'how are the ionic concentration gradients maintained?' Each of these questions is but one aspect of the more general problem of how ions are transported across cell membranes.

It is useful at the outset to recognize two distinct classes of transport phenomena in membranes:

i) passive transport, due to the diffusion of ions down their concentration gradients, that occurs spontaneously across any permeable membrane that separates two solutions of different ionic composition, and

ii) energy-dependent transport that counteracts the dissipative effects of diffusion by transporting ions against their concentration gradients. This type of transport will be called active transport.

Both active and passive transport occur in nerve membranes. The resting and action potentials are generated by ions diffusing passively between the axoplasm and the extracellular fluid (Hodgkin, 1958), and the very fact that the fibers can maintain large concentration gradients in the face of this continuous diffusion implies that some active transport must occur in the membrane to recoup the losses produced by diffusion and to establish and maintain the concentration gradients. In order to study ion transport in nerve one must develop some operational criteria for distinguishing between the two classes of transport processes, and for determining the relative contributions of each process to the transport of a given ion. Although there are no entirely unambiguous procedures for doing this, some general guidelines can be obtained if one uses the diffusion process, as it occurs in dilute aqueous solution, as a model for the diffusion process in membranes.

II. ELEMENTARY THEORY

The ions of a dilute aqueous solution are in incessant thermal motion and are continually colliding with molecules of the solvent. Each individual ion performs a chaotic, random, Brownian motion that is independent of the motion of all other ions and influenced only by the local electric field. This random, unpredictable movement of the individual ions gives rise to a migration of the population of ions as a whole that is regular and predictable and that proceeds as though each ion were driven by an external force and were moving with a uniform velocity down the concentration gradient. This spontaneous, uniform drift of the population of ions down the concentration gradient is what is recognized as diffusion (Jacobs, 1935). For ions in aqueous solution the driving force per mole of ions is the negative gradient of the electrochemical potential, and the mobility coefficient is the proportionality factor relating the mean drift velocity of the population of ions to this driving force. The value of the mobility coefficient is determined by the diameter of the ion and by the viscosity of the solution. The formal expression relating the ionic flux to the driving force is Fick's law which, for dilute solutions, may be written (Teorell, 1953)

$$\frac{dn_j}{dt} = -A\,C_j\,U_j\frac{d}{dx}\left[RT\ln(C_j)+z_j\,F\Psi\right] \tag{1}$$

where

$\frac{dn_j}{dt}$ = the net flux of ions of species j
A = area of the surface across which diffusion is occurring
C_j = concentration of ions of species j at point x
U_j = mobility coefficient of ion j
x = distance in the direction of diffusion
z_j = valence of the ion
Ψ = electric potential at point x
R = gas constant
F = Faraday constant
T = absolute temperature
and the term in brackets is the electrochemical potential.

In the light of this equation, the primary and defining characteristics of passive transport are that the net movement of an ion occurs only down a gradient of electrochemical potential, and that the electrochemical potential of a given ion is the sole source of energy for the transport of that ion. Some subsidiary characteristics of passive transport, that occur only in the simplest circumstances, are that the diffusing ion moves independently of, and does not interact directly with, other ions or molecules, and that the flux of the diffusing ions is directly proportional to the concentration gradient over a wide range of concentrations.

In contrast to passive transport, active transport is considerably more complex. It requires an input of energy, and the source of energy is usually unknown. In addition, the existence of active transport implies that the transported ions do not move across the membrane freely and independently as they do in simple diffusion, but interact with other substances, or with elements of the membrane during their passage through. Hence, some general criteria for active transport are the net movement of an ion up a gradient of electrochemical potential, a dependence of the rate of transport on cellular metabolism, a sensitivity of the transport process to specific poisons or other agents, or the dependence of the transport of one substance on the transport of other substances. A successful analysis of an active transport process consists in identifying the energy source for the transport, determining which substances interact with the ion during the transport process, and determining the molecular nature of these interactions.

In principle, it should be fairly simple to account for the passive transport properties of cell membranes since the energy source and the basic mechanism of this transport processes are assumed to be known. The membrane contributes no energy to the transport and acts only as a valve regulating the flow of material across it. What is needed are ways of expressing and measuring these passive regulatory properties and a conceptual framework for interpreting the measurements. The passive properties of the membrane are expressed quantitatively in terms of permeability coefficients, and elementary diffusion theory provides the conceptual framework for defining and measuring these coefficients and for interpreting them in molecular terms. The risk inherent in this approach is that the diffusion process within the confines of the membrane may be quite different from the diffusion process in dilute aqueous solution, which is serving as our conceptual model. Hence any interpretations cannot be regarded as definitive but must be treated as crude first approximations.

Obviously, the permeability properties of a membrane are strongly influenced by its structure and composition. The solid structural matrix may be impervious to all substances so that the membrane behaves as though only a fraction of its geometrical area is available for diffusion. The diffusion paths through the membrane may not follow straight lines perpendicular to the plane of the membrane, but may wind tortuously among the structural elements. The concentration of ions within the membrane may differ appreciably from the concentrations in the bathing solutions. Some ions may be excluded from the interior of the membrane either because they cannot dissolve in the substance of the membrane or because they are repelled by charged groups firmly fixed to the matrix of the membrane. Other ions may be concentrated within the membrane because they are attracted by these same charged groups. During their passage through the membrane, the diffusing ions may interact strongly among themselves or with elements of the membrane. Thus, the mobilities of the ions within the membrane may differ appreciably from their mobilities in aqueous solution. Furthermore, even though the energy for the diffusion of an ion within the membrane may be derived solely from its own electrochemical potential, equation (1) may not express correctly either the functional form of this potential or the relation of the ionic flux to it. Permeability coefficients are global measures of this complex array of effects, and studies of membrane permeability have long been carried out with the hope of learning something about membrane structure and about the diffusion process in the membrane.

In order to obtain an expression that defines a permeability coefficient in terms of elementary diffusion theory, equation (1) must be applied to a real membrane. In general, this is impossible, since any real structure is hopelessly complex. Hence, the real membrane will be represented by a highly simplified model whose properties are that a part of the membrane is completely impermeable to the ion under consideration and that Fick's law applies in the remaining, permeable portion. For such a membrane, the ionic flux per unit area is

$$J_j = -A'_j\, U'_j\, C'_j \frac{\mathrm{d}}{\mathrm{d}x}\left[\mathrm{R}T \ln (C'_j) + z_j\, \mathrm{F}\Psi'_j\right] \tag{2}$$

where J_j is the ionic flux into the cell per unit area of membrane and A'_j is the fraction of the membrane area that is available for the diffusion of ions of species j. The primes are used to emphasize that the various parameters are measured within the membrane, and Ψ' has a subscript to allow for the possibility that different ions permeate the membrane by different pathways.

Equation (2) can be rearranged to give

$$J_j = -\frac{A'_j\, D'_j\, \mathrm{d}/\mathrm{d}x[C'_j \exp (z_j\, \mathrm{F}\Psi'_j/\mathrm{R}T)]}{\exp (z_j\, \mathrm{F}\Psi'_j/\mathrm{R}T)} \tag{3}$$

where $D'_j = U'_j\mathrm{R}T$, is the diffusion coefficient of the ions in the membrane.

If we assume that A'_j, D'_j and J_j are independent of x, equation (3) can be integrated to give

$$J_j = P_j\, [C_{j_\mathrm{o}} - C_{j_\mathrm{i}} \exp (z_j \mathrm{F}\, V/\mathrm{R}T)] \tag{4}$$

where P_j is the permeability coefficient of the membrane for ion j, C_{j_o} is the concentration of the ion in the solution on the outside of the membrane, C_{j_i} is the concentration of the ion in the solution on the inside of the membrane and V is the electrical potential of the solution on the inside of the membrane, the outside being taken as zero.

Equation (4) defines the permeability coefficient in terms of the net flux of an ion, the membrane potential, and the concentration difference across the membrane. In addition, equation (4) specifies the value of the membrane potential at which the passive flux of a particular ion is zero. This value of the membrane potential is called the equilibrium potential for that ion, and is given by

$$V_{ej} = \left[\frac{RT}{z_j F}\right] \ln (C_{j_o}/C_{j_i}) \tag{5}$$

When the membrane potential equals the equilibrium potential for a given ion, no net diffusion of that ion occurs, and the distribution of that ion will remain unchanged indefinitely without any work being done; the ion is at equilibrium. If the membrane potential differs from the equilibrium potential, then unless the membrane is completely impermeable to that ion, the ion will diffuse and its distribution will change at a rate determined in part by the permeability coefficient. The ionic distribution can remain constant under these conditions only if some process, other than simple diffusion, also contributes to the transport of the ion.

The permeability coefficient, P_j, defined by equation (4) is given by

$$P_j = \frac{A'_j D'_j}{\int_0^{l_j} \exp(z_j F \Psi'_j / RT) dx} \tag{6}$$

where l_j is the length of the diffusion path for ion j.

Equation (6) shows that the permeability coefficient of a membrane to a particular ion is determined by the area of the membrane that is available to that ion for diffusion, by the diffusion coefficient of the ion in the membrane, and by the distribution of the electrical potential within the membrane. If the electric field in the membrane is constant and if the potential difference across the membrane is due entirely to this field [this condition should obtain when the solutions on each side of the membrane are isotonic (see MacInnes, 1961, appendix)], then the integral in equation (6) can be readily evaluated and the permeability constant becomes

$$P_j \text{ (constant field)} = P'_j \left[\frac{z_j F V/RT}{\xi_j - 1}\right] \tag{7}$$

$$\text{where } P'_j = \frac{A'_j D'_j \beta_j}{l_j} \tag{8}$$

ξ_j is equal to exp $(z_j F V/RT)$ and β_j is the partition coefficient of the ion, i.e. it is the ratio of the concentration immediately within the membrane to the concentration in the outside solution. For a membrane containing fixed charges, β_j is equal to the Donnan ratio at the membrane interfaces, and this ratio is determined by the density of the fixed charges in the membrane (Teorell, 1953).

P'_j is the permeability coefficient that would be measured if the potential gradient in the membrane were zero, and it depends only on the geometrical properties of the membrane, on the diffusion coefficient of the ion in the membrane, and on the partition coefficient or the concentration of fixed charges in the membrane (Teorell, 1953).

Since ions are electrically charged, a flux of ions can be considered as though it were an electric current whose density is given by the relation

$$I_j = Z_j \, \mathrm{F} \, J_j \tag{9}$$

If different ions diffused across a membrane at completely different rates, then ionic diffusion would generate an electric current that would eventually produce a measurable accumulation of electric charge in the bulk of the solutions on the two sides of the membrane. In practice such a macroscopic accumulation of electric charge does not occur because during ionic diffusion an electric potential develops across the membrane, and this potential assumes precisely that value necessary to insure that the ionic fluxes are exactly balanced so that the sum of the ionic currents is zero. This value of the membrane potential is determined by the condition

$$\Sigma I_j = \Sigma[\mathrm{F} \, z_j \, P_j \, (C_{j_o} - C_{j_i} \xi_j)] = 0 \tag{10}$$

In general, equation (10) cannot be solved explicitly for V, but if we assume constant field conditions within the membrane, and if we assume that only monovalent ions are diffusing, then the value of V is given by (Goldman, 1943; Hodgkin and Katz, 1949a)

$$V = \frac{\mathrm{R}T}{\mathrm{F}} \ln \left[\frac{\Sigma(P'_j \, C_{j_o})^+ + \Sigma(P'_j \, C_{j_i})^-}{\Sigma(P'_j \, C_{j_i})^+ + \Sigma(P'_j \, C_{j_o})^-} \right] \tag{11}$$

where the superscripts + and − indicate that the summations are taken over the cations, or over the anions, respectively.

This equation shows that the membrane potential is determined by those ions to which the membrane is most permeable and which are present in highest concentration. In general, the membrane potential is not equal to the equilibrium potential of any given ion, but approaches that equilibrium potential as the relative permeability of the membrane to that ion becomes progressively greater.

Equation (11) has been used to explain the origin of the resting and action potentials of squid giant axons. The equation accounts satisfactorily for the dependence of the resting membrane potential on the external concentrations of K^+, Na^+, and Cl^- ions if it is assumed that the permeability of the squid nerve to K^+ ions is five to twenty times as great as the permeability to Na^+ or Cl^- ions (Hodgkin and Katz, 1949a). During the action potential the permeability of the nerve membrane to Na^+ increases rapidly and then quickly returns to the resting value. At the peak of the action potential the permeability to Na^+ is much greater than the permeability to K^+ or Cl^-, and the membrane potential approaches the sodium equilibrium potential (Hodgkin and Katz, 1949a).

The permeability coefficients, P_j, can be used to estimate the electrical conductance of the membrane. Rewriting equation (9) in terms of these coefficients we get

$$I_j = z_j \mathrm{F} P'_j \left[\left\{ \frac{z_j \mathrm{F} V/\mathrm{R}T}{(\xi_j - 1)} \right\} (C_{j_o} - C_{j_i}\xi_j) \right] \tag{12}$$

The conductance of the membrane for the ion of species j, may be defined as

$$G_j = -\frac{\partial I_j}{\partial V} \tag{13}$$

If P'_j is assumed to be independent of V, then the ionic conductance is given by

$$G_j = -z_j \mathrm{F} P'_j \frac{\partial[\mathrm{f}(V)]}{\partial V} \tag{14}$$

where $\mathrm{f}(V)$ represents the term in brackets in equation (12). This complete expression is too complex to write out, but it may be found elsewhere (Brinley, 1965).

If the permeability coefficient increases as the nerve is depolarized, then the conductance computed from equation (14) will be less than the true conductance. The permeability of nerve fibers to Na^+ and to K^+ increases, at least transiently, as the nerve is depolarized, so that the conductances computed from equation (14) are not very reliable. Nevertheless, it is useful to make these computations to see if the fluxes can account approximately for the conductance measured by electrical means. If all our assumptions are correct, this latter conductance should equal the sum of the individual ionic conductances computed from equation (14).

III. TRACER FLUXES

Although studies of the dependence of the membrane potential of nerve fibers on the ionic concentrations of the bathing solution provide information about the relative permeabilities of the membrane to the various ions, they provide no information about the absolute values of the permeability coefficients. However, these coefficients can be readily measured with radioisotopes with the aid of equation (4).

Radioactive ions may be added to the bathing solution at concentration $C^*_{j_o}$, and the initial rate at which they enter the axon can be measured. This initial rate of entry of the radioactive ions, $M^*_{j_i}$, is called the tracer influx and is given by equation (4) as

$$M^*_{j_i} = P_j \, C^*_{j_o} \tag{15}$$

The radioactive ions can also be introduced into the axoplasm of a nerve fiber and the fiber can then be washed in a large volume of non-radioactive solution so that the external concentration of radioactive ions is always held near zero. Under these conditions the rate of loss, or efflux, of tracer ions, $M^*_{j_o}$, is

$$M^*_{j_o} = P_j \, C^*_{j_i} \xi_j \tag{16}$$

Thus when ions cross a membrane solely by passive diffusion it is a simple matter to determine permeability coefficients by using radioisotopes. However, in living cells, we believe that diffusion is not the only means by which ions cross the membrane, and it is not always safe to compute permeability coefficients directly from tracer fluxes. Thus, in order to avoid any implications about the number or the nature of the processes that may be involved in the transport of a particular species of ion across a membrane, the rate of transport is measured in terms of influxes or effluxes, which are measures of the total number of ions that are shuttling to-and-fro across the membrane in unit time. The influx of a particular species of ions is the total number of those ions that in unit time cross a unit area of membrane in the direction outside to inside. The efflux is the number crossing simultaneously in the opposite direction. The net flux of an ion, or the rate at which ionic content of the axon is changing, is equal to the difference between its influx and efflux. In terms of fluxes the problem is to determine how many processes contribute to a particular ionic flux, to determine the molecular mechanism of each process, and to determine the contribution of each process to the total flux.

Influxes and effluxes are computed directly from the tracer fluxes under the assumptions that the stable and radioactive isotopes of an ion are chemically identical and that they participate in all processes strictly according to the ratio of their concentrations. Hence the influx of stable ions, M_{j_i}, can be computed from the measured influx of radioactive ions by the equation

$$M_{j_i} = M^*_{j_i} \, C_{j_o}/C^*_{j_o} \tag{17}$$

Similarly, the efflux of stable ions is computed from the efflux of tracer ions by the formula

$$M_{j_o} = M^*_{j_o} \, C_{j_i}/C^*_{j_i} \tag{18}$$

If a single axon is in a steady state with respect to its ionic distribution, ($M_{j_i} = M_{j_o} = M_j$), then the uptake and release of radioactive ions follows an exponential time-course with a time constant, T_j^*, given by (Keynes, 1951)

$$T_j^* = C_{j_i} V / M_j A \tag{19}$$

where A is the area of axon surface and V is the volume of axon.

Thus for axons in a steady state, ionic fluxes can be readily calculated from the time constant of tracer exchange and the intracellular concentration of stable ions.

The permeability coefficient can be readily calculated from the influx or efflux of an ion, if these fluxes are due only to passive diffusion. Thus, combining equations (15) and (17) and equations (16) and (18) we get

$$M_{j_i} = P_j \, C_{j_o} \tag{20}$$

$$M_{j_o} = P_j \, C_{j_i} \xi_j \tag{21}$$

The ratio of these passive fluxes is given by

$$\frac{M_{j_i}}{M_{j_o}} = \frac{C_{j_o}}{C_{j_i} \xi_j} \tag{22}$$

This equation for the ratio of passive fluxes has sometimes been used as a test to help decide whether or not the transport of a particular species of ion is purely passive (Ussing, 1949).

Although it is relatively simple to obtain numerical values for permeability coefficients, the interpretation of the significance of these coefficients is ambiguous. One cannot say from diffusion studies alone which of the parameters in equations (6) or (8) plays the major role in determining the value of the permeability coefficient. The simplest way to interpret the permeability coefficient is to assume that ions cross the membrane by diffusing through special channels. Some of these channels may even be water-filled pores (Koefoed-Johnsen and Ussing, 1953). The selectivity of the permeability properties of the membrane is determined by the selectivity of the channels, and a membrane can be pictured as containing many specialized areas penetrated by channels that are highly selective as to which ions are permitted to pass. The absolute value of the permeability coefficient for a particular ion is determined by the number of channels which permit that ion to enter, by the partition coefficient of the channel for that ion, and by the diffusion coefficient of that ion when it is in the channel. If the channels are assumed to exist in two states, one open or functional, the other closed and impermeable, then changes in permeability to a particular ion could be achieved by varying the number of channels, selective for that ion, that are in the open state.

IV. IONIC MOVEMENTS IN NERVE FIBERS

The ionic movements across the membrane of peripheral nerve fibers may be conveniently grouped into three categories:

i) the brief, intense passive movements that occur during the action potential;
ii) the slow passive diffusion that occurs continuously in the resting axon;
iii) the active transport processes.

The ionic fluxes that flow during the action potential have been described in great detail for the giant axons of the squid (Hodgkin and Huxley, 1952) and for the nodes of myelinated nerve fibers (Dodge, 1963; Dodge and Frankenhauser, 1958, 1959; Frankenhauser and Huxley, 1964). The main results of these penetrating analyses are:

i) The action potential is caused by profound changes in the permeability of the nerve membrane first to Na^+ ions and then to K^+ ions. Large, strictly passive, fluxes of Na^+ and K^+ ions flow across the nerve membrane as a result of these permeability changes. These ionic movements cause significant changes in the axoplasmic concentrations of Na^+ and K^+ if many action potentials are conducted in rapid succession (Keynes and Lewis, 1951).
ii) The permeability coefficients are functions of the membrane potential.
iii) The Na^+ fluxes may be blocked selectively by the poisons, tetrodotoxin and saxitoxin (Nakamura and coworkers, 1965; Hille, 1968); and the potassium fluxes by tetraethylammonium ions (Hille, 1967). Thus, it seems that during an action potential Na^+ and K^+ ions cross the membrane through ion-specific, physically separate channels.
iv) The specificity of these channels is not absolute. In squid axons and nodes of Ranvier the Na^+-selective channel permits K^+ ions to pass but favors Na^+ over K^+ by a ratio of ten or twenty to one (Frankenhauser and Moore, 1963; Chandler and Meves, 1965).
v) The mechanisms of the changes in permeability are unknown, but it has been suggested that the channels are opened or closed by the movement or rearrangement of charged groups within, or at the entrance to the channels (Hodgkin and Huxley, 1952; Goldman, 1964).
vi) The basis of the ionic selectivity of the channels is also unknown, but it has been suggested that before entering the membrane the ions shed part, or all, of their water of hydration and then associate with charged groups that line the channels. The selectivity of the channel is determined by the diameters of the channel and the unhydrated ions and by the strength of the electric field around the charged groups in the channel (Mullins, 1960; Eisenman, 1962).

V. PASSIVE FLUXES IN RESTING GIANT FIBERS

The analysis of ion transport in resting nerve fibres has progressed slowly because several different processes contribute to the flux of any one ion. It has also been difficult to determine the contribution of each process to the total flux because of the ambiguities that surround the interpretation of experiments with radioisotopes. The ideal preparation for studying transport across membranes is a large single cell whose membrane is directly exposed to readily accessible, well-stirred solutions on the inside and outside. Giant nerve fibers of invertebrates, particularly squid, come close to fulfilling these requirements. The fibers are large enough so that chemical analyses can be readily performed on single axons. Relatively pure samples of axoplasm, uncontaminated by extra-cellular material, can be obtained from axons by extruding it like toothpaste from a tube. The diffusion and activity coefficients of the Na^+, K^+ and Cl^- ions in the axoplasm have been measured, and the results show that these ions are in true solution and are not bound to any great extent to other components of the axoplasm (Hodgkin and Keynes, 1953; Hinke, 1961; Keynes, 1963). Radioactive ions, and many other substances, can be injected directly into the axoplasm of these fibers (Hodgkin and Keynes, 1956; Caldwell and coworkers, 1960a) so that one can study ion movements across the membrane of a single axon, uncomplicated by the movements in any other cells. More recently, the interior of living axons have been perfused either directly (Baker and coworkers, 1962a and b; Tasaki and coworkers, 1962) or through porous glass tubes inserted into the axon (Brinley and Mullins, 1967), and the chemical composition of the interior of the axon has been varied virtually at will.

However, giant axons are not perfect preparations, for their surface membranes are not exposed directly to the bathing solution. Almost all axons are closely invested by Schwann cells which almost completely surround individual axons, or small groups of axons. The Schwann cell membrane is separated from the axon membrane by a space a few hundred Ångstroms wide, and the entire axon–Schwann cell complex is covered by an amorphous basement membrane and layers of connective tissue. The nerve fiber communicates with the bathing solution through winding, narrow clefts, about a hundred Ångstroms wide, between the membranes of adjacent Schwann cells or between an adjacent process of a single Schwann cell. Figure 1 taken from the recent review by Bunge (1968) indicates some of the ways in which Schwann cells envelop axons.

The layer of Schwann cells around the giant axons of the squid is about 1 μ thick, and the length of the diffusion path through the clefts between adjacent cells is about 5 μ (Geren and Schmitt, 1954; Villegas and Villegas, 1960). This layer of Schwann cells seriously impedes the diffusion of water

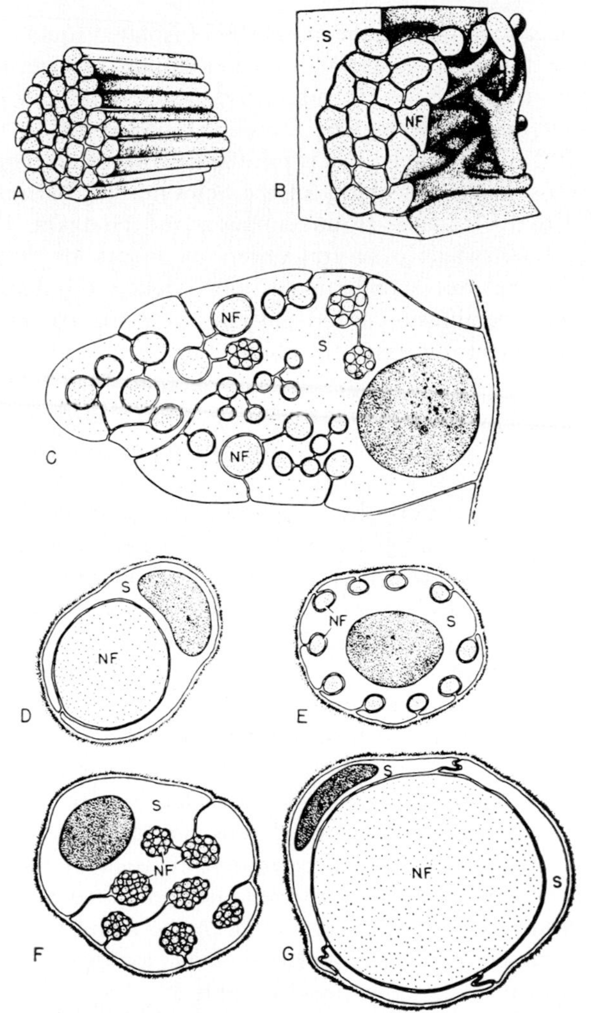

Figure 1. Forms of ensheathment known to occur in relation to nerve fibers are schematically depicted. The glial or Schwann cell contribution is emphasized rather than the connective tissue components. It is known (as in A) that in some forms (e.g. coelenterates) nerve fibers have no sheaths at all; in many instances (as in B) small nerve fibers are ensheathed as a group by some glial element interposed between them and their source of nutrient; in some forms (as in C) a giant glial cell ensheaths many axons singly or as groups (as in the leech); large nerve fibers (as in D) may be surrounded by a single Schwann cell as in insect peripheral nerves; unmyelinated fibers (as in E) may be enclosed in individual troughs of Schwann cell cytoplasm (as in vertebrate peripheral nerve); the very fine fibers of olfactory nerve (as in F) may be enclosed as groups in such troughs; in some invertebrates (as in the giant axon of squid) a large axon may have a mosaic of Schwann cells applied to its surface (as in G). (From Bunge, 1968.)

from the bathing solution into the axoplasm of isolated squid axons, and the permeability of the axon membrane to water cannot be determined directly from the rate of penetration of radioactive water molecules into the axoplasm, without first correcting for the effects of these cells (Villegas and coworkers, 1962). Fortunately, this is not the case for the penetration of ions into these axons. The permeability of the Schwann cell barrier to ions and water is 10^{-4} to 10^{-5} cm/sec (Frankenhauser and Hodgkin, 1956; Villegas and Villegas, 1960) which is several orders of magnitude greater than the permeability of the axon membrane to ions. Hence, this barrier does not seriously impede the diffusion of ions into or out of the axoplasm and tracer fluxes provide direct information about the permeability of the axon membrane to ions. All things considered, the giant fibers of the squid are particularly useful preparations for the study of ion transport and the results obtained with these axons are assumed to be representative of most nerve fibers.

The concentrations of potassium, sodium and chloride in the axoplasm of a variety of giant axons are given in table 2. If the equilibrium potentials of the various ions are compared with the resting membrane potentials of the nerve fibers, it becomes clear that none of the ions is distributed at equilibrium in the axons of squid or cuttlefish. It is also clear that Na^+ is not at equilibrium in either the lobster or crayfish axon, although K^+ and Cl^- are not far from equilibrium in these preparations.

Table 3 lists the rates at which Na^+, K^+ and Cl^- ions exchange across the surfaces of some giant axons. The time constants of the exchanges are long and vary with the ion and with the fiber diameter, but the ion traffic per unit area of membrane is quite similar for all ions and for all axons. These results show that the axon membrane is permeable to each of these ions and it follows that the observed ionic distributions cannot be maintained unless some active transport process is occurring in the membrane. The problem now is to determine the contributions of active and passive transport processes to each of the fluxes mentioned in table 3.

In general potassium and chloride occur in the axoplasm at concentrations greater than, and sodium occurs at concentrations less than, the equilibrium concentrations. One would thus expect that active transport will contribute to the influx of potassium and chloride and to the outflux of sodium, and this expectation is substantiated by the experimental evidence. Thus, the potassium influx and sodium efflux have high temperature coefficients and these fluxes are reduced by about two-thirds when the axons are treated with the cardiac glycoside, ouabain (Caldwell and Keynes, 1959; Brinley and Mullins, 1968), or alternatively when they are exposed for several hours to metabolic inhibitors such as azide, cyanide or dinitrophenol (Hodgkin and Keynes, 1955a and b; Sjodin and Beaugé, 1968). The potassium influx and

Table 2. Concentrations of inorganic ions in freshly dissected giant axons and in typical bathing solutions

Animal	Concentration in Axoplasm[1] (mM)			Concentration in Bathing Solution (mM)			Membrane Potential[1] (mV)	Equilibrium Potential[2] (mV)		
	Potassium	Sodium	Chloride	Potassium	Sodium	Chloride		Potassium	Sodium	Chloride
Squid										
Loligo forbesii	340	49	114	10·4	463	592	−60	−88	+57	−42
Loligo pealeii	360	69	157	10	425	496	−60	−90	+46	−27
Dosidicus gigas	428	85	158	10	420	555		−94	+40	−32
Sepioteuthis sepiodea	310	48	125	10	445	580	−56	−87	+46	−37
Cuttlefish										
Sepia officinalis	352	40		10·4	463	592	−62	−89	+62	
Lobster										
Homarus americanus	270	43	53	10	465	533	−75	−83	+60	−59
Crayfish										
Procambarus clarkii	265	17	12	5·4	207	242	−87	−98	+63	−61

[1]Concentrations were computed, where necessary, by assuming the density of axoplasm to be 1·05 and its water content to be 88 per cent by weight.

[2]Equilibrium potential computed from equation 5 in the text.

(From Keynes and Lewis, 1951; Koechlin, 1955; Deffner, 1961; Keynes, 1963; Brinley, 1965; Villegas and coworkers, 1965; Wallin, 1967.)

Table 3. Ionic fluxes in resting isolated giant axons[1]

	Average Diameter of axons (μ)	Temperature (°C)	Time constant tracer Exchange (hours)			Flux (pmol/cm². sec)					
			Potassium	Sodium	Chloride	Potassium In	Potassium Out	Sodium In	Sodium Out	Chloride In	Chloride Out
Loligo forbesii	800	19	53	7·7	27	18	38	42	35	22	21
			30				51				
Loligo pealeii	550	14	(71–3)	4·5	42	25	(18–440)	40	71	13	11
Sepia officinalis	200	19	14	2·2		13	35	39	22		
Homarus americanus	100	9	14	10	4·2	11	15	7	3·2	9·5	9·3

[1]Outfluxes were calculated using the time constant for tracer exchange and the intracellular concentrations mentioned in table 2.

(From Hodgkin and Keynes, 1955; Caldwell and Keynes, 1960; Caldwell and coworkers, 1960a; Tasaki and coworkers, 1961; Keynes, 1963; Brinley and Mullins, 1965; Sjodin and Beaugé, 1968.)

the sodium efflux seem to be coupled in some manner, since removal of potassium from the external solution reduces sodium efflux by 50–80 per cent in freshly dissected nerve fibers (Hodgkin and Keynes, 1955; Mullins and Brinley, 1967). External Rb^+ and Cs^+ ions can substitute for K^+ ions in maintaining the sodium efflux (Sjodin and Beaugé, 1968), but external Na^+ or Li^+ ions inhibit this effect of potassium (Baker, 1968). About half of the chloride influx may be due to active transport since dinitrophenol reduces chloride influx to this extent (Keynes, 1963). This active transport of chloride seems to be independent of the active transport of sodium or of potassium since it is unaffected by ouabain (Keynes, 1963).

On the other hand, the efflux of potassium and chloride and the influx of sodium seem to be passive fluxes. High concentrations of dinitrophenol are without effect on chloride efflux (Keynes, 1963), but they increase potassium efflux, and slightly reduce sodium influx (Hodgkin and Keynes, 1955). Furthermore, these fluxes of sodium and potassium have small temperature coefficients (Hodgkin and Keynes, 1955a). These results indicate the fluxes require no special energy sources. But even though these fluxes may be passive in an energetic sense, it is not yet clear whether the ions cross the membrane by simple diffusion. Indeed some poorly understood interaction occurs between the passive influx and passive efflux of K^+ ions. When the concentration of potassium is increased around poisoned *Sepia* axons, and the membrane potential is held constant, the potassium efflux is reduced. K^+ ions moving in opposite directions through the membrane seem to interfere with one another, and it has been proposed that the K^+ ions move in a saltatory manner along a chain of fixed sites that line a long, narrow pore through the membrane (Hodgkin and Keynes, 1955b).

In giant axons of *Loligo pealeii* a different interaction between the passive potassium fluxes is observed. When the concentration of K^+ ions is increased around these nerves, and the membrane potential held constant, the potassium efflux is increased, rather than decreased as in *Sepia* axons (Sjodin and Mullins, 1967). This phenomenon, in which an increase in the concentration of an ion on one side of a membrane produces an increase in the flux toward that side, independently of changes in membrane potential, is called exchange diffusion. Its presence implies that the movements of individual ions are not independent, as is the case for free diffusion.

In spite of these uncertainties about the details of the transport process, the efflux of potassium and chloride, and the influx of sodium will be assumed to be due to passive diffusion, and permeability coefficients and conductances will be computed from equations (7), (20), and (21). The results of the calculations are presented in table 4, together with the permeability coefficients for some non-electrolytes.

The permeability to K^+ is five to ten-fold greater than the permeability

Table 4a. Permeability coefficients and conductances of some giant axons to ions[1]

	Permeability coefficient $P'(10^{-8}cm/s)$			Conductance (millimhos/cm^2) Calculated[2]				Measured[3]
	Potassium	Sodium	Chloride	Potassium	Sodium	Chloride	Total	
Sepia officinalis	42	3·1		0·10	0·047			0·11
Loligo forbesii	45	3·5	7·3	0·13	0·03	0·05	0·21	0·5
Loligo pealeii	57	3·6	2·9	0·15	0·05	0·02	0·22	0·7
Homarus americanus	34	0·5	5·6	0·050	0·013	0·026	0·089	0·12

Table 4b. Permeability of giant fibers of squid (*Doryteuthis plei*) to water and some non-electrolytes[4]

Permeability coefficient $P'(10^{-8}$cm/s)	Water	Erythritol	Sucrose
	36,000	36	3–9

[1]Permeability coefficients were computed from potassium and chloride efflux and sodium influx using equations 7, 20, and 21 of the text.

[2]The conductances were calculated using equation 14 of the text.

[3]From Weidmann, 1951; Hodgkin and Huxley, 1952; Cole and Moore, 1960; Brinley, 1965.

[4]From Villegas and coworkers, 1962; Mullins, 1966.

to Na^+ or Cl^- ions, in rough agreement with the conclusions arrived at from measurements of the dependence of the resting potentials on the external concentrations of K^+, Na^+ and Cl^-. The absolute permeability of the nerve membrane to Na^+ or Cl^- ions is comparable to its permeability to sucrose, suggesting that all these substances pass through the membrane by the same, uncharged, non-specific channels. If the length of the diffusion pathway through the membrane is assumed to be equal to the membrane thickness, about 80Å, and if the diffusion coefficients of the ions and molecules within these channels are assumed to be equal to their diffusion coefficients in water (about 10^{-5} cm²/sec for these small ions and molecules), then, from equation (8), the fraction of the membrane area occupied by the pores is

$$A' = \frac{P'l}{D'} = \frac{5\times10^{-8}\times80\times10^{-8}}{10^{-5}} = 4\times10^{-9}$$

Osmotic studies (Villegas and coworkers, 1962; Villegas and coworkers, 1968) indicate that the radius of these non-specific channels is about 4·5 Å. Hence, the number of these non-specific channels per unit area of membrane is

$$N = \frac{A'}{\pi r^2} = \frac{4\times10^{-9}}{\pi(4{\cdot}5\times10^{-8})^2} = 6\times10^5/\text{cm}^2$$

It is possible that most of these non-specific, or leakage, channels are not present in the axon *in situ*, but represent portions of the axon surface that were damaged during the dissections.

If these calculations are repeated for K^+ ions, assuming $\beta_k = 1$, then the number of potassium channels in the resting axons is about $5\times10^6/\text{cm}^2$.

The permeability of the nerve membrane to water is many times greater than its permeability to ions, but it is comparable to the water permeability observed in artificial thin lipid membranes (Cass and Finkelstein, 1967). These membranes, which may be formed from lipid extracts of brain tissue, are about 60 Å thick and are virtually completely impermeable to inorganic ions, and seem to contain no channels. Water may permeate these membranes by dissolving between the lipid molecules and then diffusing across, and water may penetrate the membrane of squid axons by a similar mechanism.

In axons of lobster and *Sepia*, the ionic conductances calculated from the passive fluxes account fairly well for the measured membrane conductance (table 3), but in squid axons the measured conductances are several times greater than the calculated conductances. The discrepancy between the calculated and the measured conductances of axons of *L. pealeii* is largely eliminated if the calculations take into account the variation of the potassium permeability coefficient with membrane potential (Sjodin and Mullins, 1967). In lobster nerves there are about 13×10^8 sodium-specific channels/cm^2 (Moore and coworkers, 1967), and the maximum sodium conductance is about 1,200 millimhos/cm^2 (Hille, 1968). In squid (*L. forbesii*) axons the maximum conductances of the sodium, potassium and leakage channels are 120, 34, and 0·26 millimhos/cm^2, respectively (Hodgkin and Huxley, 1952). Assuming that each of these channels has a conductance equal to that of a sodium channel in lobster nerve, then the number of sodium, potassium and leakage channels per cm^2 of squid axon membrane are: $1{\cdot}3 \times 10^8$, $0{\cdot}3 \times 10^8$ and 3×10^5, respectively. The latter figure for the density of leakage channels compares well with that calculated previously from the permeability of the axon to sodium, chloride and sucrose. However, it is possible that much of the sodium influx into resting fibers occurs through the sodium-specific channels, for this flux would be accounted for if only 0·4 per cent of these channels were open. The high permeability of the resting nerve fiber to K^+ ions would be accounted for if about 17 per cent of the potassium-specific channels remain open in the resting nerve, and the conductance would be accounted for if only 1 per cent of these channels remains open.

Thus, the passive permeability properties of the resting squid axons can be accounted for by a relatively simple model. Most of the membrane is impermeable to ions and water soluble molecules, but permits the passage of water at rates characteristic of thin lipid membranes. The membrane is penetrated by relatively non-selective channels about 4·5 Å in radius that permit the diffusion of small monovalent ions and hydrophilic molecules. There are about half a million of these non-specific channels per square centimeter of membrane. In addition, each square centimeter of membrane contains about 10^8 channels that are specific for K^+ or Na^+ ions. At the normal resting potential virtually all of the sodium channels are closed and do not permit the diffusion of Na^+ ions, but 1–17 per cent of the potassium channels are open so that the membrane is about ten times more permeable to potassium than to other ions. Given the distribution of ions in the resting nerve, these permeability properties appear sufficient to account roughly for the electrical properties of the resting axon.

VI. ACTIVE FLUXES IN GIANT FIBERS

A. General Considerations

The molecular details of the active transport of potassium and sodium are still obscure, but recently preparations have been developed in which it is possible to control the ionic composition of the interior of squid giant axons. Studies of these preparations might shed new light on the molecular basis of active transport. Several years ago, Baker and coworkers (1962a) and Oikawa and coworkers (1961) discovered that the axoplasm of a giant nerve fiber could be eroded away, or squeezed out from the axon without destroying the electrical properties of the membrane. The axon could then be perfused with various solutions, and if the ionic composition of the perfusing solution resembled normal axoplasm, the electrical properties of the nerve fiber remained essentially normal. These preparations permitted additional rigorous tests of the Hodgkin–Huxley hypothesis of the ionic basis of the resting and action potentials, and the experimental results substantiated the main features of this hypothesis (Baker and coworkers, 1962b). It was also found that the intracellular concentrations of Na^+ and K^+ ions were not the only factors that were important to the function of the axon, for the survival time of the perfused preparations was markedly dependent on the anions in the perfusing solution (Tasaki and coworkers, 1965). These perfused axons also transported sodium very poorly, perhaps because they had lost some crucial substance from the axoplasm or from the inside of the membrane (Baker and coworkers, 1968; Canessa-Fischer and coworkers, 1968).

More recently, Brinley and Mullins (1967) and Mullins and Brinley (1967) inserted a thin-walled porous glass capillary down the axon, and then perfused the capillary with various solutions. These axons transported sodium and potassium in the normal manner when they were perfused with solutions that resembled the normal axoplasm. Since substances of molecular weight less than one thousand readily crossed the walls of the glass capillary, the chemical composition of the axoplasm approached that of the perfusing solution after ten to twenty minutes of perfusion. By varying the quantities of ATP, arginine phosphate and other energy-rich phosphate compounds in the perfusing solutions, these workers were able to vary the quantities of high-energy phosphates in the axoplasm and to show quite conclusively that ATP is the energy source for the active transport of sodium (Brinley and Mullins, 1968). This is illustrated in figure 2 which is taken from their work. When the axon was perfused with a solution that contained only the common ions of the axoplasm at their usual levels, and no high-energy phosphates, the sodium efflux declined almost to zero. If ATP was added to the perfusing fluid at a concentration of about 4 mM, the normal concentration in squid axoplasm, then the efflux of sodium was increased to its normal value. ATP and deoxy-ATP were the only substances which supported any substantial sodium efflux. A slight increase in the sodium efflux occurred when an axon was perfused with arginine phosphate or with ADP, but this additional

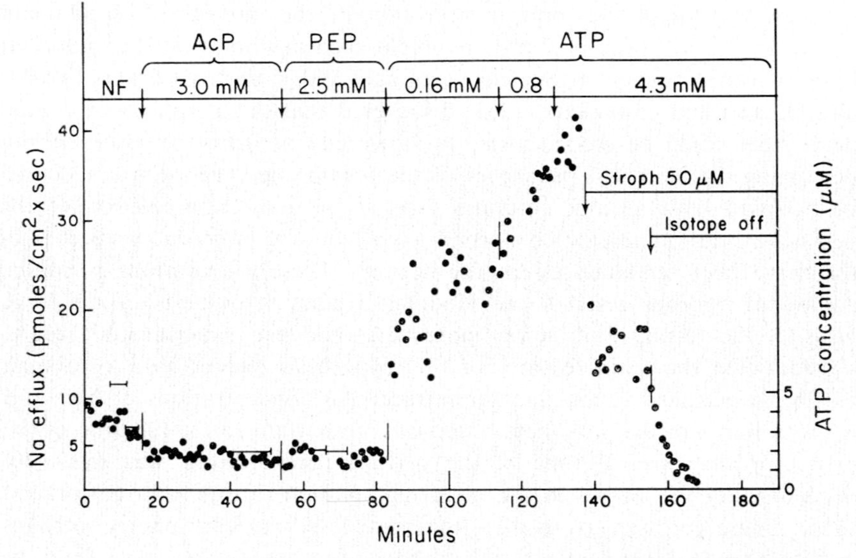

Figure 2. Effects of high energy phosphate compounds and strophanthidin on sodium efflux from perfused giant axon of *L. pealeii*. Ordinates: left and dots, sodium efflux; right and horizontal bars, ATP concentration in the effluent perfusate. Abscissa: time. The axon was in sea water and perfused initially with a solution that contained no high energy phosphates (NF) and the sodium efflux was low and still declining and the internal ATP concentration was a few micromolar. At the first arrow, the perfusing medium was switched to one that contained acetyl phosphate (AcP) and the sodium efflux, and ATP concentration, continued to fall. Next, phospho (enol) pyruvate (PEP) was found to be without effect. Three subsequent switches to solutions with increasing concentrations of ATP produced in each instance a rise in the sodium efflux until normal levels of efflux were attained with 4·3 mM ATP. Then 50 μM strophantidin was applied to the sea water around the axon and the sodium efflux was reduced by about two-thirds. (After Brinley and Mullins, 1968)

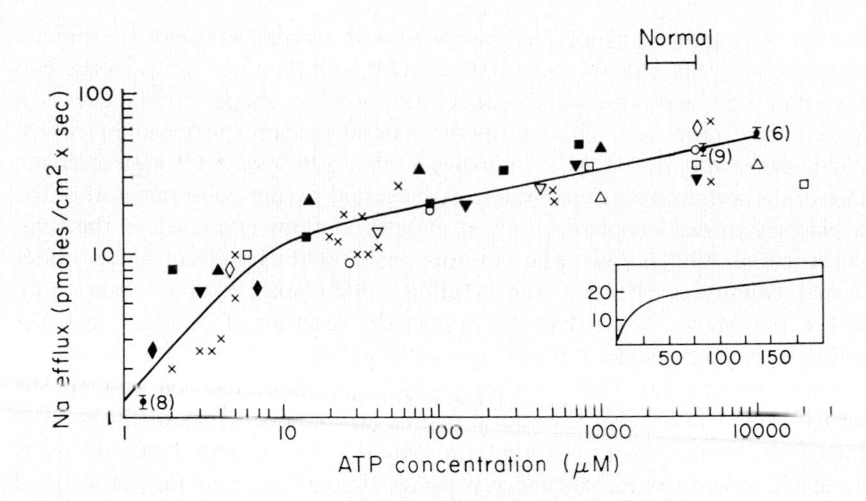

Figure 3. Effect of ATP concentration on the Na efflux from perfused giant axons of *L. pealeii*. Main graph, logarithmic scales—ordinate: sodium efflux, abscissa: ATP concentration in perfusing solution. Inset—initial portions of the main curve shown on expanded linear scales. The three points with vertical bars and numbers in parentheses represent the average results ± S.D. from a total of 20 axons, the crosses represent 18 single values from individual axons, and the other symbols represent values taken from 9 other axons in which more than one concentration of ATP was tested. The data have been normalized to 15°C, assuming a Q_{10} of 2·2 (Mullins and Brinley, 1967). The normal range of ATP concentrations in these fibers is indicated by the horizontal line in the upper right. (After Brinley and Mullins, 1968).

sodium efflux was probably not supported directly by these substances, but by the ATP that was generated from them by the phosphokinases of the axoplasm.

The dependence of the sodium efflux on the internal concentration of ATP is shown in figure 3 (Brinley and Mullins, 1968). The sodium efflux rises rapidly with increasing concentration of ATP over the range 0–10 μM, and then continues to rise slowly but steadily as the ATP concentration is further increased to 10 mM. When the internal ATP concentration is 0·1 mM, the sodium efflux is about half as great as it is when the concentration is 10 mM. Although ATP is hydrolysed by the nerve membrane and provides energy for sodium transport (Baker and Shaw, 1965; Mullins and Brinley, 1967), it is likely that the role of ATP in the transport process is not limited to serving as a fuel. Some of the sodium efflux, as measured with radioactive tracer, may require the presence of ATP without requiring that the ATP be hydrolysed. This is known to occur in red cells (Garrahan and Glynn, 1965). Thus the rate of sodium transport, as measured with tracers, may not be directly proportional to the rate of hydrolysis of ATP, and the ratio between the two rates may vary with the ATP concentration. The shape of the curve in figure 3 may reflect such changes in the state of the transport system as well as changes in the rate of ATP hydrolysis. That a lack of ATP may alter the state of the sodium transport system is suggested by the observation that the cardiac glycoside, strophanthidin, stimulates sodium efflux when the concentration of ATP is low, whereas this agent inhibits sodium efflux under normal conditions (Brinley and Mullins, 1968), and by the observation (Baker and Manil, 1968) that the rate of the reaction of ouabain with the sodium transport system depends upon the metabolic state of the axon.

When the internal ATP concentration in perfused fibers is held at the normal axoplasmic levels, the sodium efflux has the properties of the sodium efflux from normal axons (Brinley and Mullins, 1967). About two-thirds of the efflux is sensitive to cardiac glycosides (figure 2), or to the removal of potassium ions from the external solution (figure 4), and the efflux increases linearly with the internal concentration of sodium over the range 0–200 mM.

The dependence of sodium efflux on external potassium suggests that sodium efflux and potassium influx are coupled in some manner (Hodgkin and Keynes, 1955), and this coupling has been studied fairly extensively. The active movements of sodium and potassium are not tightly coupled in the sense that the ions are always transported in fixed stoichiometric amounts. Although both the active sodium efflux and the active potassium influx increase as the internal concentration of sodium rises, the active sodium efflux increases more rapidly than does the active potassium influx. Thus, when the internal concentration of sodium is 22 mM the ratio of the active sodium efflux to the active potassium influx is about 1:1, and when the internal sodium is 118 mM the ratio of the active fluxes is about 2·5:1 (Sjodin and Beaugé, 1968).

The active potassium influx and the coupling between the active fluxes are more sensitive to metabolic inhibition than is the active sodium efflux. The effect of external potassium on sodium efflux is absent, or is much reduced, in axons that are only partially poisoned with dinitrophenol or cyanide and that still have essentially normal rates of sodium efflux (Caldwell and coworkers, 1960b). Indeed, if cyanide is applied to axons in a potassium-free bathing solution, the sodium efflux first increases from the low value characteristic of the potassium-free solution and approaches values characteristic of unpoisoned axons in normal bathing solution before it starts to decline again (Caldwell and coworkers, 1960b). This increase in sodium efflux that is induced by cyanide or dinitrophenol in the potassium-free solution is blocked by adding ouabain to, or by removing sodium from, the bathing solution (Caldwell and coworkers, 1960b; Baker, 1968). Thus it seems that

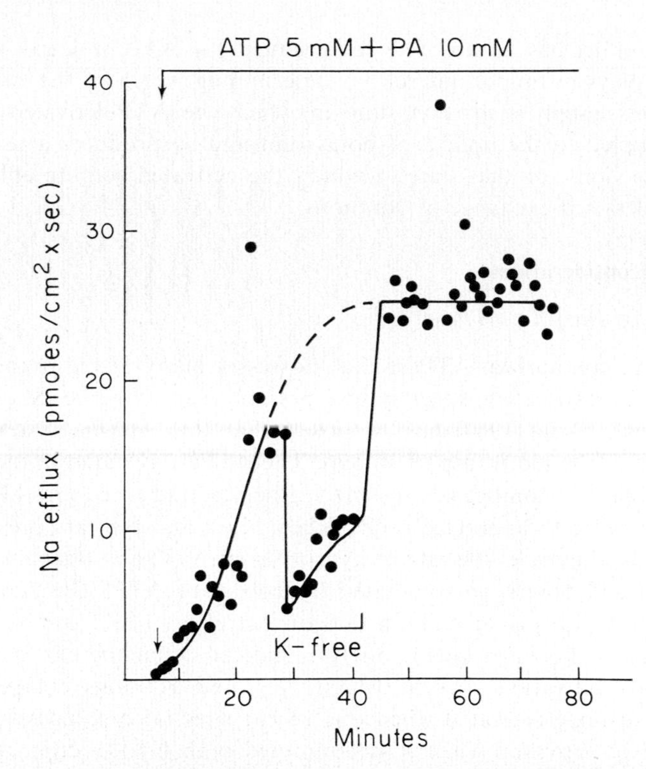

Figure 4. Effect of potassium-free solutions on sodium efflux from a perfused giant axon of *L. pealeii*. Ordinate: sodium efflux. Abscissa: time. Initially, the axon was in artificial sea water with 9 mM potassium and was perfused with a solution that contained no high energy phosphate compounds. The sodium efflux was low. At the arrow the perfusing solution was changed to one that contained 5 mM ATP and 10 mM phosphoarginine and the sodium efflux rose. During the time indicated by the bracket, the solution around the axon was replaced by a potassium-free artificial sea water. The sodium efflux was reduced by about two-thirds in the potassium-free solution and recovered rapidly when potassium was restored to the sea water. The dashed line indicates the assumed time course of the rise in sodium efflux in normal sea water. (After Mullins and Brinley, 1967.)

under these experimental conditions the state of the transport system has been altered from one that mediates an exchange of internal sodium for external potassium to one that mediates a self-exchange of sodium. This latter state of the transport system can also be induced by injecting ADP into normal axons, or by injecting ATP into fully poisoned axons whose sodium efflux has been reduced to very low levels (Caldwell and coworkers, 1960a). These axons that are carrying out this ATP-dependent self-exchange of sodium should all contain abnormally large amounts of ADP. Although the sodium efflux has an absolute requirement for ATP, it seems that the ADP level plays an important role in determining which of the two states the transport system is in; that state in which the ATP-activated sodium efflux is coupled to the uptake of potassium and so produces a useful net movement of ions, or that state in which the activated sodium efflux produces a useless self-exchange of Na^+ ions.

B. Special Considerations

1. *The role of Na^+–K^+ ATPase*

Crab nerve contains an ATPase that possesses many of the properties of the sodium–potassium transport system present in squid nerve (Skou, 1957, 1960). Indeed, Skou (1964) has proposed that this enzyme system is an integral part of the ion transport system. The ATPase is found in the microsomal fraction of a homogenate of crab leg nerves, and it requires Mg^{2+} and Na^+ ions in order to hydrolyze ATP. When Mg^{2+} and Na^+ are present, the addition of K^+ increases the rate of hydrolysis of ATP, and ouabain inhibits this sodium and potassium-stimulated hydrolysis of ATP. The dependence of the rate of hydrolysis of ATP on the concentrations of K^+ and Na^+ ions is complex, and a thorough kinetic analysis indicates that the enzyme system contains two activation sites at which Na^+ and K^+ ions compete. The ATPase is strongly activated when one of the sites is combined with Na^+, and is weakly activated when it is combined with K^+. In crab nerve the affinity of this site for sodium is about six to eight times as great as the affinity for potassium. The second site on the ATPase is activated by potassium and not by sodium, and has an affinity for potassium sixty times that for sodium (Baker and Connelly, 1966). This potassium-activated site is also activated by Li^+, NH_4^+, Rb^+, and Cs^+ ions, but the sodium-activated site is activated only by Na^+. Since the ATPase is recovered in the microsomal fraction of the nerve homogenate, it is presumed to be located in the surface membrane of the intact nerve where it is oriented with the sodium-activated site facing the axoplasm and the potassium-activated site facing the external solution.

Thus those ions that can activate the ATPase (Na^+ ions inside the axons and the other alkali cations outside) are also the ions that can be transported by the axons. The concept underlying these ATPase studies is that some of the microsomal particles are fragments of the surface membrane of the axon that are still capable of transporting sodium and potassium, that the rate of hydrolysis of ATP is a measure of the rate of ion transport in these fragments, and that the need for both potassium and sodium to achieve maximal rates of hydrolysis of ATP is an expression of the coupling between the active transport of these ions.

However, it is not clear whether activation of the ATPase is synonymous with transport, and it is not clear whether the ions that combine with the activating sites of the ATPase are the very ions that are transported. When the crab nerve ATPase is studied at high concentrations of potassium and low concentrations of sodium, conditions that normally prevail in the axoplasm, the rate of hydrolysis of ATP varies with sodium concentration along a sigmoid curve with half-maximal activation occurring at about 40 mM, and saturation occurring at about 100 mM (Skou, 1960). On the other hand, in perfused squid axons both the ATP-dependent, and the strophanthidin-sensitive, sodium efflux vary linearly with internal sodium concentration over the range from 0 to 200 mM (Brinley and Mullins, 1968). If the rate of hydrolysis of ATP by the Na^+–K^+ ATPase were proportional to the total active sodium efflux as measured with tracers, both processes should show the same dependence on the internal sodium concentration. However, it was indicated earlier that the tracer efflux of sodium was not necessarily associated with the hydrolysis of ATP and that the coupling between the active fluxes of sodium and potassium is not rigid and varies with the metabolic state of the axon and with the sodium concentration in the axoplasm. Perhaps in the axon, as in the red cell, the hydrolysis of ATP is associated only with that part of the sodium efflux that is tightly coupled to the uptake of potassium. If the potassium-coupled component of the sodium efflux were a variable fraction of the total sodium efflux, then the rate of hydrolysis of ATP would not be linearly related to the total tracer efflux.

This difference between the dependence of the ATPase and the sodium efflux on the internal sodium concentration may also indicate that more than one active transport process contributes to the sodium efflux. Thus, although virtually all of the sodium efflux from perfused axons requires ATP, only about two-thirds of the efflux is sensitive to strophanthidin or to ouabain and, hence, only two-thirds of the total sodium efflux may be accounted for by the ATPase system. The nature of the cardiac glycoside-insensitive efflux is not known. In axons from *L. pealeii* this stophanthidin resistant efflux does not require ATP, and seems not to be due to changes in the sodium permeability of the axon membrane, nor to represent exchange diffusion of sodium (Brinley and Mullins, 1968). In *L. forbesii* the ouabain-insensitive sodium efflux is inhibited by cyanide (Baker, 1968), suggesting that this efflux requires ATP in these nerves. In these latter axons the ouabain-insensitive sodium efflux increases when the sodium of the external solution is replaced by lithium and under these conditions of low external sodium, the sodium efflux is strongly dependent on the external concentration of calcium (Baker and coworkers, 1967b). Thus, the significance of the glycoside-insensitive sodium efflux is unclear.

Bearing in mind the ambiguities just mentioned, the following simple model can be used to account for some of the major properties of sodium and potassium transport in squid giant axons. About half the internal activation sites of the ATPase are occupied by Na^+ ions and half by K^+ ions, and nearly all the external sites of the ATPase are occupied by K^+. There is a continual splitting of ATP, outward transport of Na^+, and inward transport of K^+ that is just sufficient to balance the passive diffusion of these ions in the opposite directions and to maintain the ionic contents of the nerve fiber in a steady state. When the fiber is stimulated, Na^+ enters the axoplasm and further activates the ATPase so that the rate of hydrolysis of ATP and the rate of ion transport increase. The axons thus extrude the extra Na^+ from the axoplasm and take up K^+ in exchange. The ADP generated by this increased rate of ion transport should ultimately be rephosphorylated in the mitochondria and stimulate the oxygen consumption. Thus the rate of ion transport should determine, in part, the rate of oxygen consumption of the nerve fibers. The rate of respiration will be high when the rate of ion transport is high, and low when the rate of ion transport is low.

2. *Calcium Transport*

Calcium transport has also been extensively studied in squid axons. Although the total concentration of calcium in the axoplasm is 0·4–0·5 mM (Keynes and Lewis, 1956), the concentration of free calcium is only about 0·005 mM (Hodgkin and Keynes, 1957; Luxoro and Yañez, 1968), and the rest appears to be bound in an uncharged complex to proteins of the axoplasm. The axon requires a low intracellular calcium concentration to function properly, and when the calcium level rises to a few millimolar the normal gelatinuous axoplasm begins to liquify (Hodgkin and Katz, 1949b), conduction block occurs (Grundfest and coworkers, 1954; Tasaki and coworkers, 1965), and the Na^+–K^+ ATPase is inhibited (Skou, 1957). The calcium concentration of normal artificial sea water is about 10 mM and, for a membrane potential of minus sixty millivolts, the equilibrium concentration of calcium in the axoplasm is about one molar, or some five orders of magnitude greater than the observed intracellular concentration. From this it follows that calcium must be extruded from the nerve by an active process, but little is known of the nature of the process. The calcium efflux from squid giant fibers is not affected by electrical stimulation or by short exposures to cyanide or to dinitrophenol (Hodgkin and Keynes, 1957; Luxoro and Yañez, 1968), but when the nerves are exposed to cyanide for several hours, the calcium outflux increases (Blaustein and Hodgkin, 1968). This change in calcium efflux is opposite to what one would expect if calcium efflux were mediated by an active process. However, the study of calcium efflux is complicated by the fact that the rate of loss of radioactive ^{45}Ca from intact axons is controlled, not by the rate at which $^{45}Ca^{2+}$ ions cross the axon membrane, but by the rate at which the bound ^{45}Ca dissociates from the axoplasmic proteins. This is indicated by the fact that samples of axoplasm extruded into perforated dialysis tubing lose ^{45}Ca at roughly the same rates as do intact nerve fibers (Luxoro and Yañez, 1968), and by the fact that cyanide increases calcium efflux from these samples of axoplasm as it does from the intact fibers (Blaustein and Hodgkin, 1968). This state of affairs increases considerably the difficulty of studying the movements of calcium across the axon membrane.

On the other hand, the influx of calcium into the nerve fiber seems to be limited by the rate of penetration through the membrane. This is because the rate constant for the association of free calcium within the axoplasm with the binding sites on the axoplasmic proteins, is about an order of magnitude larger than the rate constant for exchange across the membrane (Luxoro and Yañez, 1968). Hence studies of calcium influx may give information about membrane events. Calcium influx is increased by stimulation of the nerve fiber, by increasing the potassium concentration of the external solution (Hodgkin and Keynes, 1957), by reducing the sodium concentration of the external solution, or by increasing the sodium concentration of the axoplasm (Baker and coworkers, 1967a). The calcium influx increases with increasing concentration of external calcium in a non-linear manner and becomes saturated when the external concentration is about 44 mM (Hodgkin and Keynes, 1957). These results suggest that Na^+ and Ca^{2+} ions cross the membrane by a process that involves some kind of competitive ionic exchange for transport sites so that neither the influx nor efflux of Ca^{2+} seems to be due to simple diffusion.

The calcium influx in giant axons under normal conditions is about 0·2 pmol/cm^2. sec, and the permeability coefficient for calcium, calculated from this flux is

$$P'_{Ca} = 0{\cdot}4 \times 10^{-8} \text{ cm/sec},$$

which is about an order of magnitude less than the permeability coefficients of the monovalent ions and the nonelectrolytes. This indicates that Ca^{2+} ions do not cross the axon membrane by the non-specific channels that are available to these other substances.

VII. ION TRANSPORT IN SMALL NERVE FIBERS

Ion transport and metabolic studies have also been carried out on bundles of small nerve fibers. Pure samples of axoplasm are not available from these preparations, so that the intracellular ionic measurements which are available have been estimated from information about the total ionic content of a nerve by correcting for the ions contained in the extracellular space. The results thus obtained give the average ionic content of all the cells of the bundles, including the Schwann cells. They also include unknown quantities of ions that may be bound and not be in true solution. The Schwann cells concentrate potassium and extrude sodium by a transport mechanism that is sensitive to cardiac glycosides (Villegas and coworkers, 1965; Villegas and coworkers, 1968). It is not known how much the Schwann cells contribute to the ionic contents and fluxes measured in nerve bundles, but the cytoplasm of these cells accounts for about 20 per cent of the total cytoplasm in mature myelinated fibers of the rat (Friede and Samorajski, 1968) and in small unmyelinated fibers of crab (Baker, 1965). Table 5 summarizes the results of such analyses for the small, unmyelinated C-fibers of crab leg nerves and rabbit vagus nerve, and for the myelinated fibers of frog sciatic nerve. The membrane potential of the vagus C-fibers has not been measured, and the value listed in this table is not reliable (Keynes and Ritchie, 1965). The membrane potential is unusually small because these fibers seem to have unusually high permeabilities to Na^+ (Armett and Ritchie, 1963) and to Cl^- ions (Rang and Ritchie, 1968b). In general the results obtained for these small nerve fibers are similar to those obtained with giant fibers in that K^+ and Cl^- ions are present at concentrations greater than the equilibrium concentrations and Na^+ is present at a concentration less than the equilibrium concentration.

When one is dealing with a bundle of axons, the interpretation of experiments with radioactive tracers is complicated by the fact that the time course of the tracer exchange seldom follows the single exponential curve expected for an ideal system divided into two compartments by a single membrane. Deviations from the ideal behavior may occur because of differences in the surface to volume ratios of the various cells in the nerve bundle, or because the tissue contains many compartments other than the axoplasm, that take up the tracer. Tracer exchange in the tissue as a whole may not be representative of tracer exchange across the axon membranes. Free access of the tracer to the axon membrane may be restricted by diffusion barriers in the extracellular space or even within the axon. This problem may be particularly acute in myelinated nerve fibers. Each of these fibers is enveloped by a highly impermeable myelin sheath composed of layers of Schwann cell membrane (Geren, 1954) that is interrupted every one or two millimeters at the nodes of Ranvier. At these nodes the axon membrane is exposed for a few microns and the myelin sheath is closely applied to the axon membrane on either side of the node (Robertson, 1959). The nerve membrane gains access to the extracellular space through channels between the interdigitating fingers of Schwann cell cytoplasm that engulf the node. The structure of these nodes is

Table 5. Estimated intracellular concentrations of K^+, Na^+ and Cl^- ions in small unmyelinated fibers and myelinated fibers.

	Internal concentration (μ moles/g cell water)			Concentration in bathing solution (mM)			Membrane potential (mV)	Equilibrium potential (mV)		
	Potassium	Sodium	Chloride	Potassium	Sodium	Chloride		Potassium	Sodium	Chloride
Crab leg (*Maia squinado*)	258	38	49	10	470	550		−82	+63	−61
Rabbit cervical vagus (desheathed)	145	63	40	5·6	154	167	−40	−82	+23	−36
Frog sciatic (desheathed) (*Rana pipiens*)	140	28	26	2	116	115	−71	−106	+34	−37

Concentrations were calculated in terms of intracellular water by assuming 1·13 g cell water/g dry wt for frog nerve, and 0·75 g cell water/g wet nerve for crab nerves, and 0·25 g cell water/g wet wt for desheathed cervical vagus nerves (Rang and Ritchie, 1968c).

(From Huxley and Stämpfli, 1951; Hurlbut, 1958, 1963a; Baker, 1965; Keynes and Ritchie, 1965.)

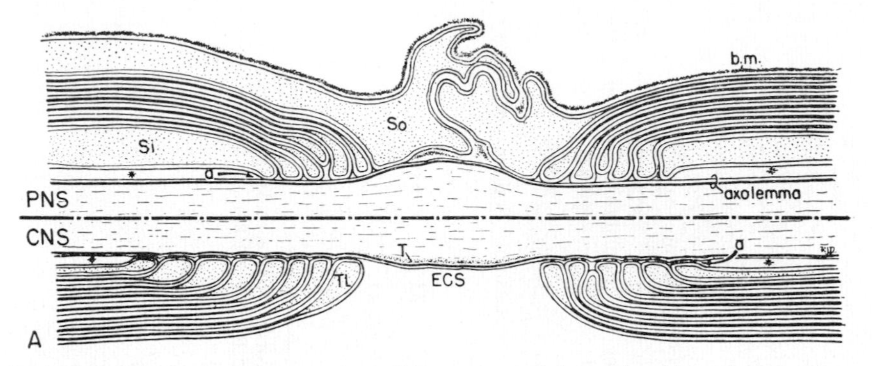

Figure 5. Schematic representation of the structure of a node of Ranvier, showing a nodal region from the peripheral nervous system (PNS) above, and from the central nervous system (CNS) below. In the PNS the Schwann cell provides both an inner collar (S_i) and an outer collar (S_o) of cytoplasm in relation to compact myelin. Outer collar (S_o) is extended into the nodal region as a series of loosely interdigitating processes. Terminating loops of the compact myelin come into close opposition to the axolemma in a region near the node, apparently providing some barrier (arrow at a) for movement of materials into or out of the periaxonal space (marked *). The Schwann cell is covered externally by a basement membrane. In the CNS the myelin ends similarly in terminal loops (tl) near the node and there are periodic thickenings of the axolemma where the glial membrane is applied in the paranodal region. These may serve as diffusion barriers and thus confine the material in the periaxonal space (marked *), so that movement in the direction of the arrow at A would be restrained. At many CNS nodes there is considerable extracellular space (ECS). (From Bunge, 1968.)

shown in figure 5 (Bunge, 1968). The ionic currents that flow during the action potential are restricted to the nodes (Stämpfli, 1954), and it may be that in the resting fibers as well, the ionic exchange between the axoplasm and the extracellular spaces occurs primarily at the nodes. If this be true, the ion traffic here will be quite intense (see below) and the rate of penetration of isotopes into the axons may be limited, not by exchange across the axon membrane at the node, but by the rate of penetration of tracer through the channels between the Schwann cells.

These preparations containing small nerve fibers are interesting because changes in the rates of active transport of sodium and potassium in them produce large changes in the intracellular sodium and potassium concentrations, large changes in the rates of oxygen consumption by the fibers, and large changes in the magnitude of their membrane potentials. Therefore, these preparations are useful for studying the relationship between active ion transport and oxidative metabolism and for studying the effects of ion transport on the membrane potential.

Radioisotopes have been used to study potassium fluxes in the small

Table 6. Estimated sodium and potassium fluxes in small diameter nerve fibers[1]

	Temperature °C	Mean fiber diameter (μ)	Membrane area (cm^2/g cell water)	Time constant tracer exchange (hours)		Resting fluxes[2] (μmoles/g cell water . h)		Resting fluxes[2] (pmol/cm^2 . as sec)	
				Potassium	Sodium	Potassium	Sodium	Potassium	Sodium
Crab leg	16–20	2	$1{\cdot}3 \times 10^4$	2		129	96	2·8	2·1
Rabbit vagus	20–25	0·75	$2{\cdot}4 \times 10^4$	1		132	50	1·5	0·6
Frog sciatic	20	10	5·3[3]	8	1·6	18	18	930	930

[1]The fibers in the nerve bundles are assumed to be in a steady state and that influxes and effluxes are equal.

[2]For crab and rabbit nerves the sodium fluxes were estimated from the changes in ionic content during a one hour immersion period in a solution containing ouabain.

[3]This figure was calculated by assuming there were 10^7 nodes/g dry nerve (Hurlbut, 1965) and that the area of a node was 60 μ^2 (Stämpfli, 1952). The area of the internodes is about 750 times larger.

(From Abbott and coworkers, 1958; Hurlbut, 1963a and b; Baker, 1965; Keynes and Ritchie, 1955.)

unmyelinated fibers of crab and rabbit nerves (Keynes and Ritchie, 1965), and both the sodium and potassium fluxes have been studied in desheathed sciatic nerves from frogs (Hurlbut, 1963a, 1963b), toads (Shanes and Berman, 1955), and cats (Dainty and Krnjevic, 1955). The isotope exchange in vertebrate nerves is taken as being representative of exchange in the myelinated axons. The results of some of these investigations are summarized in table 6. In compiling this table, the sodium fluxes in rabbit and crab nerves were estimated from the rates at which the nerves gained sodium when they were exposed to high concentrations of ouabain. It was assumed that ouabain abolished two-thirds of the sodium efflux and produced no changes in the sodium influx. Since the ionic distributions of these preparations are in a steady state, the influxes and effluxes are assumed to be equal and are not listed separately. The time constant for the exchange of radioactive potassium is considerably shorter in the small unmyelinated fibers than in giant axons, and this is due primarily to the enormous surface to volume ratios of the small fibers, for the fluxes per unit area of membrane are about an order of magnitude smaller in these fibers than in the giant fibers.

The permeabilities and conductances of the small fibers to sodium and to potassium are collected in table 7. When the computations were made for myelinated fibers, the ionic exchange was assumed to occur only at the nodes. The permeabilities of the small unmyelinated fibers are about an order of magnitude less than, and the permeability of the node is about two orders of magnitude greater than the permeabilities of the giant axons. The conductance of the nodal membrane, which was measured on isolated single fibers, is also about two orders of magnitude greater than the conductances of the giant axons. In general, the passive transport properties of nodes,

Table 7. Resting sodium and potassium permeabilities and conductances of various nerves[1]

	Permeability Coefficient, P′ (10^{-8}cm/s)		Conductance (millimho/cm^2) Calculated			Measured[2]
	Potassium	Sodium	Potassium	Sodium	Total	
Crab leg	5·9	0·15	0·0097	0·0024	0·012	
Rabbit vagus	2·7	0·19	0·0051	0·001	0·0061	0·11
Frog sciatic	3600	270	2·7	1·1	3·8	42

[1]Permeability coefficients and conductances were calculated from potassium efflux and sodium influx using the data of tables 5 and 6 and equations 7, 14, 20 and 21 of the text.

[2]From Stämpfli, 1954; Straub, 1963

Table 8. Changes in sodium and potassium content produced by electrical activity in various nerves

	Net Change per Impulse			
	(μ moles/g cell water)		(pmoles/cm^2)	
	Potassium loss	Sodium gain	Potassium loss	Sodium gain
Crab leg	$2{\cdot}3 \times 10^{-2}$	$3{\cdot}2 \times 10^{-2}$	1·8	2·5
Rabbit vagus	$2{\cdot}0 \times 10^{-2}$		0·8	
Frog sciatic	$1{\cdot}4 \times 10^{-4}$	$1{\cdot}6 \times 10^{-4}$	26	30
Giant axon[1] (*Loligo forbesii*)	$1{\cdot}6 \times 10^{-10}$	$1{\cdot}9 \times 10^{-10}$	3·0	3·5

[1]The axon diameter was taken as 800 μ and the surface to volume ratio as 54 cm^2/g cell water.

(From Keynes and Lewis, 1951; Asano and Hurlbut, 1958; Baker, 1965; Keynes and Ritchie, 1965.)

whether measured by electrical means or isotopic tracers, resemble qualitatively the passive properties of giant fibers, but quantitatively, the intensity of the ionic movements at a node is ten to a hundred times greater than in giant fibers. The maximum permeabilities of the fully-opened Na^+-specific, and K^+-specific channels at the nodes are $(4{\cdot}4\text{–}6{\cdot}5) \times 10^{-3}$ cm/sec and $1{\cdot}2 \times 10^{-3}$ cm/sec, respectively (Frankenhauser, 1960, 1962; Dodge, 1963). Hence at the nodes the resting Na^+ and K^+ permeabilities are about 0·05 per cent and 3 per cent, respectively, of these maximum ion specific permeabilities. These ratios of the resting to the peak permeabilities at the nodes are comparable to the ratios of the resting to the peak permeabilities that were estimated previously for giant axons. The ionic shifts per unit area of membrane that occur during an action potential are about ten times greater at a node than in a giant axon (table 8), and this result is consistent with electrical studies which show that during an action potential the current densities at isolated nodes are also about ten-fold greater than those in squid axons (Frankenhauser, 1960).

The conductance of the nodal membrane calculated from the permeability coefficients in table 7 is only about one-tenth the conductance measured electrically in isolated single fibers. The origin of this discrepancy is not known. It may result from the inadequacy of the theoretical treatment, or it may indicate that the rate of tracer exchange in the nerve trunk is not governed by the rate of exchange across the nodal membrane and that the true fluxes across the node are many times larger than those in table 6. It could also be that the conductance of the isolated node is increased above normal by the trauma of the dissection. However, many other results obtained from chemical and isotope studies on nerve trunks compare favorably with the results of electrical studies on isolated fibers. Thus, the sodium equilibrium potential estimated on isolated fibers is about +45 mV (Dodge, 1963) and the intracellular sodium concentration computed from this potential is 19 mM. These figures are in rough agreement with those in table 5. (However, if K^+ ions pass the sodium-selective channel, then the sodium equilibrium potential would be greater than, and the intracellular sodium concentration less than, the values just given). The net sodium gain produced by a single impulse at a single node estimated from chemical analyses of stimulated nerve trunks is $1{\cdot}8 \times 10^{-17}$ moles (Asano and Hurlbut, 1958), and this compares favorably with the net sodium gain of $2{\cdot}4 \times 10^{-17}$ moles estimated from electrical studies in single fibers (Dodge, 1963). It was pointed out above, that on a relative scale at least, the results of tracer and electrical studies done on myelinated fibers are consistent with those based on giant axons. Thus, in some respects, the results of chemical and tracer studies of passive ion movements in bundles of myelinated fibers are in rough agreement with electrical studies carried out on single fibers.

The active transport of sodium and potassium has also been studied fairly extensively in frog nerves and, in general, the results are similar to those obtained from giant fibers. About 75 per cent of the potassium influx and 50–75 per cent of the sodium efflux from nerve seems to be due to active transport, since these fluxes are abolished by azide, dinitrophenol and ouabain, or by oxygen deprivation (Shanes, 1957; Hurlbut, 1963a and b, 1965). The component of the sodium efflux that persists in the presence of these metabolic inhibitors is insensitive to changes in environmental temperature, and therefore is presumably not mediated by an active process. As in the case of sodium efflux from squid axons, the sodium efflux from frog nerve varies linearly with the intracellular sodium concentration over the range 10–100 mM. The potassium influx in frog nerve remains relatively constant as the internal sodium concentration is varied over this range, but the potassium influx falls rapidly to its passive level as the internal sodium concentration is reduced below 10 mM. Thus, as in squid axons, the ratio of the active sodium efflux to the active potassium influx is not fixed, but varies considerably with the level of intracellular sodium. At high levels of intracellular sodium, active sodium efflux may be several times greater than the active potassium influx, and at low levels of intracellular sodium the active potassium influx may exceed the active sodium efflux.

Unlike the situation in squid nerves, the sodium efflux from frog nerves is not reduced when potassium is removed from the bathing solution (Hurlbut, 1963a). Instead, sodium influx increases, the increase in sodium influx being equal to the normal potassium influx. Thus, in frog nerves it appears that the sodium efflux normally is not tightly coupled to the potassium influx. When potassium is absent from the external solution, the ion transport system mediates a self-exchange of sodium. Another possible interpretation of this observation is that because the node is shielded by the Schwann cells, the potassium concentration at the nodal membrane changes very little when potassium is removed from the external solution, so that the sodium efflux can persist unmodified in a nominally potassium-free solution.

It is evident from the data in tables 6 and 8, that because of the great surface to volume ratio of these small fibers, changes in the sodium and potassium fluxes can produce profound changes in the intracellular concentrations of these ions. The contrast between the effects of stimulation on the sodium contents of C-fibers and of giant axons is especially striking (see table 8). In crab nerve, conduction of a hundred action potentials increases the sodium concentration of the axoplasm by 3·2 mM, or about 8 per cent, whereas in giant axons the same activity increases the sodium concentration by $1{\cdot}9 \times 10^{-8}$ mM, about eight orders of magnitude less. The large increase in sodium concentration that accompanies electrical activity in small fibers should stimulate the transport ATPase, so that Na^+ ions are extruded, K^+ ions are taken up, ADP and inorganic phosphate are produced, and the rate of respiration is increased.

Stimulation of crab nerves produces a large ouabain-sensitive increase in the inorganic phosphate content of these nerves (Baker, 1965), and this increase in inorganic phosphate seems to be due to the hydrolysis of ATP by the Na^+–K^+ ATPase. Although little net hydrolysis of ATP or creatine phosphate occurs in stimulated rabbit vagus nerves (Greengard and Straub, 1959) or in frog sciatic nerve (Cheng, 1961), the rate of respiration of these nerves increases when they are stimulated, and the smaller the fibers of the nerve, the greater the increase in the rate of respiration. The rate of respiration of rabbit vagus nerve doubles when the nerve is stimulated at the rate of 3/sec for 5 minutes (Rang and Ritchie, 1968a); the rate of respiration of crab nerve increases by about 50 per cent after conducting as few as 720 impulses (Baker and Connelly, 1966); the rate of respiration of frog nerve increases by about 30 per cent when the A-fibers are stimulated for 30 minutes at a rate of 50/sec (Brink and coworkers, 1952); and squid giant axons increase their rate of respiration by about 10 per cent when stimulated at a rate of 100/sec (Connelly, 1952). There is good evidence that this increase in the rate of respiration is triggered by the increase in the intracellular sodium concentration that accompanies electrical activity in the nerves and that the rate of 'activity respiration' is determined by the rate of hydrolysis of ATP by the Na^+–K^+ ATPase.

Nerves can be stimulated when they are bathed in solutions which contain Li^+ ions in place of the normal Na^+ ions. Under these conditions the nerves conduct action potentials, but cannot gain sodium, and the 'activity respiration' is abolished (Connelly, 1959; Ritchie, 1966). The 'activity respiration' is also blocked, and the resting rate of respiration is reduced, if ouabain is added to the bathing solution at concentrations sufficient to block sodium extrusion (Connelly, 1962; Baker and Connelly, 1966; Ritchie, 1966; Rang and Ritchie, 1968a). After stimulation has stopped, the rate of respiration of a nerve returns slowly to the resting level. Complete recovery requires from a few minutes to an hour, depending on the nerve, the frequency and duration of stimulation, and the ionic composition of the bathing solution. When the tetanus is brief, the rate of respiration relaxes along a curve that is exponential in time. The rate constant of this relaxation increases as the external potassium concentration increases and, in crab nerve, the rate constant increases as the external sodium concentration decreases. These observations are consistent with the idea that the rate of the 'activity respiration' is controlled by the transport ATPase whose external site is activated by potassium and inhibited by sodium. When the ATPase is strongly activated, sodium and potassium are pumped rapidly, and the rate of respiration returns quickly to its resting level.

The active ion fluxes and the resting rates of oxygen consumption of the nerves under discussion are shown in table 9, together with the total extra oxygen consumed when a single impulse passes along these nerves. If it is assumed that the extra oxygen is used only to transport the Na^+ and K^+ ions that entered during the impulse, then the stoichiometry of the transport process can be estimated by comparing the extra oxygen consumed per impulse with the net ionic exchange per impulse. The stoichiometry can also be estimated in resting nerves by comparing the size of the ouabain-sensitive component of the resting respiration with the size of the ouabain-sensitive component of the ionic fluxes. The results of these calculations are presented in table 10.

The activity of the Na^+–K^+ ATPase has been estimated directly in resting crab and squid nerves and these results may be combined with the estimates of the active fluxes to determine the number of ions that are transported per molecule of ATP that is hydrolyzed. These Ion : P ratios are also shown in table 10. In rabbit and frog nerves, the Ion : P ratios were calculated from the Ion : O_2 ratios by assuming a P : O ratio of 3. In the resting nerves the Na : P ratios vary from 1·2 to 3·2 and the spread of the K : P ratios is even greater, a situation that is to be expected if the ratio of the active sodium efflux to active potassium influx is not constant. If Na : P ratios are computed from the oxygen consumption data for stimulated nerves, the ratio varies from 0·5 for frog sciatic nerve, to 3·5 for crab leg nerve. It is not clear whether the low Na : P ratio in frog nerve indicates that the 'activity respiration' in frog nerve provides energy for processes other than sodium and potassium transport, or whether the estimates of the ionic changes produced

Table 9. Active fluxes and ouabain-sensitive oxygen consumption of various nerves

	Temperature (°C)	Active flux[1] in resting nerve (μ moles/g cell water.h)		Oxygen consumption in resting nerve (μ moles/g cell water.h)		Extra O_2/impulse (μ moles/g cell water.imp)
		Potassium in	Sodium out	Total	Ouabain sensitive	
Crab leg	16	80	64	4·3	2·5	$1{\cdot}5\times10^{-3}$
Rabbit vagus	20	44–33	33	24	4·8	$2{\cdot}8\times10^{-3}$
Frog sciatic	20	14	12	4·4	1·3	$5{\cdot}3\times10^{-5}$
Giant axon (*Loligo pealeii*)	14–16	7	11	3·4		

[1]Active fluxes were estimated in the following ways: crab nerve, from net changes produced by ouabain; rabbit nerve, from net changes in potassium-free bathing solution or from ouabain-sensitive potassium influx; frog nerve, from ouabain or azide-sensitive fluxes; giant axons, from cyanide or strophanthidin-sensitive fluxes.

(From Brink and coworkers, 1952; Connelly, 1952, 1962; Hurlbut, 1963a and b Baker, 1965; Baker and Connelly, 1965; Keynes and Ritchie, 1965; Mullins and Brinley, 1967; Rang and Ritchie, 1968a and b; Sjodin and Beaugé, 1968.)

Table 10. Stoichiometry of ion transport in nerve

	Ion: O_2 Ratio[1] Rest		Ion: O_2 Ratio[1] Stimulated		Na^+–K^+ ATPase[2] resting nerve	Ion: P Ratio[3] resting nerve	
	Potassium	Sodium	Potassium	Sodium	(μ moles/g cell water.h)	Potassium	Sodium
Crab leg	32	26	15	21	20	4	3·2
Rabbit vagus	10–7	7	7			1·2–1·7	1·2
Frog sciatic	11	9	3	3		1·8	1·5
Giant axon (*Loligo pealeii*)					6	1·2	1·8

[1]Computed from data in tables 8 and 9.

[2]Na^+–K^+ ATPase in crab leg nerve is the ouabain-sensitive ATPase measured in cyanide-poisoned tissue (Baker, 1965). In giant axons it is membrane ATPase defined as ATPase activity of whole axon minus ATPase activity of extruded axoplasm (Mullins and Brinley, 1967).

[3]For rabbit and frog nerves these ratios were calculated from the Ion:O_2 ratios assuming a P:O ratio of 3.

by activity in these nerves are in error because of diffusion limitations around the node.

The concentrations of some high energy phosphate compounds have been determined for each of the nerves under discussion and the results are summarized in table 11. The free energy of hydrolysis of ATP, under the conditions in the axoplasm, is about 11 kcal/mole, and this is sufficient to permit the exchange of up to 3 moles of sodium and potassium per mole of ATP hydrolysed. Thus with a Na : P and K : P ratios of 3 : 1 the transport system is operating at nearly 100 per cent efficiency.

The membrane potential of myelinated fibers and mammalian C-fibers is increased above normal following a period of intensive stimulation or following a period of anoxia (Gasser, 1935; Lorente de No, 1947; Ritchie and Straub, 1957). The recovery of the fibers from these hyperpolarized states proceeds slowly and may require many minutes to over an hour to be complete. These post-tetanic and post-anoxic hyperpolarizations seem to be caused by the increased rates of ion transport that prevail at these times. The post-tetanic hyperpolarization is abolished by ouabain or by replacing the sodium of the external solution by lithium, and the rate of relaxation of this hyperpolarization increases as the potassium concentration of the bathing solution is increased (Ritchie and Straub, 1957; Connelly, 1959, 1962; Rang and Ritchie, 1968b). Two mechanisms have been proposed to account for this hyperpolarization and its dependence on active ion transport. According to one point of view, the nerve membrane removes K^+ ions from the space between the nerve fiber and the Schwann cell faster than these ions diffuse in from the extracellular space. The concentration of potassium at the nerve surface is reduced, and the membrane potential is increased (Ritchie and Straub, 1957). According to the second view, the active transport process generates an outward going electrical current in the membrane (this current contributes an additional term to the summation in equation (10)), and the magnitude of the hyperpolarization is determined by the magnitude of this current and by the value of membrane resistance (Connelly, 1959). The current may arise whenever the number of Na^+ ions extruded by the transport system exceeds the number of K^+ ions taken up. A transport process that generates an electric current within the membrane is called an electrogenic pump. It is usually difficult to decide whether a particular hyperpolarization is due to an electrogenic pump or to potassium depletion at the nerve surface, and in general both effects may occur simultaneously. Isolated myelinated nerve fibers, whose nodes may be in direct contact with the bathing solution (Meves, 1960) do not show post-tetanic or post-anoxic hyperpolarizations (Meves, 1961; Maruhashi and Wright, 1967), and this suggests that the potassium-depletion effect may be an important source of the hyperpolarization observed with bundles of these fibers.

Tabel 11. High energy phosphates in various nerves[1]

Nerve	Concentration[1] (μ moles/g cell water)				Free energy of Hydrolysis of ATP, ΔF_H[2] (kcal/mole)	Free energy of Transport, ΔF_T[3] (kcal/mole)	$-\frac{\Delta F_H}{\Delta F_T}$
	ATP	ADP	P_i	Arg. P or Cr. P			
Giant axon (*Loligo forbesii*)	4·6	0·8	4·3	3·9	−11·2	+3·33	3·4
Giant Axon (*Loligo pealeii*)	5·0	0·46	(4)	5	−11·5	+3·24	3·6
Crab Leg (*Maia squinado*)	5·7	(1)	3·9	7·8	−11·3	+3·35	3·4
Rabbit vagus	8·4	1·1	(4)	7·8	−11·3	+2·42	4·7
Frog sciatic (*Rana pipiens*)	3·2	1·2	4·4	6·8	−10·8	+3·30	3·3

[1]Numbers in parentheses are assumed values

[2] $\Delta F_H = -7 + (RT/1000) \ln \left[\frac{[ADP]\ [P_i]}{[ATP]}\right]$

[3] $\Delta F_T = (RT/1000) \ln \left[\frac{[K_i]}{[K_o]} \times \frac{[Na_o]}{[Na_i]}\right]$

(From Greengard and Straub, 1959; Caldwell, 1960; Cheng, 1961; Baker, 1965; Mullins and Brinley, 1967).

However, the potassium-depletion effect may not account entirely for the post-tetanic hyperpolarization in these bundles (Connelly, 1959).

There is good evidence, however, that the electrogenic pump produces the post-tetanic hyperpolarization in rabbit vagus nerve. After these nerves have been stimulated in a K^+-free and Cl^--free solution, the membrane potential of the nerve fibers first increases and then rapidly returns to normal. If K^+ is then added to the external solution, the membrane potential *increases* again, and then declines exponentially with time to a value appropriate to the resting nerve. It is highly unlikely that this initial increase in membrane potential is due to a depletion of potassium at the nerve membrane, for it is difficult to imagine how the addition of K^+ to the bathing medium could reduce the K^+ concentration at the nerve surface. This phenomenon may be explained qualitatively in terms of an electrogenic pump as follows: after tetanization in a potassium-free solution the nerve fibers contain much sodium that cannot be extruded because the activity of the pump is depressed by the lack of potassium at the nerve surface. When K^+ is added to the bath, the potassium concentration of the nerve surface increases, sodium is rapidly extruded, a current is generated, and the membrane potential increases and then returns to normal as the concentration of sodium in the axoplasm falls. This view is supported by the fact that ouabain blocks this post-tetanic hyperpolarization and by the fact that those ions that activate the external site of the transport ATPase also elicit the hyperpolarization when they are added to the potassium-free solution.

In frog myelinated fibers and vagus C-fibers models have been proposed in which the magnitude of the post-tetanic hyperpolarization is proportional to the rate of active sodium extrusion (Connelly, 1959; Rang and Ritchie, 1968b). In both of these models relatively small changes in internal sodium concentration saturate the extrusion process. In the myelinated fibers the amplitude of post-tetanic hyperpolarization, and therefore, the rate of sodium extrusion, is nearly completely saturated after a tetanus of 2 minutes at a rate of 50/sec. In rabbit vagus nerve the hyperpolarization is fully developed after stimulation for 5 seconds at a rate of 30/sec. According to table 8, the average internal concentration of sodium in these fibers should have increased by 1 mM and 3 mM, respectively, as a result of this activity, assuming that no sodium was extruded during the tetanus. At the node it is very likely that the sodium that enters the axons during such a short burst of activity is not uniformly distributed throughout the axoplasm, but is confined to the vicinity of the nodes. Conventional diffusion theory predicts that after 2 minutes of stimulation at 50/sec, the intracellular sodium concentration at the nodes should have increased by 4 to 16 mM, depending on whether the diffusion coefficient of sodium in the axoplasm is assumed to be 10^{-5} cm^2/sec, the approximate value in aqueous solution, or 10^{-6} cm^2/sec.

The apparent ease with which the hyperpolarization saturates with increasing levels of internal sodium does not tally with the tracer studies that indicate that the sodium efflux is proportional to the intracellular sodium concentration over a wide range of concentrations. This lack of concordance between the results of experiments with tracers and the interpretation of the studies on the origin of the post-tetanic hyperpolarization implies that both are not direct measures of the rate of sodium extrusion.

One possible origin of the discrepancy between these two sets of observations would be that sodium in the axoplasm is compartmentalized so that sodium entering during activity does not readily mix with the rest of the sodium in the axon, but is confined to restricted regions near the membrane. Under these circumstances tracer exchange measured in bundles of nerve fibers would not be a valid index of ionic fluxes across the fiber membranes. Another explanation of the discrepancy would be that the electrogenic component of the sodium efflux is so small a fraction of the total efflux that it cannot be readily detected in measurements with radioactive tracers or in measurements of ionic content. This possibility has been examined in stimulated rabbit vagus nerves and in frog sciatic nerves recovering from anoxia.

In the case of rabbit nerve, Rang and Ritchie (1968a) postulated that the electrogenic extrusion of sodium would produce an outward movement of chloride from the fibers (because the chloride permeability of these nerves seems to be high). The loss of NaCl from the axons would induce a loss of water so that the tonicity of the axoplasm remains constant. Since significant changes in chloride content or water content could not be measured in nerves stimulated in potassium-free solution, these workers concluded that the electrogenic component of the sodium pump accounts for only a small fraction of the total sodium transport and that the major portion of the sodium extrusion is coupled to the uptake of potassium. In frog sciatic nerve, the active extrusion of sodium and uptake of potassium as measured by movements of radioactive tracers or by net changes in the ionic content of the nerves, may be blocked without blocking the post-anoxic hyperpolarization (Hurlbut, 1965). These results also indicate that if the electrogenic extrusion of sodium is responsible for this hyperpolarization, then the electrogenic component of the sodium efflux accounts for only a small fraction (about 12 per cent) of the total efflux. Thus, although there is good evidence that active transport may be the basis of the post-anoxic or post-tetanic hyperpolarizations in peripheral nerve fibers, many details of the relationship between the rate of transport and the degree of hyperpolarization remain to be worked out.

Biochemical studies indicate that a site on the transport ATPase is phosphorylated during the hydrolysis of ATP. At 24 °C, the catalytic activity of the enzyme system isolated from eel electroplaques is $2{\cdot}4 \times 10^3$ moles ATP hydrolysed per minute per mole of enzyme bound phosphate (Albers and coworkers, 1968). Assuming a Q_{10} of 2·3 for the ATPase, this is equivalent to catalytic activity at 18 °C of 27 moles of ATP hydrolysed per mole of enzyme bound phosphate per second. This temperature was chosen because it is representative of the temperatures in table 9. If we assume that each molecule of the transport ATPase binds a molecule of phosphate, then we can estimate the number of ATPase molecules from the maximum rate of ATP hydrolysis. The maximum rate of ATP hydrolysis can be estimated from the maximum rate of ouabain-sensitive respiration, assuming a P : O ratio of 3. The maximum rate of ouabain-sensitive respiration is given by the sum of the ouabain-sensitive respiration in resting nerve plus the maximum increase in respiration that occurs when the nerve is being stimulated or when it is extruding sodium taken up during an exposure to a potassium-free solution. In frog nerve the ouabain-sensitive component of the resting respiration is about 30 per cent of the total, and the maximum increment produced by stimulation is also about 30 per cent of the total (Brink and coworkers, 1952). The maximum rate of ouabain-sensitive respiration is, therefore, about 2·6 μ mols/g cell water . hr. This is equivalent to a rate of hydrolysis of ATP per unit area of nodal membrane of 840 pmol/cm^2. sec. Since the activity of the ATPase is 27 moles ATP hydrolyzed per mole of ATPase per second, the density of the ATPase molecules at the node is about 1900×10^{10} molecules/cm^2. In crab and rabbit nerves the maximum increase in the rate of ouabain-sensitive respiration is about equal to the resting rate of respiration (Baker and Connelly, 1965; Rang and Ritchie, 1968a). If similar calculations are made to estimate the densities of the ATPase sites in crab and rabbit nerves, the results are $1·5 \times 10^{10}$ ATPase molecules/cm^2 and $4·5 \times 10^{10}$ ATPase molecules/cm^2, respectively. Giant axons of squid increase their rate of oxygen consumption by about 10 per cent when they are stimulated at 100/sec. According to table 9 this amounts to 0·34 μmoles 0_2/g cell water. hr which is equivalent to a rate of hydrolysis of ATP of 7 pmol/cm^2 . sec for fibres 550 μ in diameter. The resting rate of ATP hydrolysis, measured in perfused axons, is 6 μmols/g cell water . hr or 17 pmol/cm^2. sec (Mullins and Brinley, 1967). The maximum rate of ATP hydrolysis is, then, 24 pmol/cm^2. sec, so that the density of ATPase molecules in the giant axons is 54×10^{10} molecules/cm^2. These results are collected in table 12 together with some estimates of the number of passive ionic channels in the various nerves under discussion. The number of channels were estimated from sodium, potassium and leakage conductances assuming that the conductance of each of the channel was $0·9 \times 10^{-9}$ mho in sea water and $0·2 \times 10^{-9}$ mho in frog Ringer's solution (Hille, 1968). For the crab and rabbit fibers, for which no conductance data are available, the number of leakage channels was estimated from the resting sodium permeability in the manner used for giant axons.

Table 12. Densities of ion transport sites and permeability channels in various nerves

	Passive channels (Number/cm²)			ATPase sites/cm²
	Sodium	Potassium	Leakage	
Giant Axon (*Loligo pealeii*)	$1{\cdot}3\times 10^8$	$0{\cdot}3\times 10^8$	3×10^5	54×10^{10}
Crab leg (*Maia squinado*)			$0{\cdot}19\times 10^5$	$1{\cdot}5\times 10^{10}$
Rabbit vagus			$0{\cdot}24\times 10^5$	$4{\cdot}5\times 10^{10}$
Frog sciatic (*Rana pipiens*)	80×10^8	15×10^8	3000×10^5	1900×10^{10}

(From Hille, 1968; other values taken from text.)

The number of densities of the ionic channels at a node can also be estimated from the sodium permeability coefficients given on page 129 and in table 7. The radius of the leakage channel is assumed to be 4·5 Å, and the radii of the sodium and the potassium-specific channels are assumed to be 3Å (Hille, 1968). The number of densities of the sodium, potassium and leakage channels estimated in this way are $1300\times 10^8/\text{cm}^2$, $360\times 10^8/\text{cm}^2$ and $360\times 10^5/\text{cm}^2$. These estimates of the number of densities of the sodium and the potassium-specific channels are about twenty-fold greater than the estimates based on conductance data. This difference between the two estimates probably is a reflection of the fact that the estimated conductance of a single sodium channel at a node is equivalent to the conductance of a Ringer's-filled pore only 5 Å long, or one-eighteenth of the membrane thickness (Hille, 1968). Perhaps this indicates that the partition coefficient of the sodium pore is about twenty. It will be remembered that in a 0·1 M solution of monovalent ions, the average distance between ions of a given type is about 25 Å. If there were no concentrating of ions in a channel, then a channel would contain only about three ions along its length.

To summarize, all these nerves seem to have similar ion transport properties. The density of the specialized membrane regions is found to be about fifty times greater at a node than at the surface membrane of a giant fiber, and about one thousand-fold greater at the node than at the membrane of a C-fiber. Most of the other quantitative differences in the transport properties of the different fibers are due mainly to differences in their surface to volume ratios.

REFERENCES

Abbott, B. C., A V. Hill and J. V. Howarth (1958) *Proc. Roy. Soc. Ser. B*, **148**, 149
Albers, R. W., G. J. Koval and G. J. Siegel (1968) *Mol. Pharmacol.*, **4**, 324
Armett, C. J. and J. M. Ritchie (1963) *J. Physiol.*, **165**, 130
Asano, T. and W. P. Hurlbut (1958) *J. Gen. Physiol.*, **41**, 1187
Baker, P. F. (1965) *J. Physiol.*, **180**, 383
Baker, P. F. (1968) *J. Gen. Physiol.*, **51**, No. 5, part 2, 172S
Baker, P. F. and C. M. Connelly (1965) *J. Physiol.*, **185**, 270
Baker, P. F., A. L. Hodgkin and T. I. Shaw (1962a) *J. Physiol.*, **164**, 330
Baker, P. F., A. L. Hodgkin and T. I. Shaw (1962b) *J. Physiol.*, **164**, 355
Baker, P. F. and J. Manil (1968) *Biochim. Biophys. Acta.*, **150**, 328
Baker, P. F. and T. I. Shaw (1965) *J. Physiol.*, **180**, 424
Baker, P. F., M. P. Blaustein, A. L. Hodgkin and R. A. Steinhardt (1967a) *J. Physiol.*, **192**, 43P
Baker, P. F., M. P. Blaustein J. Manil and R. A. Steinhardt (1967b) *J. Physiol.*, **191**, 100P
Baker, P. F., R. F. Foster, D. S. Gilbert and T. I. Shaw (1968) *Biochim. Biophys. Acta.*, **163**, 560
Blaustein, M. P. and A. L. Hodgkin (1968) *J. Physiol.*, **198**, 46P
Brink, F. J., Jr., D. W., Bronk, F. D. Carlson and C. M. Connelly (1952) *Cold Spring Harbor Symp. Quant. Biol.*, **17**, 53
Brinley, F. J., Jr. (1965) *J. Neurophysiol.*, **28**, 742
Brinley, F. J., Jr. and L. J. Mullins (1965) *J. Neurophysiol.*, **28**, 526
Brinley, F. J., Jr. and L. J. Mullins (1967) *J. Gen. Physiol.*, **50**, 2303
Brinley, F. J., Jr. and L. J. Mullins (1968) *J. Gen. Physiol.*, **52**, 181
Bunge, R. P. (1968) *Physiol. Rev.*, **48**, 197
Caldwell, P. C. (1960) *J. Physiol.*, **152**, 545
Caldwell, P. C., A. L. Hodgkin, R. D. Keynes and T. I. Shaw (1960a) *J. Physiol.*, **152**, 561
Caldwell, P. C., A. L. Hodgkin, R. D. Keynes and T. I. Shaw (1960b) *J. Physiol.*, **152**, 591
Caldwell, P. C. and R. D. Keynes (1959) *J. Physiol.*, **148**, 8P
Caldwell, P. C. and R. D. Keynes (1960) *J. Physiol.*, **154**, 177
Canessa-Fischer, M., F. Zambrano and E. Rojas (1968) *J. Gen. Physiol.*, **51**, no. 5, part 2, 162S
Cass, A. and A. Finkelstein (1967) *J. Gen. Physiol.*, **50**, 1765
Chandler, W. K. and H. Meves (1965) *J. Physiol.*, **180**, 788
Cheng, S. C. (1961) *J. Neurochem.*, **7**, 278
Cole, K. S. and J. W. Moore (1960) *J. Gen. Physiol.*, **44**, 123
Connelly, C. M. (1952) *Biol. Bull.*, **103**, 315
Connelly, C. M. (1959) *Rev. Mod. Phys.*, **31**, 475
Connelly, C. M. (1962) *Proc. Intern. Physiol. Cong.*, P600
Dainty, J. and K. Krnjevic (1955) *J. Physiol.*, **128**, 489
Deffner, G. G. J. (1961) *Biochim. Biophys. Acta.*, **47**, 378
Dodge, F. A., Jr. (1963) *Ph.D. Thesis*, Rockefeller University, New York
Dodge, F. A., Jr. and B. Frankenhauser (1958) *J. Physiol.*, **143**, 76
Dodge, F. A., Jr. and B. Frankenhauser (1959) *J. Physiol.*, **148**, 74
Eisenman, G. (1962) *Biophys. J.*, **2**, no. 2, part 2, 259
Frankenhauser, B. (1960) *J. Physiol.*, **152**, 159
Frankenhauser, B. (1962) *J. Physiol.*, **160**, 54
Frankenhauser, B. and A. L. Hodgkin (1956) *J. Physiol.*, **131**, 341
Frankenhauser, B. and A. F. Huxley (1964) *J. Physiol.*, **171**, 302
Frankenhauser, B. and L. E. Moore (1963) *J. Physiol.*, **169**, 438
Friede, R. L. and T. Samorajski (1968) *J. Neuropathol. and Exptl. Neurol.*, **27**, 546
Garrahan, P. J. and I. M. Glynn (1965) *Nature*, **207**, 1098
Gasser, H. S. (1935) *Amer. J. Physiol.*, **111**, 35
Geren, B. B. (1954) *Exptl. Cell Res.*, **7**, 558
Geren, B. B. and F. O. Schmitt (1954) *Proc. Natl. Acad. Sci. (U.S.)*, **40**, 863

Goldman, D. E. (1943) *J. Gen. Physiol.*, **27**, 37
Goldman, D. E. (1964) *Biophys. J.*, **4**, 167
Greengard, P. and R. W. Straub (1959) *J. Physiol.*, **148**, 353
Grundfest, H., C. Y. Kao and M. Altamirana (1954) *J. Gen. Physiol.*, **38**, 245
Hille, B. (1967) *J. Gen. Physiol.*, **50**, 1287
Hille, B. (1968) *J. Gen, Physiol.*, **51**, 199
Hinke, J. A. M. (1961) *J. Physiol.*, **156**, 314
Hodgkin, A. L. (1958) *Proc. Roy. Soc. Ser. B*, **148**, 1
Hodgkin, A. L. and A. F. Huxley (1952) *J. Physiol.*, **117**, 500
Hodgkin, A. L. and B. Katz (1949a) *J. Physiol.*, **108**, 37
Hodgkin, A. L. and B. Katz (1949b) *J. Exptl. Biol.*, **26**, 292
Hodgkin, A. L. and R. D. Keynes (1953) *J. Physiol.*, **119**, 513
Hodgkin, A. L. and R. D. Keynes (1955a) *J. Physiol.*, **128**, 28
Hodgkin, A. L. and R. D. Keynes (1955b) *J. Physiol.*, **128**, 61
Hodgkin, A. L. and R. D. Keynes (1956) *J. Physiol.*, **131**, 592
Hodgkin, A. L. and R. D. Keynes (1957) *J. Physiol.*, **138**, 253
Hurlbut, W. P. (1958) *J. Gen. Physiol.*, **41**, 959
Hurlbut, W. P. (1963a) *J. Gen. Physiol.*, **46**, 1191
Hurlbut, W. P. (1963b) *J. Gen. Physiol.*, **46**, 1223
Hurlbut, W. P. (1965) *Amer. J. Physiol.*, **209**, 1295
Huxley, A. F. and R. Stämpfli (1951) *J. Physiol.*, **112**, 476
Jacobs, M. H. (1935) *Ergeb. Biol.*, **12**, 1
Keynes, R. D. (1951) *J. Physiol.*, **113**, 99
Keynes, R. D. (1963) *J. Physiol.*, **169**, 690
Keynes, R. D. and P. R. Lewis (1951) *J. Physiol.*, **114**, 151
Keynes, R. D. and P. R. Lewis (1956) *J. Physiol.*, **134**, 399
Keynes, R. D. and J. M. Ritchie (1965) *J. Physiol.*, **179**, 333
Koechlin, B. A. (1955) *J. Biophys. Biochem. Cytol.*, **1**, 511
Koefoed-Johnsen, V. and H. H. Ussing (1953) *Acta Physiol. Scand.*, **28**, 60
Lorente de No, R. (1947) *Studies from the Rockefeller Institute for Medical Research*, **131**
Luxoro, M. and E. Yañez (1968) *J. Gen. Physiol.*, **51**, no. 5, part 2, 115S
MacInnes, D. A. (1961) *The Principles of Electrochemistry*, Dover Publications, New York
Manery, J. F. (1939) *J. Cell. Comp. Physiol.*, **14**, 365
Maruhashi, J. and E. B. Wright (1967) *J. Neurophysiol.*, **30**, 434
Meves, H. (1960) *Pflüg. Arch. Ges. Physiol.*, **271**, 655
Meves, H. (1961) *Pflüg. Arch. Ges. Physiol.*, **272**, 336
Moore, J. W., T. Narahashi and T. I. Shaw (1967) *J. Physiol.*, **188**, 99
Mullins, L. J. (1960) *J. Gen. Physiol.*, **43**, no. 5, part 2, 105S
Mullins, L. J. (1966) *Ann. N.Y. Acad. Sci.*, **137**, 830
Mullins, L. J. and F. J. Brinley, Jr. (1967) *J. Gen. Physiol.*, **50**, 2333
Nakamura, Y., S. Nakagima and H. Grundfest (1965) *J. Gen. Physiol.*, **48**, 985
Oikawa, T., C. S. Spyropoulos, I. Tasaki and T. Teorell (1961) *Acta Physiol. Scand.*, **52**, 195
Rang, H. P. and J. M. Ritchie (1968a) *J. Physiol.*, **196**, 163
Rang, H. P. and J. M. Ritchie (1968b) *J. Physiol.*, **196**, 183
Rang, H. P. and J. M. Ritchie (1968c) *J. Physiol.*, **196**, 223
Ritchie, J. M. (1966) *J. Physiol.*, **188**, 309
Ritchie, J. M. and R. W. Straub (1957) *J. Physiol.*, **136**, 80
Robertson, J. D. (1959) *Z. Zellforsch. Mikroskop. Anat.*, **50**, 553
Shanes, A. M. (1957) In Q. R. Murphy (Ed.), *Metabolic Aspects of Transport across Cell Membranes*. University of Wisconsin Press, Madison, p 127
Shanes, A. M. and M. D. Berman (1955) *J. Cell. Comp. Physiol.*, **45**, 199
Sjodin, R. A. and L. A. Beaugé (1968) *J. Gen. Physiol.*, **51**, no. 5, part 2, 152S
Sjodin, R. A. and L. J. Mullins (1967) *J. Gen. Physiol.*, **50**, 533
Skou, J. C. (1957) *Biochim. Biophys. Acta.*, **23**, 394
Skou, J. C. (1960) *Biochim. Biophys. Acta.*, **42**, 6

Skou, J. C. (1964) *Prog. Biophys. Mol. Biol.*, **14**, 133
Stämpfli, R. (1954) *Physiol. Rev.*, **34**, 101
Straub, R. W. (1963) *Pflüg. Arch. Ges. Physiol.*, **278**, 108
Tasaki, I., I. Singer and T. Takenaka (1965) *J. Gen. Physiol.*, **48**, 1095
Tasaki, I., T. Teorell and C. S. Spyropoulos (1961) *Amer. J. Physiol.*, **200**, 11
Tasaki, I., A. Watanabe and T. Takenaka (1962) *Proc. Natl. Acad. Sci., U.S.* **48**, 1177
Teorell, T. (1953) *Prog. Biophys. Biophys. Chem.*, **3**, 305
Ussing, H. H. (1949) *Acta Physiol. Scand.*, **19**, 43
Villegas, R., I. B. Bruzual and G. M. Villegas (1968) *J. Gen. Physiol.*, **51**, no. 5, part 2, 81S
Villegas, R., C. Caputo and L. Villegas (1962) *J. Gen. Physiol.*, **46**, 245
Villegas, R. and G. M. Villegas (1960) *J. Gen. Physiol.*, **43**, no. 5, part 2, 73S
Villegas, J., R. Villegas and M. Giménez (1968) *J. Gen. Physiol.*, **51**, 47
Villegas, J., L. Villegas and R. Villegas (1965) *J. Gen. Physiol.*, **49**, 1
Wallin, B. G. (1967) *Acta Physiol. Scand.*, **70**, 419
Weidmann, S. (1951) *J. Physiol.*, **114**, 372

CHAPTER 5

Ion transport and metabolism in brain

R. M. Marchbanks
Department of Biochemistry,
University of Cambridge,
Cambridge, England

I. INTRODUCTION

The functional activity of the brain depends, as it does in other excitable tissues on the permeability properties of the membranes of its constituent cells and the processes which bring about an asymmetric distribution of ions and other substances across them. Three major problem areas can be discerned in attempting to relate ion transport and the metabolism and functional activity of brain. First, what permeability barriers exist in brain tissue, and what is the relationship of these to the gross anatomy of the brain as a whole, and to the membranes and other fine structure of its constituent cells? Secondly, what are the characteristics of the permeability barriers which account for the development of an asymmetric distribution of ions and do these characteristics differ for the permeability barriers associated with the membranes of the various cell types comprising brain tissue? Thirdly, what explanations at the molecular level can be offered to account for the properties of the permeability barriers?

Both in its gross anatomy, and in the fine structure of its cells brain tissue is exceedingly complex. This has impeded investigations of the physiological and biochemical aspects of transport in brain. In particular, elucidation of the intimate mechanisms of transport in brain has been hampered. Most investigators have been content to try to establish the general principles of transport in brain, to indicate their similarity to transport processes in other tissues and to identify morphologically the membranes responsible; all sufficient problems in themselves. This review will attempt an appraisal of the progress to date along these lines.

A. Anatomical Considerations

Anatomical inspection of the brain suggests that there are at least four gross fluid compartments separated by permeability barriers. The blood plasma is separated by the capillary walls from the extracellular fluid of brain tissue, and by the choroidal epithelium from the cerebrospinal fluid (CSF). The CSF is separated from the extracellular fluid by the ependymal layer of cells. The extracellular fluid bathes the neuronal and glial cells and is separated from the intracellular compartment by the membranes of these cells themselves. Some consequences of the gross fluid compartmentation are discussed in Section II.

Most interest is focussed on the relationship of the extracellular to intracellular fluid compartment because it is the properties of the permeability barrier between these two compartments which cause the asymmetric distribution of ions between the inside and outside of brain cells that is essential for the functional activity of the brain. The brain slice preparation provides the most useful tool for investigating the properties of the cell membranes of brain tissue. Results from this preparation are discussed in Sections III and IV.

Brain tissue is composed mainly of two general cell types; neurons, excitable cells of diverse morphology whose membrane propagates the action potential, and glial cells of two main types, astroglia and oligodendroglia. A supportive metabolic role is generally ascribed to glial cells (De Robertis and Gerschenfeld, 1961), though a more fundamental role in impulse transmission in brain has sometimes been suggested (see Section V C). It is to be noted that the permeability and properties of the membranes of cells as complicated as neurons and glia may vary at different points on the cell surface. This is particularly likely to be the case where morphological and functional specialization occurs, for instance at the synaptic terminals of neurons, at the end feet of astroglia and at the myelinic exfoliations of oligodendroglia.

The cell itself contains discrete organelles, bounded by membranes. Obvious examples are the nuclei and mitochondria, and in neurons the synaptic vesicles. These subcompartments of cells are discussed in other chapters of this volume (cf. Chapters 8 and 9). Even though information on this subject in brain is scant, it is clear that in their local environment subcellular organelles will modify the intracellular concentrations of ions, and they may well affect the behaviour of the cell as a whole in its ability to concentrate them.

A full description of transport processes in the brain will necessitate simplification of the system by isolation of different cell types and subcellular organelles. Progress along these lines has been small (Section V) and studies on artificially separated components of the brain are always subject to the criticism that the trauma resulting from the separation procedure renders the experimental results invalid. This objection can only be countered by a careful evaluation of the physiological state of the separated preparations. Nevertheless, the effort seems merited since some simplification of the systems under study is necessary if a successful resolution of the nature of the transport processes between the various fluid compartments is to be achieved.

B. Diffusion and Transport

The notion of a fluid compartment implies an impediment to the movement of water and solutes into or out of it. The impediment to free movement in biological systems can usually be recognized as one or a series of membranes of characteristic fine structure. The existence of a compartment can also be inferred from metabolic studies, without any reference to a particular membrane structure. This however is rare, the only example of interest in brain tissue being perhaps the metabolic compartmentation of glutamate (Berl and coworkers, 1968).

Biological membranes exhibit varying degrees of impediment to the movement of substances by passive diffusion which is usually expressed as the permeability of the membrane to the substances in question. They also exhibit varying capabilities to transport substances between the compartments they separate. The discussion of this section is only intended to provide a frame of reference for the description of these processes in brain. The reader may consult the excellent book by Stein (1967) for further details.

The rate at which a substance crosses a membrane (net flux) when only passive diffusion forces operate is described by Fick's law. The rate is linearly dependent on the concentration gradient (or in the case of ions the electrochemical potential gradient), and the permeability constant of the membrane for that substance. Linear dependency of the rate of movement on the concentration gradient is diagnostic of passive diffusion processes. From this it follows that if only passive diffusion forces operate the net movement of a substance ceases when the concentration is equal on either side of the membrane because the unidirectional fluxes (i.e. influx and efflux) are equal. Therefore, if there are no losses of the substance from the system by movements into other compartments, passive diffusion forces will tend to equalize the concentration on either side of the membrane as the system approaches equilibrium.

However, maintenance of an inequality of concentrations on either side of a membrane does not necessarily indicate processes other than passive diffusion. If the free concentration of a substance is reduced by binding to some non-diffusible substance this will cause an apparent deviation from concentration equality. For charged molecules the equilibrium distribution is subject to further conditions. An ion can only move across a membrane accompanied by the passage of an ion with opposite charge, or in exchange for an ion with similar charge. If, as is usually the case in biological systems, there are ions which are unable to traverse the membrane these will affect the distribution of ions which can diffuse resulting in the well known Donnan distribution.

If there is a transmembrane potential then the equilibrium condition for ions across the membrane is that their electrochemical potentials should be equal. The difference in electrochemical potential of an ion on either side of a membrane is given by the Nernst equation and is dependent on the transmembrane potential and the concentration, or, more strictly, the thermodynamic activity of the ion. In neurons of brain the transmembrane potential is about −60 mV (inside negative) and glial cells in tissue culture maintain a similar membrane potential (Hild and Tasaki, 1962). These potentials will impose substantial concentration differences of ions at equilibrium on either side of the membranes.

It is clear therefore that inequalities of concentration on opposite sides of a membrane though probably a necessary condition are not a sufficient condition for the diagnosis of processes other than passive diffusion across the membrane. This situation with respect to ions and charged membranes has been formalized by Ussing's (1952) flux ratio test. If the ratio of influx to efflux is equal to the ratio of electrochemical potentials on either side of the membrane, then there is no proof that any process other than simple diffusion across the membrane is taking place.

The term transport implies that substances are traversing the membrane at a rate faster than would be expected by passive diffusion. The term facilitated diffusion embodies this minimal description. Studies over the last decade (reviewed by Stein, 1967) suggest that facilitated diffusion is carried out by combination of the substrate being transported with a carrier on one side of the membrane, the carrier–substrate complex then translocating to the other side of the membrane and dissociating to release the substrate.

Carrier-mediated transport does not necessarily require the input of chemical free energy, nor does it necessarily result in the accumulation of a substance against its concentration or electrochemical gradient. However, metabolic inhibitors are often found to reduce carrier-mediated transport. The flux is not linearly proportional to the concentration of the substrate, but reaches a limiting value (saturation) as the concentration is increased. The situation is formally analogous to the limiting of the rate of an enzyme reaction at high substrate concentration, hence the term 'Michaelis–Menten kinetics' is often used to describe it. As a consequence of this, the unidirectional flux is higher than the net flux, the discrepancy increasing as the concentrations of substrate exceed the dissociation constant of the carrier-substrate complex. Structural analogues of the substrate often reduce the flux, and by analogy with enzymes can be regarded as competitively inhibiting the formation of the carrier–substrate complex.

Carrier-mediated transport is often stimulated by the movement of sodium ions in the same direction (co-transport), or by structural analogues of the substrate in the opposite direction (counter-transport). In practice co-transport with sodium ions has not been shown in brain tissue very frequently, but dependency of flux on sodium ions has been demonstrated for many substances. The characteristics of saturability and inhibition by structural analogues have been most commonly used to distinguish carrier-mediated transport from passive diffusion in studies on brain systems. The flux of a substance across a membrane is usually composed of both a linear and a saturable component, a well known case being the flux of potassium across the erythrocyte membrane (Glynn, 1956).

It is to be noted that the criteria of carrier-mediated transport are all expressed in terms of flux, and not steady state concentrations. The steady state concentration (usually expressed as the ratio of internal concentration to external concentration) will result from a balance between carrier-mediated influx and efflux, and flux due to passive diffusion, the former saturating as the concentration increases while the latter is linearly proportional to concentration. Such systems, are frequently described as 'pump and leak' and result in a complicated relationship between the steady state concentration ratio and the external concentration of substrate. In the simplifying case where the affinity constant for efflux is very high or efflux is not carrier mediated the concentration ratio tends to unity as the external concentration of substrate is raised. However, in the case of amino acids (see Section III E) there is evidence that efflux is carrier mediated with affinity constants similar to that of influx. Most of the data on transport in brain has assumed the simplified case, and the constants derived should therefore be regarded as phenomenological rather than having a strict relationship to the affinity and kinetics of the carrier.

Active transport is a special case of carrier-mediated transport in which the transport of a substance against its electrochemical gradient takes place. The situation is clearly identifiable from Ussing's flux ratio test. The energy required to drive it may be derived directly from the high energy phosphates produced by metabolism, or it may be derived from the electrochemical gradient directed into the cell of sodium or the concentration gradients directed out of the cell of other substances. Active transport systems employing these forms of energy supply have been termed primary, secondary and tertiary, respectively. The problem of identifying the energy source for active transport of any substance is rather formidable in a tissue as complex as brain; furthermore, in only a few cases is the membrane potential known in order to determine whether transport is actually against its electrochemical gradient. For this reason the many cases in brain where there appears to be concentration will be described as concentrative accumulation unless the evidence is very clear that transport is thermodynamically uphill, with a clearly distinguishable energy source.

C. Molecular Mechanism of Ion Transport

The sodium and potassium activated ATPase discovered by Skou (1957) is involved in the active transport of sodium in erythrocytes (Post and coworkers, 1960; Dunham and Glynn, 1961) and in crab nerve (Skou, 1960). The central importance of sodium and potassium ions to nervous system function has stimulated investigators to examine the properties of this enzyme in brain. Brain is very rich in Na^+–K^+-ATPase (Bonting and coworkers, 1962) and the enzyme is localized in microsomal fractions from cerebral cortex (Deul and McIlwain, 1961; Schwartz and coworkers, 1962). Using morphologically characterized subcellular fractions from cerebral cortex, Hosie (1965) was able to show that the enzyme is located on the external membrane of nerve cells. Some possible models of the relationship of the enzyme action to ion transport in brain are discussed by McIlwain (1963).

The hydrolysis of ATP which is a concomitant of ion transport has prompted attempts at identification of phosphorylated intermediates of membrane components that might be concerned with the transport of sodium and potassium. The incorporation of phosphate into phosphatidic acid, triphosphoinositide or any other phospholipid does not seem to correlate with the activity of Na^+–K^+-ATPase (Yoshida and coworkers, 1961; Kirschner and Barker, 1963; Glynn and coworkers, 1965; De Graeff and coworkers, 1965).

Electrical stimulation of brain tissue promotes the incorporation of inorganic phosphate into phosphoryl serine (Heald, 1956, 1958). Ahmed and Judah (1965) report a sodium-stimulated, ouabain-inhibited incorporation of ^{32}P into phosphoryl serine which by centrifugation methods could be shown to be derived principally from the microsomal or membrane rich fraction (Trevor and Rodnight, 1965). However, the phosphate group appears to be transferred to the serine residue during the extraction procedure (Bader and coworkers, 1965; Rodnight and Lavin, 1966).

Despite these difficulties in identification, it is clear that cerebral microsomes which are rich in membrane fragments and Na^+–K^+-ATPase bind ^{32}P when incubated with γ-^{32}P ATP. The binding is activated by sodium ions; potassium ions added before or during the reaction reduce the amount of bound phosphate. Its formation is inhibited by ouabain in the presence of sodium (Rodnight and coworkers, 1966).

Incorporation with similar characteristics of ^{32}P from ATP into acid insoluble material has also been observed with kidney membrane Na^+–K^+-ATPase (Post and coworkers, 1965) and with a small particle fraction from the electric organ of *Electrophorus* (Albers and coworkers, 1963). The ^{32}P incorporating activities and Na^+–K^+-ATPase activities of various preparations are summarized by Albers (1967). The bound phosphate is not found in phosphoryl serine, and it seems to be a protein-bound acyl phosphate of a high energy type, easily hydrolysable and capable of phosphorylating methanol. (Hokin and coworkers, 1965; Bader and coworkers, 1966; Hems and Rodnight, 1966; Rodnight and coworkers, 1966).

The Na^+–K^+-ATPase is oriented across the membrane of the erythrocyte with its potassium-activated and ouabain-inhibited sites facing outwards, and the sodium-activated and ATP-hydrolytic sites facing inwards (Glynn, 1962; Whittam, 1962b; Whittam and Ager, 1964). For each molecule of ATP split, three ions of sodium are transported outwards and less than three, probably two, ions of potassium transported inwards (Glynn, 1962). Medzihradsky and coworkers (1967) have recently reported the separation of this important enzyme from cerebral microsomes using detergent with addition of ATP to stabilize it. The molecular weight estimated by gel filtration is about 670,000. Alkylation of the enzyme at the site where ouabain binds has been achieved with haloacetate derivatives of strophanthidin (Hokin and coworkers, 1966).

Investigation into the general properties of transport systems and their molecular mechanisms are at an exciting phase. The rest of this review will be an attempt to see how far the principles and phenomena noted so far apply to the heterogeneous collection of transport systems that constitutes the brain.

II. THE BRAIN *IN SITU*

A. Fluid Compartments

The plasma comes into close contiguity with the CSF at the choroid plexuses but is separated from it by two layers of cells, the pial layer and the ependymal cells. The CSF occupies the ventricles, subarachnoid spaces and central canal of the spinal cord all of which are interconnected. Ependymal cells line the walls of the ventricles and beneath them lie a layer of astroglial processes. Astroglial processes (end feet) also enfold the capillaries as they pass through the brain.

The extracellular fluid bathing the cells is therefore in principle accessible from the plasma via the capillary walls, and from the CSF via the ependymal layer. The morphological relationship of these compartments is discussed in greater detail by Davson (1967) and the reader is referred to reviews by Lajtha (1962), Johnstone and Scholefield (1962) and the symposium edited by Lajtha and Ford (1968) for a more complete discussion of the problems connected with the morphological identity of blood brain barrier and movement of substances between these fluid compartments.

There is divergence of views about the volume of the extracellular space. Direct measurements of the space between cells as shown in electron micrographs gives a value for the extracellular space as 5 per cent of total (Horstmann and Meves, 1959). Studies by Van Harreveld and coworkers (1965) suggest that this is because anoxic swelling of the cells occurs very rapidly, and when the tissue is rapidly frozen extracellular spaces of 15–20 per cent of the total are found. Artefacts of fixation also contribute to the low extracellular volume measured microscopically (Bahr and coworkers, 1957; Pappius and coworkers, 1962; see also discussion in Tower, 1968).

By measuring the electrical resistance of tissue Van Harreveld and Ochs (1956) conclude that the extracellular space is about 25 per cent of the total. The distribution of extracellular space markers administered *in vivo* is complicated by the permeability barriers to free diffusion from plasma to CSF and extracellular fluid, so that true equilibrium is not reached, and also there is doubt about the validity of some markers being truly extracellular (see Section III. B; Davson, 1967, p. 104 *et seq*). Studies with markers *in vitro* yield in general much higher values (30–40 per cent of total) for the extracellular space, but here the problem is complicated by the swelling of brain slices (see Section III B).

Rall and coworkers (1962), by infusing the CSF with inulin, compute a value of 12 per cent of total for the inulin space which they equate with the extra-cellular space. In a comprehensive discussion of these complexities, Davson (1967) concludes that the volume of the extracellular space is measured most accurately by techniques of this kind and that it is 12–15 per cent of

Table 1. Concentrations of some ions in compartments of rabbit brain[1]

Ion	Plasma	Cerebrospinal fluid	Extracellular fluid	Brain tissue (intracellular fluid compartment[2])
Na^+	148[6]	149[6]	150·4[5]	51[8]
K^+	4·3[3]	2·9[3]	4·71[5]	90[8]
Ca^{2+}	5·9[4]	2·38[4]		1·31[8]
Mg^{2+}	2·02[7]	1·74[7]		5·59[8]
HCO_3^-	25[6]	22[6]		
Cl^-	106[6]	130[6]		40[8]

[1]Concentrations are expressed as mmoles/kg and are not corrected for solid content of tissue.

[2]An estimate of the intracellular concentration can be made by assuming 15 per cent w/w is the extracellular fluid compartment (see text p. 153). Thus intracellular concentrations of Na^+ and K^+ are 42 and 104 mmoles/kg, respectively. Where estimates of the concentration in the extracellular fluid are not available they can probably be regarded as approximately equal to concentrations in the CSF. (see text p. 154).

[3]Bradbury and Davson (1965). [4]Bradbury (1965). [5]Bito and coworkers (1966), values obtained from an *in vivo* dialysate of dog brain. [6]Davson (1967). [7]Bradbury, quoted by Davson (1967). [8]Ames and Nesbett (1958).

the total volume of the brain tissue. The concentrations of some ions in the fluid spaces of brain are drawn together in table 1. Plasma, CSF and extracellular fluid are approximately similar, but the intracellular fluid has substantial differences of composition from the others.

B. Movements of Substances between the Fluid Compartments

Impediments to the movement of dyes from the plasma to brain tissue led to the notion of the blood–brain barrier. Since then restrictions of the movements of many other substances have been observed, in general lipid-soluble substances penetrate brain tissue from the plasma more rapidly. It will be apparent from morphological considerations that the permeability barrier is composed of not just one but many membranes.

The CSF is 5 mV positive with respect to plasma in the jugular vein (Held and coworkers, 1964). From this it is possible to calculate whether the distribution of ions between the plasma and CSF can be accounted for by passive diffusion. Held and coworkers (1964) conclude that sodium and magnesium ions are at higher concentrations in the CSF than those in electrochemical equilibrium with the plasma, potassium and bicarbonate ions are at a lower concentration, whereas chloride and calcium ions equilibrate according to their electrochemical potential. It seems therefore that sodium and magnesium are secreted into the CSF and potassium and bicarbonate removed from it by active mechanisms. Other anions such as iodide and perchlorate are transported out of the CSF (Woodbury, 1968). Secretion of sodium into the CSF occurs at the choroid plexus since the ventricles receive radioactivity more rapidly than the subarachnoid space when $^{24}Na^{+}$ is injected intravenously (Sweet and coworkers, 1949). The role of the choroid plexus in the case of the other ion movements is less well established.

Intraventricular–cisternal perfusion experiments suggest that movements between CSF and extracellular fluid of potassium (Bradbury and Davson, 1965; Cserr, 1965), sodium (Davson and Pollay, 1963), calcium (Bradbury, 1965; Graziani and coworkers, 1965) and magnesium (Bradbury, 1965) take place readily by passive diffusion. Indeed Rall (1968) has suggested that the ependyma is no barrier to the diffusion of large molecules such as inulin. However, analysis of *in vivo* dialysates of extracellular fluid (Bito and coworkers, 1966) suggests that it has slightly higher concentrations of potassium and amino acids than the CSF.

Experimentally, it is difficult to distinguish between movements from plasma to extracellular fluid and plasma to CSF, but since present evidence suggests that the extracellular fluid and CSF readily equilibrate by passive diffusion this problem has little bearing on the origin of substances taken up intracellularly. However, ion movements direct from the plasma to brain tissue do take place but at a slower rate than movements from CSF to brain tissue. Bakay (1960) has shown that the flux of $^{24}Na^{+}$ from CSF to brain is three times greater than that from plasma to brain; allowing for the greater area of the capillary bed this suggests that the $^{24}Na^{+}$ flux per unit area of CSF into brain is twenty times that from plasma to brain.

In general, the CSF contains lower concentrations of amino acids than the plasma (Bito and coworkers, 1966). Intracisternal injection of lysine and leucine increased the rate of penetration of radioactively labelled lysine and leucine from the plasma, suggesting that carrier-mediated transport into the CSF was taking place (Lajtha and Mela, 1961). The efflux of amino acids from the cerebrospinal fluid appears to be carrier mediated (Lajtha and

Mela, 1961), and can be carried out against a concentration gradient (Lajtha and Toth, 1961). In the brain as a whole the concentrations of the non-essential amino acids are substantially higher than in the plasma (Carver, 1965). It is clear that concentrative mechanisms must exist for some amino acids; these are present on the cell membranes of neurons and glia and are the subject of discussion in Section III E.

Studies on the brain *in situ* suggest that the CSF and extracellular fluid are separated from the plasma by permeability barriers and that transport processes do mediate the concentration of various ions in these compartments to a certain extent. However, the permeability barrier with the most dramatic facility for concentrative processes exists between these two fluid compartments and the intracellular compartment. The transport processes affecting the composition of the CSF appear to regulate its ion concentrations so that it provides an ideal extracellular fluid for the functioning of nerve cells.

Transport between the intracellular and extracellular compartments can be most conveniently explored using the brain slice preparations, and the results of such investigations are discussed in Sections III and IV.

III. TRANSPORT IN BRAIN SLICES

A. Properties of the Preparation

Brain slices are a more simplified system for the study of cerebral transport phenomena than the brain *in situ*. The permeability barriers between the plasma and CSF and the extracellular fluid have been removed and the incubation medium has direct access to the membranes of the constituent cells of the slice. This simplification has made the brain slice an extremely attractive preparation for the study of metabolism and transport in brain. Properly handled the cells of the slice can be shown to maintain *in vitro* the ionic gradients characteristic of excitable tissue. The membrane potential of individual cells in the slice can be measured and it can be shown that the cells from certain regions are electrically excitable. The environment of the slices can be readily manipulated and the effects of chemical and electrical stimuli investigated.

The development of the preparation to its current usefulness has been in great measure due to the work of McIlwain and his colleagues (see McIlwain, 1963, 1966) whose investigations have been distinguished by the careful appraisal given to the physiological state of the isolated preparation. It can be argued that the isolation of any organ or tissue from its environment produces more artefacts to confuse the understanding than simplifications to enlighten it. Occasionally this view has been put forward with respect to brain slices (Ruščák and Whittam, 1967). The brain slice certainly contains damaged cells and certain characteristic changes on incubation, such as swelling (Section III B), are pathological. However, McIlwain (1966, p. 50) has estimated that in a 35 mm thick cerebral cortex slice the cut surfaces constitute about 0·2 per cent of the cell surfaces. Although it is difficult to be certain that all the phenomena observed in a slice correspond to the behaviour of brain tissue *in situ*, there is an impressive degree of correlation between transport and metabolic characteristics observed *in vivo* and in slices *in vitro*. A further problem is that the constituent cells of the brain slice are heterogeneous in function and morphology. Both neurons and glia of various types are represented and it is always to be remembered that all the cell types contribute to the behaviour of the slice. Consequently, it is unwise to attribute a property of slices to any particular cell type unless further evidence is available (cf. Section V).

The preparation of brain slices of reproducible thickness less than 0·4 mm with minimum damage to the tissue has been the object of a number of critical studies. Where few slices of large surface area are required as for studies involving electrical stimulation and recording the slices can be cut by hand using the cutting edge of a razor moistened with saline and a template to maintain constant thickness. Mechanical chopping provides larger quantities of tissue but it is probably more damaged than hand-cut slices. A suitable apparatus which is commercially available has been described by McIlwain and Buddle (1953). Practical details of these and other methods are critically and comprehensively reviewed by McIlwain and Rodnight (1962). Frank and coworkers (1968) have described a novel cutting table in which the slices are cut by a vibrating nylon thread.

The criteria used to evaluate the optimal, or adequate conditions of incubation include:

i) the absence of excessive swelling of the incubated slices (see Section III B);

ii) the maintenance of gradients of sodium and potassium across the membranes of the cells of the slice similar to those found *in vivo* (see Section III C);

iii) the maintenance of the characteristic electrical properties of the cells in the slice (Section III C) and

iv) the maintenance of appropriate respiratory and metabolic response to various forms of stimulation (Section IV A).

The suspension media for slices are usually Krebs–Ringer's solutions or variants of these (Krebs, 1950). McIlwain and Rodnight (1962) discuss details of these and mention a number of alternative buffer systems such as phosphate or tris-HCl. The advantages of reducing the calcium ion concentration of the medium from that of the normal saline to 0·75 mM have been noted by Keesey and coworkers (1965) (see also Section III B,C).

The incubation medium is usually fortified by the addition of a source of metabolic energy. Glucose is most frequently used for this purpose at a final concentration of 10 mM. Many other energy supplying substrates have been investigated (Krebs, 1950). L-glutamate has been used by Terner and coworkers (1950) but it causes substantial swelling and depolarizes the membranes of cells in the slice (see Section IV D). Some amino acids and glycolytic and citric acid cycle intermediates have been investigated as energy sources by Joanny and coworkers (1966) who conclude that they have no advantages over glucose in this respect. Difficulties in the use of pyruvate have been discussed by Cremer (1967). Further comments on the use of these and other substrates are to be found in reviews by McIlwain and Rodnight (1962), Elliott and Wolfe (1962), and McIlwain (1966).

For metabolic experiments where respiration is to be measured conventional manometric methods can be used. Arrangements for applying electrical stimulation to the slices in a manometer vessel are described by Ayres and McIlwain (1953), McIlwain (1954), and McIlwain (1960). It is frequently advantageous to clamp the slice between the stimulating electrodes, this is particularly so when it is necessary to transfer the slice rapidly to another medium in order to monitor the very rapid metabolic changes which result from electrical stimulation. A suitable apparatus for stimulation and quick transfer is described by Heald and McIlwain (1956), and McIlwain (1960).

A substantial advance in the technology of slice investigation was the description (Li and McIlwain, 1957), and subsequent improvements (Hillman and McIlwain, 1961; Gibson and McIlwain, 1965) of an apparatus enabling the membrane potential of the cells in a slice to be measured. The slice is maintained on a grid at the surface of the incubation medium. Microelectrodes are advanced from above by a micromanipulator and potential changes from these amplified and recorded on an oscilloscope. Regions of stable negative potential from —40 to —90 mV can be recorded as the electrode penetrates the slice and these are thought to represent the successful impalement of a neuron (Section III C). The slice and medium are oxygenated by gas flow over the surface, arrangements are included for additions to and withdrawal of the medium and the temperature of the whole apparatus is maintained by a thermostatically regulated bath. Yamamoto and McIlwain (1966) have shown that action potentials can be evoked and postsynaptic events recorded in pyriform cortex slices *in vitro*.

B. The Intracellular Volume of Brain Slices and Swelling

In order to establish the concentration of ions and other substances within the various types of cell in brain slices and thus the gradients of these substances developed across the membrane of the cell, it is necessary to know the intracellular volume of the cells in the slice. The use of glass electrodes specific for sodium or potassium ions (McLaughlin and Hinke, 1966) may offer an alternative method for measurements of the concentrations of these ions, but for studies on other substances the intracellular volume must be known. It is not sufficient to measure the volume of the slice and assume that this represents the intracellular volume, since there is a substantial interstitial volume into which the medium can readily diffuse.

The total volume of the slice is estimated by measuring the wet or dry weight of the slice and calculating the total volume from a knowledge of the water content and specific gravity of cerebral tissue. A characteristic value for the water content of cerebral tissue is about 800 mg/g and a specific gravity of 1·00 is usually assumed.

Alternatively, the slices may be equilibrated with a substance which penetrates all the cells and the total volume calculated by assuming that the concentration of this substance is the same as it is in the external medium. Tritiated water has been used to estimate the total water space of slices; equilibration takes place with the total water content of the slices within two minutes (Cohen and coworkers, 1968).

The volume of the interstitial space of a slice has been estimated by equilibrating the slices with a known concentration of a substance to which the cells are not permeable. By subtracting the volume of the interstitial space from the total volume, a measure of the intracellular volume can be obtained. Various markers for the interstitial space have been used of which ^{14}C-inulin is probably the most satisfactory. Protein, and therefore presumably other macromolecules like inulin, do not penetrate the membrane of cells (Pappius and coworkers, 1962).

The use of chloride as an indicator for interstitial space depends on the assumption that the membrane potential is maintained near its normal value in the slice. In muscle cells the chloride ion distributes to maintain its electrochemical potential equal on both sides of the membrane (Adrian, 1961). If the membrane potential of the cells in the slice drops then chloride, being freely diffusible across the cell membrane (Thomas and McIlwain, 1956), will enter the cell to an extent predictable from the change in membrane potential (Gibson and McIlwain, 1965). Increases in the chloride space therefore reflect decreases in the membrane potential rather than increases in the interstitial space, and the same considerations apply to the use of thiocyanate as an indicator of the interstitial space in slices. It is noteworthy that a wide variety of conditions expected to decrease the membrane potential of the cells in the slice caused much larger increases in the chloride and thiocyanate spaces than in the sucrose or inulin spaces (Pappius and Elliott, 1956; Bourke and Tower, 1966a). Whether sucrose can be used as an indicator of interstitial space is controversial; Pappius and Elliott (1956) found that the sucrose space was some 50 per cent greater than the inulin space, thereby implying that sucrose can penetrate the cells. Bourke and Tower (1966a), on the other hand, found that the sucrose and inulin space were equivalent.

Numerous studies have been carried out on these lines, and a rather complex situation has been revealed. It is clear that the total volume of the slice is very sensitive to changes in its environment. When cerebral cortex slices are incubated in a glucose saline they swell by imbibition of water (Elliott, 1946; Pappius and Elliot, 1956; Varon and McIlwain, 1961). During the first thirty minutes of incubation the slices swell approximately 20 per cent of their original weight, and thereafter at about 0·1 per cent of their original weight per minute (Varon and McIlwain, 1961). The extent of swelling is increased by delays during the preparation and transfer of the slices to the medium (Varon and McIlwain, 1961; Bachelard and coworkers, 1962) and also if the subsequent incubation is carried out under anaerobic conditions (McIlwain, 1952; Pappius and Elliot, 1956; Bourke and Tower, 1966a; Franck and coworkers, 1968). Addition of glutamate (10 mM) or high potassium concentrations to the incubation medium causes increased swelling (Pappius and coworkers, 1958), although replacement of Cl^- by the less diffusible isethionate as counter-ion largely prevents K^+-induced swelling (Bourke and Tower, 1966a). Ouabain (10 μM) in the medium causes further swelling (Bourke and Tower, 1966a).

Reducing the calcium concentration of the medium to 0·75 mM rather than the usual 2·3 mM reduces the extent of swelling (Keesey and coworkers, 1965) but complete lack of calcium increases swelling (Dawson and Bone, 1965; Ames and coworkers, 1967). Addition of basic proteins to the medium prevents inulin accessible swelling (Varon and McIlwain, 1961), and blood plasma fractions IV and V are also reported to reduce swelling (Davrainville and Gayet, 1965).

During the initial swelling there is a loss of potassium from the slice and an uptake of sodium and chloride (Leaf, 1956; Bachelard and coworkers, 1962; Keesey and Wallgren, 1965; Bourke and Tower, 1966b; Franck and coworkers, 1968). The loss of potassium has been estimated to be about 40 μ equiv./g tissue. About 80 μ equiv./g of sodium is taken up along with about 40 μ equiv./g of chloride. The loss of potassium is balanced by the uptake of sodium, and the additional sodium is taken up along with chloride as counter-ion (Leaf, 1956; Bachelard and coworkers, 1962). Movement of potassium out of the tissue and sodium and water in has also been observed in retinal tissue incubated under conditions of osmotic stress (Ames and coworkers, 1965). Although hypertonic media depress swelling (Pappius and Elliott, 1956), the phenomena cannot be entirely explained as a movement of water in accommodation to differences in osmotic pressure between the inside and outside of the cell because during the swelling the slice takes up the salt in the incubation medium (Leaf, 1956).

The nature of the compartments into which the water is taken up has been investigated by determining the changes in the volume of the various spaces in the slice. The initial rapid swelling and the swelling associated with metabolic disturbances and high potassium concentrations are mainly in the non-inulin space, i.e. intracellular (Varon and McIlwain, 1961; Franck and coworkers, 1968). However, Pappius and coworkers (1962) find that 80 per cent of the initial swelling is in the inulin accessible space. The slow rate of swelling after the initial disturbance is mainly in the inulin accessible space (Varon and McIlwain, 1961). In a recent study Cohen and coworkers (1968) have observed that at 37°C there is an inulin space not present at 0°C. The 37°C inulin space is about 40 per cent of the 0°C inulin space and is insensitive to the physiological state of the slice.

The inulin space represents about 30–50 per cent of the total water space of the slice depending on conditions. About 10–20 per cent of the inulin space is estimated by Keesey and coworkers (1965) to be due to damaged cells at the surface of a singly cut slice. However, Pappius and coworkers (1962) by microscopic observation estimate that most of the inulin space is due to damaged cells. All estimates of the volume of the interstitial space by the inulin method in slices yield values much higher than that calculated by electron microscopic observation for reasons discussed in Section II A.

What these spaces represent in terms of structure is not well understood. By studying the swelling characteristics of cerebral cortex from cats at various stages of development, Tower and Bourke (1966) have shown that potassium-induced swelling only occurs after the physiological and morphological maturation of cortical neurons. At a later stage (three months postnatal) a slice swelling accessible to chloride becomes observable which these authors associate with the proliferation of cortical glial cells. Electron microscopic examination of swollen slices (discussed by De Robertis and Gerschenfeld, 1961) suggests that astrocytic glia contribute most to swelling.

It seems clear that one of the causes of swelling is due to metabolic disturbances which result in the failure of the membrane pump to extrude sodium. The increased intracellular sodium is accompanied by an increased intracellular chloride, which can be shown to result in an increased osmotic pressure difference between cell and medium (Leaf, 1956).

Regardless of the mechanism of swelling and the interpretations to be placed on the spaces measured by the various markers, it is clear that the phenomena as a whole must be taken into account in the conduct and interpretation of experiments measuring ion transport in brain slices. The movement of a substance into the inulin-inaccessible space is the best criteria of movement of that substance across the membranes of the cells in the slice. This criteria has been adopted by McIlwain and his colleagues (Varon and McIlwain, 1961; Bachelard and coworkers, 1962; Gibson and McIlwain, 1965; Keesey and Wallgren, 1965; Keesey and coworkers, 1965); however, the cautionary remarks on the use of inulin by Bourke and Tower (1966a) deserve attention. The sensitivity of swelling phenomena to metabolic disturbance, and in particular the effect of temperature on the inulin space noted by Cohen and coworkers (1968), make it advisable to estimate it under the exact conditions which are used to investigate the uptake of a substance into the cells of a slice.

C. Movements of Potassium, Sodium and Chloride Ions

Following the initial disturbance potassium is taken up by the slices (Krebs and coworkers, 1951) and sodium expelled from them (Bachelard and coworkers, 1962; Bourke and Tower, 1966b). In experiments described by Terner and coworkers (1950) potassium uptake was dependent on the presence of both glucose and L-glutamate in the medium, and the uptake of potassium was approximately equivalent to the uptake of L-glutamate. The apparent effect of glutamate in this case was probably due to the increase in swelling which it causes (Section IV D). Sodium in the incubation medium is necessary for potassium uptake (Pappius and coworkers, 1958), oxidizable substrates and aerobic conditions are also necessary (Bourke and Tower, 1966b).

Sodium extrusion from slices appears to be more sensitive to the trauma resulting from their preparation, and the concentrations of sodium in them do not approach the *in vivo* value as nearly as do the potassium concentrations (Bachelard and coworkers, 1962). Reducing the calcium concentration of the medium so that the tissue concentration of calcium approximates more closely to that found *in vivo* results in increases in the slice concentration of potassium and decreases in the concentration of sodium. (Lolley and McIlwain, 1964; Keesey and coworkers, 1965; Bourke and Tower, 1966b). Bourke and Tower (1966b) recommend a higher potassium concentration in the medium (16 mM) and the substitution of isethionate for chloride in order to produce concentrations of potassium and sodium in the slice more nearly approaching those found *in vivo*. Some representative ratios of the potassium, sodium and chloride concentrations between the intracellular compartment of slices and the medium are shown in table 2.

When slices are electrically stimulated there is a loss of potassium from the non-inulin space and an uptake of sodium but little change in the chloride distribution (Bachelard and coworkers, 1962; Cummins and McIlwain, 1961; Keesey and coworkers, 1965; Keesey and Wallgren, 1965). On the cessation of stimulation there is a re-uptake of potassium, and extrusion of sodium.

Table 2. Distribution of sodium, potassium and chloride between medium and intracellular compartment in incubated slices

Ion	Tissue concn/ CSF concn. (from table 1)	Concn. in slice inulin inaccessible space/concn. in medium				
		Initial		After incubation		
Na^+	0·34	0·23[2]	0·22[3]	0·69[1]	0·31[2]	0·34[3]
K^+	31	45[2]	43·6[3]	20·9[1]	22[2]	24·3[3]
Cl^-	0·31	0·107[2]	0·123[3]	0·49[1]	0·107[2]	0·286[3]

[1]Varon and McIlwain (1961)
[2]Gibson and McIlwain (1965), figure for chloride calculated assuming equilibrium with a membrane potential of —60 mV
[3]Bourke and Tower (1966)

Measurements of the flux of potassium and sodium during stimulation showed impressive increases, ×2·5 for potassium (Cummins and McIlwain, 1961); ×5 for sodium (Keesey and Wallgren, 1965).

The membrane potential of the cells of a slice can be investigated and changes as a result of various conditions of incubation examined. Potassium concentrations in the media of 6 mM and sodium concentrations of 25–300 mM are necessary for the maintenance of a membrane potential of approximately −60 mV (Hillman and McIlwain, 1961). Addition of high potassium concentrations to the incubation medium causes depolarization of the cell membrane (Li and McIlwain, 1957; Hillman and McIlwain, 1961). From a knowledge of the membrane potential of the cells in the slice, it is possible to calculate from the Nernst equation that the electrochemical gradient of potassium is slightly greater than that corresponding to the membrane potential, while the sodium gradient is opposite to that predicted by the membrane potential. Chloride ions distribute according to their electrochemical potential (Gibson and McIlwain, 1965; Keesey and coworkers, 1965).

The permeability of the cell membrane to sodium relative to its permeability to potassium can be estimated from measurements of the membrane potential (Hodgkin, 1958). The permeability to sodium is about 0·01–0·08 of that to potassium (Gibson and McIlwain, 1965). On stimulation these workers found that the sodium permeability of the membrane rose about ten-fold.

In summary, the cells of brain slices when properly prepared and maintained can be shown to exhibit the following properties, similar in general to those of other excitable cells.

i) A resting membrane potential of approximately −60 mV.

ii) Maintenance of potassium concentration inside the cell in excess of that predicted by the membrane potential; and a sodium gradient in the opposite sense to that predicted by the membrane potential.

iii) On stimulation, a loss of potassium, and an increased membrane permeability to sodium, resulting in entry of sodium to the cells.

iv) After stimulation, uptake of potassium and extrusion of sodium.

The unequivocal ascription of these properties to neurons is difficult because of the possibility that glial cells are involved. Some studies in this connection on sodium and potassium efflux from slices are discussed in Section V. It seems unlikely that the convoluted structure of glial cells would allow successful impalement by the micro-electrode but the possibility must be borne in mind.

D. Calcium Movements

Recent studies on calcium transport in squid axon have disclosed the existence of an ouabain-insensitive calcium-stimulated component of sodium efflux. Calcium influx is stimulated by high sodium concentrations inside the axon, and inhibited by high external sodium concentrations (Baker and coworkers, 1969). It seems, therefore, that calcium can cross the axon membrane by counter-transport with sodium, and high internal sodium concentrations would be expected to increase calcium influx. Furthermore, there is evidence (Blaustein and Hodgkin, 1969) that calcium efflux can be coupled to some component of sodium influx.

Studies by Lolley (1963) and Charnock (1963) on calcium in brain slices suggest that a proportion of the calcium of the tissue is bound. The amount bound can be varied by changing the calcium concentration in the medium, but is not susceptible to influence by metabolic inhibitors. A small increase in the calcium flux was observed by Lolley (1963) when the slices were electrically stimulated. However, Fujisawa and coworkers (1965) have shown that ^{45}Ca influx is increased by metabolic inhibition, high potassium concentrations or ouabain. This observation would be consistent with those of Baker and coworkers (1969) on squid axon, since the treatment of the slices would raise their internal sodium content and thus increase the calcium influx by the sodium–calcium counter-transport mechanism. Fujisawa and coworkers (1965) also found that efflux of calcium was increased by metabolic poisons, possibly because these also cause increased sodium influx which can be coupled to calcium efflux, as proposed by Blaustein and Hodgkin (1969).

A further complication in explaining calcium movement in nervous tissue is the existence of metabolically dependent calcium uptake into mitochondria (Brierley, 1963) and into microsomes from brain (Otsuka and coworkers,

1966; Robinson, 1968; see also Section V D). These would have the effect of reducing the free intracellular concentration of calcium in normally metabolising tissue.

E. Uptake of Amino Acids and Other Substances

Brain slices in common with other tissues are capable of concentrative accumulation of many amino acids. The phenomena was noticed first by Stern and coworkers (1949) in the case of L-glutamate and has been the subject of many subsequent investigations. The results of several of these have been drawn together in table 3.

In general, concentrative accumulation of amino acids by brain slices requires an energy source in the incubation medium (Stern and coworkers, 1949; Elliott and van Gelder, 1958; Takagaki and coworkers, 1959; Abadom and Scholefield, 1962a,c; Nakamura and Nagayama, 1966), and is depressed by metabolic inhibitors such as 2,4-dinitrophenol (DNP). Accumulation is usually mediated by two mechanisms, one which exhibits a Michaelis–Menten-type dependency on the concentration of the accumulated amino acid, and the other which has a linear dependency on the concentration. Almost invariably sodium ions are necessary for concentrative accumulation (Takagaki and coworkers, 1959) and they cannot be replaced by lithium, choline or potassium ions (Lahiri and Lajtha, 1964). High potassium concentrations inhibit the concentrative accumulation of amino acids. It is interesting that acetylcholine specifically reverses potassium inhibition of amino acid accumulation (Nakazawa and Quastel, 1968b). Counter-transport of methionine by histidine has been shown by Nakamura (1963).

The extent of accumulation of different amino acids varies, and tissue to medium ratios relative to that for L-glutamate (from the data of Blasberg and Lajtha, 1965) are shown in table 3. It is interesting that γ-amino butyric acid (GABA), which is unique to nervous system and is an inhibitory transmitter, is accumulated to the greatest extent and has the highest affinity for its carrier. However, the tissue to medium ratios shown in table 3 do not necessarily represent the relative number and activity of the carrier sites because of the complications caused by carrier-mediated efflux (Blasberg, 1968; see also Section I B).

Blasberg and Lajtha (1965) have also investigated the carrier specificity for different amino acids by examining mutual inhibition of accumulation by different combinations of amino acids. (see also Abadom and Scholefield, 1962b; Tsukada and coworkers, 1963; Neame, 1964a, 1968). They conclude that the carriers have rather a wide specificity based principally on size and charge. They identify six groups of amino acids each group having a common carrier; small neutral, large neutral, small basic, large basic, acidic and a

Table 3. Some properties of amino acid uptake by brain slices

Amino acid	Requirement for energy source	Requirement for Na^+	Tissue: medium ratio as % of that for L-glutamate[1]	Influx as % of that for L-glutamate[19]	Saturability of uptake system	K_M(mM)[2]	V_{max} (μmoles/min . g)	Inhibitors
L-alanine[1,5]			69		+[5]	0·44[5]	0·42[5]	
D-alanine[1,5]			155		+[5]	1·47[5]	0·44[5]	
α-amino isobutyrate[1,5,6]	+[6]	+[6]	114	67	+[5]	0·43[5]	0·25[5]	DNP[3]
γ-amino butyrate[1,7,8,9,10]	+[7,9,10]	+[8,9]	225	28	+[7,8,9]	0·022[8]	0·115[8]	K^+[9], ouabain,[8,9] DNP[8,10]
L-glutamate[1,4,9,11,12]	+[4,9,11,12]	+[9,12]	100	100				K^+[9], ouabain[4,9]
D-glutamate[1,9,12]	+[9,12]	+[9,12]	187	51	+[9]			K^+[9], ouabain[9]
glycine[1,3,4,5,18]	+[3,4]	+[18]	131	39	+[5]	0·44[5]	0·58[5]	ouabain[4,18] DNP[4]
L-histidine[1,13,14,15]	+[14]	+[13]	94		+[15]			K^+[13]
L-leucine[1,6]		+[6]	20	24				
D-leucine[1,6]		+[6]	16					
Cyclo-leucine[1,6]		+[6]	53		+[6]			
L-lysine[1,6,15]		+[6]	24	9				
D-lysine[1,6]		+[6]	11					
L-methionine[1,15]			23					
D-methionine[1]			27					
L-ornithine [15]								
L-phenylalanine[1]			15					
L-proline[1,15]			85					
Dihydroxyphenylalanine[17]	+[17]	+[17]						ouabain[17], protoveratrine[17]
5-hydroxytryptophan[16]		+[17]						

Positive signs in the table indicate a demonstration of the effect in the reference denoted by the superscript. Lack of positive sign does not indicate a negative finding.

[1]Tissue:medium ratio at an amino acid concentration of 1·5 mM expressed as a percentage of that found for L-glutamate, which was 14·1. From Blasberg and Lajtha (1965). [2]Determined from kinetic measurements. [3]Abadom and Scholefield (1962a,c). [4]Gonda and Quastel (1962). [5]Smith (1967). [6]Lahiri and Lajtha (1964). [7]Elliott and Van Gelder (1958). [8]Iversen and Neal (1968). [9]Tsukada and coworkers (1963). [10]Nakamura and Nagayama (1966). [11]Stern and coworkers (1949). [12]Takagaki and coworkers (1959). [13]De Almeida and coworkers (1965). [14]Nakamura (1963). [15]Neame (1961). [16]Schanberg and Giarman (1960). [17]Yoshida and coworkers (1963). [18]Nakazawa and Quastel (1968). [19]Blasberg and Lajtha (1966). Influx was measured at amino acid concentrations of 0·1 mM and results are expressed as a percentage of that found for L-glutamate, which was 0·568 μmole/min . g

carrier probably specific for GABA. There does seem to be some interaction between an amino acid and the carrier of a different group, thus the carrier for acidic amino acids is partially inhibited by small basic amino acids. Blasberg and Lajtha (1965) suggest that this is because any particular amino acid has an affinity for its primary carrier site and a lesser affinity for a secondary carrier site. The carriers are not in general specific for particular optical isomers, although there are frequently differences in the uptake of the D and the L form (Neame, 1964b). Somewhat surprisingly, D-glutamate, D-alanine and D-methionine are all accumulated more than their L-isomers (Blasberg and Lajtha, 1965). In the case of alanine, this has been ascribed to a more rapid efflux of the L-isomer (Neame and Smith, 1965). The carrier for GABA is separate from that of glutamate (Tsukada and coworkers, 1963; Blasberg and Lajtha, 1965; Nakamura and Nagayama, 1966), but appears to be that which also transports β-alanine (Nakamura and Nagayama, 1966; Iversen and Neal, 1968).

The relationship of the parameters of uptake of amino acids in slices to their *in vivo* distribution has been the subject of a number of investigations. Comparison of the uptake ability for lysine and glutamate by slices from different parts of the brain did not indicate any parallelism between uptake and *in vivo* concentrations (Levi and Lajtha, 1965). A similar result for histidine and methionine is reported by Nakamura (1963). However, GABA uptake ability in slices from cortex, diencephalon and mesencephalon and subcortical white matter is correlated with the GABA concentrations in these parts of the brain *in vivo* (Nakamura and Nagayama, 1966). Some correlations between uptake ability of slices from various species and *in vivo* amino acid concentrations were observed by Levi and coworkers (1967).

In a study of the uptake of several amino acids by brain slices and other tissues, Neame (1962, 1968) concludes that in general brain slices take up amino acids more actively than other tissues. The specificity of the amino acid transport systems in brain resembles those in tumour tissue more closely than those in kidney or intestine.

Concentrative accumulation by slices of various other substances have been investigated and the relevant details of some of these are collected in table 4. Chemical transmitter substances and their precursors have been a particular subject of attention in this respect. The concentrative accumulation of choline (Schuberth and coworkers, 1966; Schuberth and coworkers, 1967) is similar in general characteristics, i.e. saturability and sodium dependency, to the uptake of amino acids and to the uptake of choline in erythrocytes (Martin, 1968) and squid axon (Hodgkin and Martin, 1965). The existence of a carrier-mediated transport mechanism for choline suggests that this substance should not be regarded as a suitably inert replacement for sodium in experiments on permeability and transport. Schuberth and

Sundwall (1967) report that acetylcholine is also accumulated against a concentration gradient. It seems possible however that concentrative acetylcholine uptake in nervous system simply reflects the rather broad specificity of the choline uptake system (Marchbanks, 1969).

Other putative transmitter substances such as noradrenaline, 5-hydroxytryptamine and histamine are taken up by sodium-dependent, saturable transport mechanisms (see table 4).

IV. RELATIONSHIP OF METABOLISM TO TRANSPORT IN BRAIN SLICES

A. Respiration, ATP and Phosphocreatine

The rate of oxygen uptake of cerebral cortex slices in glucose saline is about half the rate of brain tissue *in vivo*. When electrical stimulation is applied to the slices *in vitro* the rate of respiration is substantially increased (McIlwain, 1951; McIlwain and Joanny, 1963). The rate of aerobic glycolysis of slices can also be shown to be increased by electrical stimulation (McIlwain and Tresize, 1956). High concentrations of potassium ions (30–50 mM) and omission of calcium ions cause increases in the respiration of slices (Ashford and Dixon, 1935; Dickens and Greville, 1935; Gore and McIlwain, 1952; Hertz and Schou, 1962; Hertz, 1966).

The concentrations of ATP and creatine phosphate in the slice have been found to fluctuate according to the functional state of the slices. During preparation of the slices the concentration of phosphocreatine falls, and the concentration of inorganic phosphate rises (McIlwain and coworkers, 1951). During incubation phosphocreatine is synthesized and inorganic phosphate removed, and when electrical stimulation is applied there is a transitory drop in the concentration of ATP, followed by a much more prolonged and severe drop in the levels of phosphocreatine (Heald and McIlwain, 1956). After the cessation of stimulation the level of phosphocreatine rises to normal values (Heald, 1954). Qualitatively similar changes have been observed in the high energy phosphate levels when other excitable tissues such as mammalian C-fibres are stimulated (Greengard and Straub, 1959).

The requirement of the sodium pump for ATP, demonstrable enzymically as the Na^+–K^+-ATPase (Section I C), is the connecting link between the effect of electrical stimulation and consequent changes in high energy phosphate levels and respiration. Stimulation causes depolarization of the cell membranes with a consequent rise of their permeability to sodium ions. Sodium influx occurs as well as potassium efflux and the raised intracellular concentration of sodium activates the sodium pump. The extrusion of sodium against its electrochemical potential gradient in the squid axon

Table 4. Uptake of biogenic amines and other substances by brain slices

Substance	Requirement for energy source	Requirement for Na^+	Tissue/medium ratio	Saturability	K_M(mM)[1]	V_{max} μmoles/min/g	Inhibitors
Choline[2,3]	+[2]	+[3]	3 at 100 μM[2]	+[2]	0·22[2]	0·013[2]	Ouabain[3], K^+[3] Hemicholinium[2] DNP[2]
Acetylcholine[4]	+[4]	+[4]	2·5 at 100 μM[4]	+[4]	0·008[4]	0·08[4]	K^+[4]
Carbachol[7]	+[7]		9·2 at 1 μM[7]	+[7]	0·041[7]	0·0026[7]	d-tubo-curarine[4] K^+[4]
Carnosine[5]	+[5]		9 at 7 μM[5]	+[5]			
Histamine[6]	+[6]		1·8 at 2 mM[6]				DNP[6]
Noradrenaline[10]			4 at 50 μM[10]	+[10]			ouabain[10], DNP[10]
5-hydroxytryptamine[8]			15 at 0·011 μM[8]	+[8]	0·00057[8]	0·0009[8]	ouabain[8], DNP[8]
Thiamine[9]		+[9]	1·6 at 0·2 μM[9]	+[9]			ouabain[9]

Positive signs in the table indicate the demonstration of the effect in the reference denoted by the superscript. Lack of positive sign does not indicate a negative finding.

[1] K_M determined from steady state concentration ratios (see text p. 149). [2] Schuberth and coworkers (1966). [3] Schuberth and coworkers (1967). [4] Schuberth and Sundwall (1967). [5] Abraham and coworkers (1964). [6] Neame (1964b). [7] Creese and Taylor (1967). [8] Blackburn and coworkers (1967). [9] Sharma and Quastel (1965). [10] Dengler and coworkers (1961).

requires ATP (Caldwell and Keynes, 1957). Replenishment of the small intracellular stores of this substance is accomplished by the breakdown of phosphocreatine and there is an overall liberation of inorganic phosphate. High levels of inorganic phosphate stimulate the oxidative metabolism of glucose to form more ATP.

Overall quantitative aspects of the processes of metabolism and transport have been discussed by McIlwain and Joanny (1963). For each pulse applied 5·3 mμequiv.Na/g enters the tissue and an increment of 1·56 mμmoles O_2/g of respiration is observed. Assuming that the extra respiration is concerned with providing energy to pump this sodium out, a ratio of sodium transported to high energy phosphate hydrolysed of 0·57 can be calculated. This is much lower than the same ratio calculated for erythrocytes by Glynn (1962) (Section I C); possibly there are other processes competing for the use of phosphate in nervous tissue. On the basis of the respiration in a salt free medium, Whittam (1962a) estimates that about 40 per cent of the respiration of brain is concerned with providing energy for the salt pumping mechanisms.

The kinetics of the process are puzzling because, as McIlwain (1963) has observed, the influx of sodium on electrical stimulation is not fast enough to account satisfactorily for the very rapid breakdown of high energy phosphates which takes place almost immediately on stimulation.

Because of the similarities between the electrical and potassium ion stimulation of slice metabolism, i.e. increased respiration and decreased phosphocreatine levels (McIlwain, 1952; Biesold, 1967), it has been argued that high potassium concentrations exert these metabolic effects because they depolarize the nerve cell membranes of the slice (Quastel, 1962). It is, however, clear that high potassium concentrations do not exactly mimic the effect of electrical stimulation because glycine uptake which is inhibited by high potassium concentrations is unaffected by electrical stimulation of the slice (Nakazawa and Quastel, 1968a). Furthermore, tetrodotoxin (Section IV B) prevents the respiratory response to electrical stimulation but not to high potassium concentration (McIlwain, 1967).

B. Inhibitors of Ion Movements

Recent studies on the mechanisms involved in ion movements in slices have utilized inhibitors of these processes. One of the most important of these for the understanding of active transport has been the cardiac glycoside ouabain. Ouabain is an inhibitor of the accumulation of potassium and extrusion of sodium in erythrocytes (Schatzmann, 1953; Glynn, 1957) and squid axon (Caldwell and Keynes, 1959). Ouabain is also an inhibitor of the Na^+–K^+-ATPase from crab nerve (Skou, 1960) and from erythrocytes

(Dunham and Glynn, 1961). Experiments on erythrocytes (Whittam and Ager, 1964) and squid axon (Caldwell and Keynes, 1959) indicate that ouabain competes with the potassium binding site of the transport system on the outside of the cell, and that it competes with the potassium activating site of the Na^+–K^+-ATPase.

In incubated cortex slices ouabain at μM concentrations causes inhibition of potassium accumulation and sodium extrusion (Yoshida and coworkers, 1961; Schwartz, 1962; Whittam, 1962a; Swanson and McIlwain, 1965; Bourke and Tower, 1966b). Similar effects occur *in vivo* (Cserr, 1965). Ouabain also inhibits the Na^+–K^+-ATPase of cerebral tissue (Deul and McIlwain, 1961; Aldridge, 1962; Bonting and coworkers, 1962; Skou, 1962; Whittam and Blond, 1964).

Early reports on the effects of ouabain on the respiration of slices gave conflicting results (Wollenberger, 1947; Yoshida and coworkers, 1961; Schwartz 1962; Whittam, 1962a; Gonda and Quastel, 1962). Schwartz (1962) suggested that differing concentrations of calcium in the incubation medium might account for these discrepancies. It has been found that at low (0·75 mM) concentrations of calcium in the incubation medium ouabain decreases oxygen uptake, while at higher concentrations of calcium (1·5–3 mM) respiration is increased by ouabain (Bourke and Tower, 1966b; Swanson and Ullis, 1966; Ruščák and Whittam, 1967). Ouabain inhibits the potassium stimulation of respiration (Gonda and Quastel 1962; Ruščák and Whittam, 1967).

The effects of ouabain and calcium on the metabolism of high energy phosphates has been examined by Rose (1965) and Swanson (1968a). Rose (1965) found that ouabain prevented incorporation of inorganic phosphate into ATP and acid-insoluble phosphates. In calcium-free medium ouabain had a very marked effect in decreasing phosphate incorporation into ATP. Swanson (1968a) observed that under conditions of low calcium concentration the intracellular potassium of the slice dropped slightly, but there was a more substantial drop in phosphocreatine concentrations. Ouabain accentuated further the changes found in calcium-low media and in the presence of calcium reduced intracellular potassium and phosphocreatine concentrations. It is noteworthy that ouabain not only prevents the re-accumulation of potassium and extrusion of sodium after stimulation, but also prevents the recovery of phosphocreatine levels after stimulation (Swanson, 1968b). While it is clear that an important effect of ouabain is to block the sodium pump, and the cation shifts can be readily explained on this basis, it seems likely that ouabain also has an effect on the processes supplying high energy phosphates; otherwise it is difficult to explain the effects of ouabain on the phosphocreatine concentrations (Swanson, 1968a,b). However, Swanson (1968b) was not able to demonstrate any effect on the

P:O ratios or respiratory control ratios of mitochondria isolated from slices incubated with ouabain.

Tetrodotoxin, a poison isolated from puffer fish, has been shown to inhibit the ion movements consequent upon stimulation of giant axons of lobster and squid (see Kao, 1966). In incubated cortex slices tetrodotoxin has little effect unless the tissue is electrically stimulated. But it prevents the loss of potassium from the slice, and the increase of respiration which results from electrical stimulation (McIlwain, 1967); prevention of sodium influx during stimulation was not so evident. Swanson (1968b) reported that tetrodotoxin prevents potassium loss on stimulation, and observed as well that it prevented the sodium influx and lowering of phosphocreatine concentrations that usually result from stimulation. Prevention by tetrodotoxin of the characteristic sodium movements on stimulation has also been observed by McIlwain and coworkers (1969). These results suggest that the principal action of tetrodotoxin at concentrations of 1–5 μM in cerebral tissues is to interfere with the passive downhill movements of sodium that normally occur upon stimulation. This is similar to its action in other excitable preparations where it prevents the increased permeability to sodium which takes place during depolarization (Kao, 1966).

Protamines and other basic proteins inhibit the respiratory response of slices to high potassium concentrations and electrical stimulation (McIlwain, 1959). These actions of protamine on the slices are prevented by gangliosides (McIlwain, 1961). Chlorambucil prevents the restoration of slice concentrations of potassium depleted by stimulation, without affecting the Na^+–K^+-ATPase (Evans and McIlwain, 1967). It is suggested that a nucleophilic group concerned with cation transport has been alkylated by chlorambucil.

C. Effects of Calcium

An attempt to rationalize all the complex actions of calcium on the behaviour of slices would be foolhardy given the present state of knowledge. However, certain effects can be discerned. Lowered calcium levels in the bathing medium cause nerve cell membranes to become more unstable and more susceptible to depolarization (Curtis and coworkers, 1960). This would account for the leakiness of slices when incubated in calcium-free media (Section III B). Potassium loss on electrical stimulation is decreased when the calcium concentration in the medium is high (Joanny and Hillman, 1964), and the increased respiration in low-calcium medium could be due to greater activity of the sodium pump resulting from membrane leakiness.

High intracellular levels of calcium would be expected to uncouple mitochondrial oxidative phosphorylation (Brierley, 1963), and to cause competition for ATP supplies by the microsomal calcium-uptake system (Section

V D). At the same time calcium is an inhibitor of the Na^+–K^+-ATPase (Deul and McIlwain, 1961; Whittam and Blond, 1964) and in this respect would be expected to conserve intracellular ATP levels. Ruščák and Whittam (1967) conclude that this is the reason for the abolition of the metabolic effects of calcium when the sodium pump is inactivated. The balance between these two opposing possibilities is difficult to predict until further information is available about the properties and energy requirements of the sodium–calcium counter transport mechanism discussed in Section III D.

D. Effects of Glutamate and Depolarizing Amino Acids

The effects of L-glutamate on the behaviour of cerebral cortex slices have interested investigators not only because this substance is important metabolically but also because it has the property of decreasing the membrane potential of excitable cells (Curtis and coworkers, 1960).

Glutamate causes increases in the potassium content of cerebral cortex slices (Terner and coworkers, 1950; Takagaki and coworkers, 1959). However, it also causes substantial swelling of the slices (Pappius and Elliott, 1956; Tsukada and coworkers, 1963) and the actual intracellular concentration of potassium was not found to be increased by glutamate in the studies by Bradford and McIlwain (1966) and Harvey and McIlwain (1968). Glutamate causes a fall in the slice concentrations of phosphocreatine (McIlwain, 1952; Bradford and McIlwain, 1966; Harvey and McIlwain, 1968), and in addition, causes an increase in the concentration of sodium in the inulin-inaccessible space (Bradford and McIlwain, 1966; Harvey and McIlwain, 1968). Similar effects have been observed in incubated retinal tissue (Ames, 1956; Ames and coworkers, 1967). Glutamate promptly diminishes the resting potential of the cells of the slice (Hillman and McIlwain 1961; Gibson and McIlwain, 1965; Bradford and McIlwain, 1966). Characteristically the depolarization observed is about −30 mV and it can be calculated that permeability of the cell membrane to sodium ions has increased five-fold (Bradford and McIlwain, 1966). The sodium influx increases about three-fold (Harvey and McIlwain, 1968).

Several other amino acids, and related compounds have actions on the membrane potential similar to L-glutamate; L-aspartate, L-cysteate, DL-homocysteate and L-α-aminoadipate (Bradford and McIlwain, 1966). In general, the membrane potential of the cells in the cerebral cortex slice seems less sensitive to depolarization by N-alkylated amino acids than are other excitable cells. A particularly anomalous case is N-methyl-D-aspartate which is more active than L-glutamate in increasing the frequency of firing of cerebral cortical cells of the cat (Crawford and Curtis, 1964) but is hardly active in depolarizing the slice cells *in vitro* at concentrations of 5 mM or lower (Bradford and McIlwain, 1966). In the study by Bradford and McIlwain (1966) N-methyl-DL-aspartate (5 mM) had little effect on the phosphocreatine and sodium and potassium concentrations of the slice, but Harvey and McIlwain (1968) found that 0·1 mM N-methyl-DL-aspartate caused changes in the slice concentrations of sodium and potassium comparable to those produced by 5 mM L-glutamate.

V. TRANSPORT AND METABOLISM IN SEPARATED CELLS AND SUBCELLULAR PARTICLES FROM BRAIN

A. Neurons and Glia

The ratio of glial cells to neuronal cells in nervous tissue can vary rather widely. In rat cortex Friede and Van Houten (1962) found ratios from 0·06 to 0·96, and observed that as the length of the neuronal axon increased so did the number of associated glia. Ratios of glial cells to neurons as high as 10 have been claimed by other authors (Hydén, 1960).

Glial cells are of two main types. Astrocytic glia have processes which ensheath the brain capillaries almost completely and also ramify among the oligodendroglia. Oligodendroglia have processes which closely encapsulate the nerve cell bodies, dendrites and axons and contribute towards the formation of the myelin sheath. Oligodendroglial processes also make contact with astrocytes (De Robertis and Gerschenfeld, 1961). Since glial cells isolate nerve cells from the blood supply, Hydén (1960) has argued that they mediate the transport of substances from the capillaries to the neurons possibly by pinocytosis. The ubiquity of glial cells and their close topological relationship to neurons make it important to study their metabolic and transport properties in order to establish their role in the functioning of nervous tissue.

There have been three main approaches to identifying the characteristics and properties of glial cells.

i) Tissue rich in glial cells such as subcortical white matter or gliomas can be studied and then an assessment made of the glial contribution to the properties of the cerebral grey matter by estimating the ratio of glial cells to neurons in that tissue (Elliott and Heller, 1957; Tower and coworkers, 1961; Hess, 1961; Tower and Bourke, 1966).

ii) Neuronal and glial cells may be isolated by hand microdissection and metabolic and transport studies conducted on the isolated cells (Lowry, 1957; Hydén and Pigon, 1960; Hamburger and Rockert, 1964; Roots and Johnston, 1964; Hydén and Lange, 1965; Hertz, 1966).

iii) Nervous tissue may be gently dispersed by mild homogenization or straining procedures and then submitted to density gradient centrifugation to isolate glial cells (Chu, 1954; Korey, 1957) or to separate neurons and glia (Rose, 1967). None of these methods are entirely satisfactory. Extrapolation of the properties of glial cells from gliomas and white matter to those of cortex involves assumptions about similarities of metabolism in the two environments (see for instance Hess, 1961); and demands data on cell type numbers. Microdissection methods yield very small amounts of tissue for which special methods of investigation must be devised. With structures as convoluted and fragile as neurons and glia the possibility exists of gross cellular damage occurring during dissection, which may affect the metabolic properties differentially. However, the membrane potential of neurons dissected from Deiters' nucleus is about −40 mV (Hillman and Hydén, 1965) so membrane damage cannot be extensive in this case. The homogenization procedures are liable to produce even more cellular damage.

B. Neuronal and Glial Compartments of Ions

Tower and coworkers (1961) calculate that the chloride-inaccessible space of cerebral cortex slices incubated in a physiological medium (27 mM K) for 1 hour is about 43 per cent of the total, some 11 per cent of which is contributed by glial cells. They also calculate that during the incubation sodium is expelled and potassium taken up into the non-chloride space of neurons which is also reduced in volume while the potassium and sodium concentrations and volume of the glial non-chloride space remains constant. The comments in Section III B on using chloride as a marker for extracellular space should be borne in mind in evaluating this data, particularly since the high concentration of potassium used would tend to depolarize neurons. The inability of glial cells to concentrate potassium implied by this result is at variance with the results of Hamburger and Rockert (1964) who using microdissected glial lumps found that they could accumulate potassium on incubation at 37°C, whereas dissected neurons did not. The glial accumulation of potassium could be prevented by metabolic poisons. For technical reasons chloride was replaced by bromide in their incubation medium; this unusual substitution may have some bearing on the results.

An analysis of the kinetics of efflux of radioactive sodium and potassium from slices pre-incubated with these isotopes has been conducted by Hertz (1968). Efflux showed two components, a rapidly exchanging fraction and a slowly exchanging fraction. The rapidly exchanging fraction comprises some 70 per cent of the tissue radioactivity and although the inulin space was not measured it is argued that this cannot all be accounted for by the extracellular space of the slice. Hertz (1968) proposes that glial cells are responsible for at least part of the rapidly exchanging compartment and since under aerobic conditions a high potassium concentration stimulates only the rapid component of the efflux, it is suggested that the high potassium concentration stimulates active uptake of potassium into glial cells.

Fractions of enriched glia obtained by centrifugation procedures appear to take up potassium and histidine (Bradford and Rose, 1967), the process being inhibited by metabolic poisons. The neuronally enriched fraction did not take up potassium or histidine and had no measurable membrane potential, but recent results (Rose, personal communication) indicate that in a high phosphate medium the neuronally enriched fraction can take up potassium. The evaluation of these results is complicated by the possibility of more extreme damage to the neuron than the glia and possible contamination of the glial fraction with nerve ending particles (Rose, 1967; see also Cremer and coworkers, 1968).

Neuroglial cells of the central nervous system of the leach contain high concentrations of potassium (Nicholls and Kuffler, 1965); their membrane potential of about −80 mV may be depolarized by addition of potassium (Orkand and coworkers, 1966). Mammalian glial cells in tissue culture can also be shown to have a membrane potential, sensitive to potassium, of about −60 mV (Hild and Tasaki, 1962).

C. Respiration in Neurons and Glia

Elliott and Heller (1957) estimate that on a wet weight and whole cell basis oligodendroglia respire 4–6 times as fast as astroglia; from this and using a glial:neuronal ratio of 1:1, they calculate that about 85 per cent of the respiration of cortex is due to neurons, and that a neuron respires at about 5–6 times the rate of a glial cell. Hess (1961) has also presented data of this kind and finds that the average cortical neuron with its processes respires at between 16–50 times the rate of the average cortical glial cell, and that 90–95 per cent of the respiration of the cortical tissue may be attributed to neurons. Comparison of the respiratory activity of Betz cells with that of an equivalent volume of neuropile suggests that the respiration rate of glia is at least an order of magnitude lower on a volume basis than that of neurons (Epstein and O'Connor, 1965).

The respiration of hand dissected neuronal perikarya from Deiters' nucleus and lumps of neuroglia were investigated by Hertz (1966) using Cartesian divers. Neuronal respiration was fairly steady for 3 hours and on a cellular basis about 100 times the rate of neuroglial respiration; on a fresh weight basis it was about 6–7 times faster than glial respiration. Neuroglial respiration declined in rate after about 1–2 hours, was more sensitive to lack of sodium in the medium, and showed a transient increase in respiration when potassium was added. The neuronal perikarya did not react to potassium in this way. Hydén and Lange (1965) estimate that on a cellular basis the rate of neuronal respiration is about 10 times that of the glia.

Results from centrifugal fractions enriched in neurons and glia suggested that on a cellular basis neurons respire about 10 times faster than non-neuronal cells (Korey and Orchen, 1959). Neuronal and glial fractions prepared by Rose (1967) respired at much the same rate, and both showed potassium stimulation of respiration (Bradford and Rose, 1967), though the stimulation effect was about half of that usually observed in slices.

The weight of evidence suggests that the average neuronal cell respires about ten times faster than a glial cell. However, the application of this result to estimates of the respiratory rate on a volume basis, or of the contribution each cell type makes to the total respiratory activity of the cortex, is controversial. It depends on the glial:neuronal ratio which, if Friede and Van Houten's (1962) figures are used, indicate that most of the respiration of the cortex is due to neurons. Other authors (see Hertz, 1966) have used a much higher glial:neuronal ratio and conclude that the respiration of cortex tissue is mainly due to glial cells. Further discussion seems profitless unless agreement or more data can be obtained concerning the glial:neuronal ratio, and the relative volume of glial and neuronal cells in the cortex.

Hertz (1966) argues from his results on the potassium stimulation of glial cell respiration that the well known effect of potassium on the respiration of slices is due to their content of neuroglia. If potassium stimulation of respiration can be regarded as diagnostic of the depolarization of an excitable membrane (see Section IV A), then this suggests that glial cells form the major population of cells having excitable membranes in cortical tissue. However, it should be noted that the removal of the dendritic tree from the neuronal perikarya during dissection would cause injury depolarization, thus masking the potassium stimulation of respiration. This evidence is at present too fragmentary and contradictory to allow a radical re-appraisal of the role of glia, and in particular to support the notion of neuronal–neuroglial–neuronal transmission proposed by Hertz (1965). But it is clear that glial cells have a number of interesting transport and metabolic properties which should engage the attention of investigators.

D. Subcellular Particles

The isolation of subcellular organelles from brain by the conventional methods of homogenization and differential centrifugation presents certain difficulties because of the morphological heterogeneity of the tissue. It is difficult to produce preparations of nuclei without extensive contamination from cell debris. Mitochondria produced by conventional procedures are usually heavily contaminated by presynaptic nerve terminals (synaptosomes), a circumstance which has led to erroneous ascription of glycolytic capabilities to brain mitochondria. As a consequence, studies on transport in the subcellular organelles of brain have been limited.

The isolation of synaptosomes complete with their complement of synaptic vesicles, intraterminal cytoplasm and mitochondria (Whittaker, 1959; De Robertis and coworkers, 1962; for reviews see Whittaker, 1965, and De Robertis, 1967) has made possible the biochemical investigation of this remote, but functionally important part of the nerve cell (see Marchbanks and Whittaker, 1969). During the course of these investigations it has become apparent that synaptosomes are capable of metabolic activity of some complexity, and there are indications that a surprising amount of structural and functional integrity is retained despite an extensive, and one would have thought, traumatic process of separation and isolation.

By gel filtration and centrifugation methods (Marchbanks, 1966, 1967), it is possible to separate synaptosomes from their incubation medium, and by equilibration with suitable space markers to measure their internal volume (Marchbanks, 1966). The permeability of the limiting membrane to potassium and sodium has been measured and found to be similar to that of other biological membranes suggesting that the limiting membrane is substantially intact (Marchbanks, 1967).

Synaptosomes respire with glucose as substrate when incubated in a suitable physiological medium. They are capable of phosphorylative activity (Abdel-Latif, 1966), and intrasynaptosomal concentrations of ATP and phosphocreatine rise on incubation; high potassium concentrations in the incubation medium diminish the phosphocreatine levels, and cause a small increase in the rate of respiration (Marchbanks and Whittaker, 1969; Bradford, 1969).

During incubation in physiological saline synaptosomes concentrate potassium about seven-fold as compared with a thirty-fold concentration by cortex slices (Bradford, 1969). Studies on radioactive sodium uptake (Ling and Abdel-Latif, 1968) suggest that there is a potassium-stimulated, ouabain and iodoacetate-inhibited sodium extrusion process demonstrable in synaptosomes.

The uptake of radioactive choline into synaptosomes has been investigated (Potter, 1968; Marchbanks, 1968) and sodium-dependent, hemicholinium-inhibited, carrier-mediated transport of choline across the synaptosome membrane can be demonstrated. The choline concentration required for half saturation of the carrier is similar to that found in cerebral cortex slices and squid axon. Noradrenaline and 5-hydroxytryptamine are also taken up into synaptosomes by a sodium-dependent, ouabain-inhibited process (Bogdanski and coworkers, 1968). Tryptophan uptake into synaptosomes is carrier-mediated, and stimulation of its uptake by preloading with phenylalanine, an example of amino acid counter-transport (see Section I B), can be demonstrated (Grahame-Smith, personal communication).

An evaluation of the synaptosome as a preparation for the study of transport in a specialized part of the nerve cell axon is premature at the moment. Much more needs to be known about the movements of sodium and potassium, and estimation of the potential across the synaptosome limiting membrane presents a considerable difficulty yet to be overcome. Nevertheless, some interesting membrane functions are clearly demonstrable in this preparation, and these merit further investigation in order to try and exploit the possibility of using the preparation to study transport in a specialized part of the nerve cell.

An ATP-dependent binding of calcium in brain microsomes has been shown by Otsuka and coworkers (1966), which might play a role in keeping the intracellular concentration of free calcium ions low. In a further study Robinson (1968) showed that efflux of calcium from the microsomes was increased specifically by NaCl, but reduced by ATP, which suggests that a sodium–calcium counter transport system is operating similar to that discussed in Section III D. It is to be hoped that these interesting observations will be extended, in particular, to a morphological examination of the fractions to determine whether synaptosomes (Yoshida and coworkers, 1966; Lust and Robinson, 1968), or microsomes, or both types of particle can take up calcium.

VI. CONCLUSIONS

The brain slice preparation has yielded the most information concerning the nature of transport processes in brain. The technological requirements for the successful use of this preparation have been identified and it is clear that a substantial parallelism exists between the transport processes observable in slices *in vitro* and between these processes in other tissues as well as in the brain *in situ.*

The processes which maintain sodium and potassium gradients in brain tissue, and the effects of electrical stimulation on these are similar to those found in other excitable tissues. The requirements for ATP as a primary and phosphocreatine as a secondary energy source for the sodium pump correlate well with other observations. However, details of the mechanism of coupling between the permeability changes consequent upon stimulation and the operation of the sodium pump remain obscure. Recent studies on the mechanism of action of the Na^{+}–K^{+}-ATPase and its relation to the cation sensitive phosphate binding discussed in Section I C are of considerable interest in this connection, and can be expected to throw light on the details of transport–metabolic coupling. Certain other aspects, and in particular, the role of calcium in energy-linked transport functions, remain uncertain.

The demonstration in brain tissue of sodium-dependent carrier-mediated transport systems for amino acids, and of counter-transport processes indicates again the general similarity of transport mechanisms in brain to those found in other tissues. Indeed, it is surprising, considering the functional specialization of brain that more clear cut differences in transport mechanism from those of other tissues have not emerged.

The interpretation of transport studies in brain slices is constantly bedevilled by the morphological heterogeneity of the constituent cells of the slice. This problem arises most acutely when attempts are made to identify either neuronal or glial cells as responsible for the various transport and metabolic phenomena encountered in the slice. The resolution of these complexities can be partially undertaken by developmental and comparative studies, but in principle the most promising approach lies in simplifying the system under study by separating the cell types. Investigations along these lines have not yielded extensive or easily interpretable results so far, but it remains a potentially important method of attacking the problem, particularly if the procedures used to separate the cells can be made more efficient and less liable to damage them. The possibilities of neuronal and glial cell types grown in tissue culture have barely been exploited for transport studies in this connection.

Methods exist in principle for the isolation of subcellular organelles from brain tissue and there are signs that interest is being aroused in the hitherto neglected problems of intracellular compartmentation in brain. Uptake into such subcellular organelles as mitochondria is quite possibly an important factor in the regulation of the intracellular cytoplasmic concentrations of various substances.

The general features of the transport processes of brain can be discerned; the most pressing problems concern the relationship of these transport processes to the various cell types of brain tissue and their functional activity.

Acknowledgement

I thank Professor F. G. Young for his interest.

REFERENCES

Abadom, P. N. and P. G. Scholefield (1962a) *Can. J. Biochem. Physiol.*, **40**, 1575
Abadom, P. N. and P. G. Scholefield (1962b) *Can. J. Biochem. Physiol.*, **40**, 1591
Abadom, P. N. and P. G. Scholefield (1962c) *Can. J. Biochem. Physiol.*, **40**, 1603
Abdel-Latif, A. A. (1966) *Biochim. Biophys. Acta.*, **121**, 403
Abraham, D., J. J. Pisano and S. Udenfriend (1964) *Arch. Biochem. Biophys.*, **104**, 160
Adrian, R. H. (1961) *J. Physiol.* (*London*), **156**, 623
Ahmed, K. and J. D. Judah (1965) *Biochim. Biophys. Acta.*, **104**, 112
Albers, R. W. (1967) *Ann. Rev. Biochem.*, **36**, 727
Albers, R. W., S. Fahn and G. J. Koval (1963) *Proc. Natl. Acad. Sci. U.S.*, **50**, 474
Aldridge, W. N. (1962) *Biochem. J.*, **83**, 527

Ames, A. III. (1956) *J. Neurophysiol.*, **19**, 213
Ames, A. III., J. B. Isom and F. B. Nesbett (1965) *J. Physiol.* (*London*), **177**, 246
Ames, A. III. and F. B. Nesbett (1958) *J. Neurochem.*, **3**, 116
Ames, A. III., Y. Tsukada and F. B. Nesbett (1967) *J. Neurochem.*, **14**, 145
Ashford, C. A. and K. C. Dixon (1935) *Biochem. J.*, **29**, 157
Ayres, P. J. W. and H. McIlwain (1953) *Biochem. J.*, **55**, 607
Bachelard, H. S., W. J. Campbell and H. McIlwain (1962) *Biochem. J.*, **84**, 225
Bader, H., A. K. Sen and R. L. Post (1966) *Biochim. Biophys. Acta.*, **118**, 106
Bahr, G. F., G. Bloom and U. Friberg (1957) *Exptl. Cell Res.*, **12**, 342
Bakay, L. (1960) *Neurology*, **10**, 564
Baker, P. F., M. P. Blaustein, A. L. Hodgkin and R. A. Steinhardt (1969) *J. Physiol.* (*London*), **200**, 431
Berl, S., W. J. Nicklas and D. D. Clarke (1968) *J. Neurochem.*, **15**, 131
Biesold, D. (1967) *Biochem. J.*, **102**, 20P
Bito, L., H. Davson, E. Levin, M. Murray and N. Snider (1966) *J. Neurochem.*, **13**, 1057
Blackburn, K. J., P. C. French and R. J. Merrills (1967) *Life Sci.*, **6**, 1653
Blasberg, R. G. (1968) In A. Lajtha and D. H. Ford (Eds.), *Progress in Brain Research*, *Vol.* 29, p. 245, Elsevier, Amsterdam
Blasberg, R. and A. Lajtha (1965) *Arch. Biochem. Biophys.*, **112**, 361
Blasberg, R. and A. Lajtha (1966) *Brain Res.*, **1**, 86
Blaustein, M. P. and A. L. Hodgkin (1969) *J. Physiol.* (*London*), **200**, 497
Bogdanski, D. F., A. Tissari and B. B. Brodie (1968) *Life Sci.*, **7**, 419
Bonting, S. L., L. L. Caravaggio and N. M. Hawkins (1962) *Arch. Biochem. Biophys.*, **98** 413
Bourke, R. S. and D. B. Tower (1966a) *J. Neurochem.*, **13**, 1071
Bourke, R. S. and D. B. Tower (1966b) *J. Neurochem.*, **13**, 1099
Bradbury, M. W. B. (1965) *J. Physiol.* (*London*), **179**, 67P
Bradbury, M. W. B. and H. Davson (1965) *J. Physiol.* (*London*), **181**, 151
Bradford, H. F. (1969) *J. Neurochem.*, **16**, 675
Bradford, H. F. and H. McIlwain (1966) *J. Neurochem.*, **13**, 1163
Bradford, H. F. and S. P. R. Rose (1967) *J. Neurochem.*, **14**, 373
Brierley, G. P. (1963) In B. Chance (Ed.), *Energy-linked Functions of Mitochondria*, Academic Press, New York, p. 237
Caldwell, P. C. and R. D. Keynes (1957) *J. Physiol.* (*London*), **137**, 12P
Caldwell, P. C. and R. D. Keynes (1959) *J. Physiol* (*London*), **148**, 8P
Carver, M. J. (1965) *J. Neurochem.*, **12**, 45
Charnock, J. S. (1963) *J. Neurochem.*, **10**, 219
Chu, L-W. (1954) *J. Comp. Neurol.*, **100**, 381
Cohen, S. R., R. Blasberg, G. L. Levi and A. Lajtha (1968) *J. Neurochem.*, **15**, 707
Crawford, J. M. and D. R. Curtis (1964) *Brit. J. Pharmacol.*, **23**, 313
Creese, R. and D. B. Taylor (1967) *J. Pharmacol. Exptl. Therap.*, **157**, 406
Cremer, J. E. (1967) *Biochem. J.*, **104**, 223
Cremer, J. E., P. V. Johnston, B. I. Roots and A. J. Trevor (1968) *J. Neurochem.*, **15**, 1361
Cserr, H. (1965) *Amer. J. Physiol.*, **209**, 1219
Cummins, J. T. and H. McIlwain (1961) *Biochem.*, *J.* **79**, 330
Curtis, D. R., D. D. Perrin and J. C. Watkins (1960) *J. Neurochem.*, **6**, 1
Curtis, D. R., J. W. Phillis and J. C. Watkins (1960) *J. Physiol.* (*London*), **150**, 656
Davrainville, J. L. and J. Gayet (1965) *J. Neurochem.*, **12**, 771
Davson, H. (1967) *Physiology of Cerebrospinal Fluid*, J. and A. Churchill, London
Davson, H. and M. Pollay (1963) *J. Physiol.* (*London*), **167**, 247
Dawson, J. and A. Bone (1965) *J. Neurochem.*, **12**, 167
Dengler, H. J., H. E. Spiegel and E. O. Titus (1961) *Science*, *N.Y.*, **133**, 1072
De Almeida, D. F., E. B. Chain and F. Pocchiari (1965) *Biochem. J.*, **25**, 793
De Graeff, J., E. F. Dempsey, L. D. F. Lameyer and A. Leaf (1965) *Biochim. Biophys. Acta.*, **106**, 155
De Robertis, E. (1967) *Science*, *N.Y.*, **156**, 907

De Robertis, E. and H. M. Gerschenfeld (1961) *Intern. Rev. Neurobiol.*, **3**, 1
De Robertis, E., A. P. de Iraldi, G. R. de L. Arnaiz and L. Salganicoff (1962) *J. Neurochem.*, **9**, 23
Deul, D. H. and H. McIlwain (1961) *J. Neurochem.*, **8**, 246
Dickens, F. and G. D. Greville (1935) *Biochem. J.*, **29**, 1468
Dunham, E. T. and I. M. Glynn (1961) *J. Physiol.* (*London*), **156**, 274
Elliott, K. A. C. (1946) *Proc. Soc. Exptl. Biol., Med.* **63**, 234
Elliott, K. A. C. and N. M. van Gelder (1958) *J. Neurochem.*, **3**, 28
Elliott, K. A. C. and I. H. Heller (1957) in D. Richter (Ed.), *Metabolism of the Nervous System*, Pergamon Press, London. p. 286
Elliott, K. A. C. and L. S. Wolfe (1962) In K. A. C. Elliott, I. H. Page and J. H. Quastel (Eds.), *Neurochemistry*, 2nd ed. C. C. Thomas, Springfield, Illinois. p. 177
Epstein, M. H. and J. S. O'Connor (1965) *J. Neurochem.*, **12**, 389
Evans, W. H. and H. McIlwain (1967) *J. Neurochem.*, **14**, 35
Franck, G., M. Cornette and E. Schoffeniels (1968) *J. Neurochem.*, **15**, 843
Friede, R. L. and W. H. van Houten (1962) *Proc. Natl. Acad. Sci. U.S.*, **48**, 817
Fujisawa, H., K. Kajikawa, Y. Ohi, Y. Hashimoto and H. Yoshida (1965) *Japan J. Pharmacol.*, **15**, 327
Gibson, I. M. and H. McIlwain (1965) *J. Physiol.* (*London*), **176**, 261
Glynn, I. M. (1956) *J. Physiol.* (*London*), **134**, 278
Glynn, I. M. (1957) *J. Physiol.* (*London*), **136**, 148
Glynn, I. M. (1962) *J. Physiol.* (*London*), **160**, 18P
Glynn, I. M., C. W. Slayman, J. Eichberg and R. M. C. Dawson (1965) *Biochem. J.*, **94**, 692
Gonda, O. and J. H. Quastel (1962) *Biochem. J.*, **84**, 394
Gore, M. B. R. and H. McIlwain (1952) *J. Physiol.* (*London*), **117**, 471
Graziani, L., A. Escriva and R. Katzman (1965) *Amer. J. Physiol.*, **208**, 1058
Greengard, P. and R. W. Straub (1959) *J. Physiol.* (*London*), **148**, 353
Hamburger, A. and H. Rockert (1964) *J. Neurochem.*, **11**, 757
Harvey, J. A. and H. McIlwain (1968) *Biochem. J.*, **108**, 269
Heald, P. J. (1954) *Biochem. J.*, **57**, 673
Heald, P. J. (1956) *Biochem. J.*, **63**, 242
Heald, P. J. (1958) *Biochem. J.*, **68**, 580
Heald, P. J. and H. McIlwain (1956) *Biochem. J.*, **63**, 231
Held, D., V. Fencl and J. R. Pappenheimer (1964) *J. Neurosphysiol.*, **27**, 942
Hems, D. A. and R. Rodnight (1966) *Biochem. J.*, **101**, 516
Hertz, L. (1965) *Nature*, **206**, 1091
Hertz, L. (1966) *J. Neurochem.*, **13**, 1373
Hertz, L. (1968) *J. Neurochem.*, **15**, 1
Hertz, L. and M. Schou (1962) *Biochem. J.*, **85**, 93
Hess, H. H. (1961) In S. S. Kety and J. Elkes (Eds.), *Regional Neurochemistry*, Pergamon, Oxford. p. 200
Hild, W. and I. Tasaki (1962) *J. Neurophysiol.*, **25**, 277
Hillman, H. and H. Hydén (1965) *J. Physiol.* (*London*), **177**, 398
Hillman, H. H. and H. McIlwain (1961) *J. Physiol.* (*London*), **157**, 263
Hodgkin, A. L. (1958) *Proc. Roy. Soc.* (*London*), *Ser. B.*, **148**, 1
Hodgkin, A. L. and K. Martin (1965) *J. Physiol.* (*London*), **179**, 26P
Hokin, L. E., M. Mokotoff and S. M. Kupchan (1966) *Proc. Natl. Acad. Sci. U.S.*, **55**, 797
Hokin, L. E., P. S. Sastry, P. R. Galsworthy and A. Yoda (1965) *Proc. Natl. Acad. Sci. U.S.*, **54**, 177
Horstmann, E. and H. Meves (1959) *Z. Zellforsch.*, **49**, 569
Hosie, R. J. A. (1965) *Biochem. J.*, **96**, 404
Hydén, H. (1960) In J. Brachet and A. E. Mirsky (Eds.), *The Cell*, Vol. 4, Academic Press, New York, p. 215
Hydén, H. and P. W. Lange (1965) *Acta Physiol. Scand.*, **64**, 6
Hydén, H. and A. Pigon (1960) *J. Neurochem.*, **6**, 57
Iversen, L. L. and M. J. Neal (1968) *J. Neurochem.*, **15**, 1141

Joanny, P. and H. Hillman (1964) *J. Neurochem.*, **11**, 413

Joanny, P., H. Hillman and J. Corriol (1966) *J. Neurochem.*, **13**, 371

Johnstone, R. M. and P. G. Scholefield (1962) In K. A. C. Elliott, J. H. Page and J. H. Quastel (Eds.), *Neurochemistry*, 2nd ed. C. C. Thomas, Springfield, Illinois, p. 376

Kao, C. Y. (1966) *Pharmacol. Rev.*, **18**, 997

Keesey, J. C. and H. Wallgren (1965) *Biochem. J.*, **95**, 301

Keesey, J. C., H. Wallgren and H. McIlwain (1965) *Biochem. J.*, **95**, 289

Kirschner, L. B. and J. Barker (1963) *J. Gen. Physiol.*, **47**, 1061

Korey, S. R. (1957) In D. Richter (Ed.), *Metabolism of the Nervous System*, Pergamon Press, London. p. 87

Korey, S. S. and M. Orchen (1959) *J. Neurochem.*, **3**, 277

Krebs, H. A. (1950) *Biochim. Biophys. Acta.*, **4**, 249

Krebs, H. A., L. V. Eggleston and C. Terner (1951) *Biochem. J.*, **48**, 530

Lahiri, S. and A. Lajtha (1964) *J. Neurochem.*, **11**, 77

Lajtha, A. (1962) In K. A. C. Elliott, I. H. Page and J. H. Quastel (Eds.), *Neurochemistry*, 2nd ed. C. C. Thomas, Springfield, Illinois, p. 399

Lajtha, A. and D. H. Ford (Eds.) (1968) *Progress in Brain Research Vol.* 29, Elsevier, Amsterdam

Lajtha, A. and P. Mela (1961) *J. Neurochem.*, **7**, 210

Lajtha, A. and J. Toth (1961) *J. Neurochem.*, **8**, 216

Leaf, A. (1956) *Biochem. J.*, **62**, 241

Levi, G., J. Kandera and A. Lajtha (1967) *Arch. Biochem. Biophys.*, **119**, 303

Levi, G. and A. Lajtha (1965) *J. Neurochem.*, **12**, 639

Li, C. L. and H. McIlwain (1957) *J. Physiol.* (*London*), **139**, 178

Ling, C-M. and A. A. Abdel-Latif (1968) *J. Neurochem.*, **15**, 721

Lolley, R. N. (1963) *J. Neurochem.*, **10**, 665

Lolley, R. N. and H. McIlwain (1964) *Biochem. J.*, **93**, 12P

Lowry, O. H. (1957) In D. Richter (Ed.), *Metabolism of the Nervous System*, Pergamon Press, London. p. 323

Lust, W. D. and J. D. Robinson (1968) *Federation Proc.*, **25**, 752

Marchbanks, R. M. (1966) *Biochem. J.*, **100**, 65P

Marchbanks, R. M. (1967) *Biochem. J.*, **104**, 148

Marchbanks, R. M. (1968) *Biochem. J.*, **110**, 533

Marchbanks, R. M. (1969) In S. H. Barondes (Ed.), *Cellular Dynamics of the Neuron*, Academic Press, New York, (p. 115)

Marchbanks, R. M. and V. P. Whittaker (1969) In E. E. Bittar and N. Bittar (Eds.), *The Biological Basis of Medicine*, *Vol.* 5 Academic Press, New York

Martin, K. (1968) *J. Gen. Physiol.*, **51**, 497

McIlwain, H. (1951) *Biochem. J.*, **49**, 382

McIlwain, H. (1952) *Biochem. J.*, **52**, 289

McIlwain, H. (1954) *J. Physiol.* (*London*), **124**, 117

McIlwain, H. (1959) *Biochem. J.*, **73**, 514

McIlwain, H. (1960) *J. Neurochem.*, **6**, 244

McIlwain, H. (1961) *Biochem. J.*, **78**, 24

McIlwain, H. (1963) *Chemical Exploration of the Brain*, Elsevier, Amsterdam

McIlwain, H. (1966) *Biochemistry and the Central Nervous System*, J. and A. Churchill, London

McIlwain, H. (1967) *Biochem. Pharmacol.*, **16**, 1389

McIlwain, H. and H. L. Buddle (1953) *Biochem. J.*, **53**, 412

McIlwain, H. and P. Joanny (1963) *J. Neurochem.*, **10**, 313

McIlwain, H. and R. Rodnight (1962) *Practical Neurochemistry*, Little, Brown and Co., Boston

McIlwain, H. and M. A. Tresize (1956) *Biochem. J.*, **63**, 250

McIlwain H., L. Buchel and J. D. Cheshire (1951) *Biochem. J.*, **48**, 12

McIlwain, H., J. A. Harvey and G. Rodriguez (1969) *J. Neurochem.*, **16**, 361

McLaughlin, S. G. A. and J. A. M. Hinke (1966) *Can. J. Physiol. Pharmacol.*, **44**, 837

Medzihradsky, F., M. H. Kline and L. E. Hokin (1967) *Arch. Biochem. Biophys.*, **121**, 311
Nakamura, R. (1963) *J. Biochem.* (*Tokyo*), **53**, 314
Nakamura, R. and M. Nagayama (1966) *J. Neurochem.*, **13**, 305
Nakazawa, S. and J. H. Quastel (1968a) *Can. J. Biochem. Physiol.*, **46**, 355
Nakazawa, S. and J. H. Quastel (1968b) *Can. J. Biochem. Physiol.*, **46**, 363
Neame, K. D. (1961) *J. Neurochem.*, **6**, 358
Neame, K. D. (1962) *J. Physiol.* (*London*), **162**, 1
Neame, K. D. (1964a) *J. Neurochem.*, **11**, 67
Neame, K. D. (1964b) *J. Neurochem.*, **11**, 655
Neame, K. D. (1968) In A. Lajtha and D. H. Ford (Eds.), *Progress in Brain Research, Vol.* 29, Elsevier, Amsterdam, p. 185
Neame, K. D. and S. E. Smith (1965) *J. Neurochem.*, **12**, 87
Nicholls, J. G. and S. W. Kuffler (1965) *J. Neurophysiol.*, **28**, 519
Orkand, R. J., J. G. Nicholls and S. W. Kuffler (1966) *J. Neurophysiol.*, **29**, 788
Otsuka, M., I. Ohtsuki and S. Ebashi (1966) *J. Biochem.* (*Tokyo*), **58**, 188
Pappius, H. M. and K. A. C. Elliott (1956) *Can. J. Biochem. Physiol.*, **34**, 1007
Pappius, H. M., I. Klatzo and K. A. C. Elliott (1962) *Can. J. Biochem. Physiol.*, **40**, 885
Pappius, H. M., M. Rosenfeld, D. M. Johnson and K. A. C. Elliott (1958) *Can. J. Biochem. Physiol.*, **36**, 217
Post, R. L., C. R. Merritt, C. R. Kinsolving and C. D. Albright (1960) *J. Biol. Chem.*, **235**, 1796
Post, R. L., A. K. Sen and A. S. Rosenthal (1965) *J. Biol. Chem.*, **240**, 1437
Potter, L. T. (1968) In P. N. Campbell (Ed.), *The Interaction of Drugs and Subcellular Components on Animal Cells*, J. and A. Churchill, London, p. 293
Quastel, J. H. (1962) In K. A. C. Elliott, I. H. Page and J. H. Quastel (Eds.), *Neurochemistry*, C. C. Thomas, Springfield, Illinois. p. 226
Rall, D. P. (1968) In A. Lajtha and D. H. Ford (Eds.), *Progress in Brain Research*, *Vol.* 29, Elsevier, Amsterdam, p. 159
Rall, D. P., W. W. Oppelt and C. S. Patlak (1962) *Life Sci.*, **2**, 43
Robinson, J. D. (1968) *J. Neurochem.*, **15**, 1225
Rodnight, R., D. A. Hems and B. E. Lavin (1966) *Biochem. J.*, **101**, 502
Rodnight, R. and B. Lavin (1966) *Biochem. J.*, **101**, 495
Roots, B. I. and P. V. Johnston (1964) *J. Ultrastruct. Res.*, **10**, 350
Rose, S. P. R. (1965) *Biochem. Pharmacol.*, **14**, 589
Rose, S. P. R. (1967) *Biochem. J.*, **102**, 33
Ruščák, M. and R. Whittam (1967) *J. Physiol.* (*London*), **190**, 595
Schanberg, S. and N. J. Giarman (1960) *Biochim. Biophys. Acta.*, **41**, 556
Schatzmann, H-J. (1953) *Helv. Physiol. Pharmacol. Acta.*, **11**, 346
Schuberth, J. and A. Sundwall (1967) *J. Neurochem.*, **14**, 807
Schuberth, J., A. Sundwall and B. Sorbo (1967) *Life Sci.*, **6**, 293
Schuberth, J., A. Sundwall, B. Sorbo and J-O. Lindell (1966) *J. Neurochem.*, **13**, 347
Schwartz, A. (1962) *Biochem. Pharmacol.*, **11**, 389
Schwartz, A., H. S. Bachelard and H. McIlwain (1962) *Biochem. J.*, **84**, 626
Sharma, S. K. and J. H. Quastel (1965) *Biochem. J.*, **94**, 790
Skou, J. C. (1957) *Biochim. Biophys. Acta.*, **23**, 394
Skou, J. C. (1960) *Biochim. Biophys. Acta.*, **42**, 6
Skou, J. C. (1962) *Biochim. Biophys. Acta.*, **58**, 314
Smith, S. E. (1967) *J. Neurochem.*, **14**, 291
Stein, W. D. (1967) *The Movement of Molecules across Cell Membranes*, Academic Press, New York
Stern, J. R., L. V. Eggleston, R. Hems and H. A. Krebs (1949) *Biochem. J.*, **44**, 410
Swanson, P. D. (1968a) *J. Neurochem.*, **15**, 57
Swanson, P. D. (1968b) *Biochem. Pharmacol.*, **17**, 129
Swanson, P. D. and H. McIlwain (1965) *J. Neurochem.*, **12**, 877
Swanson, P. D. and K. Ullis (1966) *J. Pharmacol. Exptl. Therap.*, **153**, 321
Sweet, W. H., B. Selverston, A. Solomon and L. Bakay (1949) *J. Clin. Invest.*, **28**, 1949

Takagaki, G., S. Hirano and Y. Nagata (1959) *J. Neurochem.*, **4**, 124
Terner, C., L. V. Eggleston and H. A. Krebs (1950) *Biochem. J.*, **47**, 139
Thomas, J. and H. McIlwain (1956) *J. Neurochem.*, **1**, 1
Tower, D. B. (1968) In A. Lajtha and D. H. Ford (Eds.), *Progress in Brain Research, Vol.* 29, Elsevier, Amsterdam. p. 465
Tower, D. B. and R. S. Bourke (1966) *J. Neurochem.*, **13**, 1119
Tower, D. B., J. R. Wherrett and G. M. McKhann (1961) In S. S. Kety and J. Elkes (Eds.), *Regional Neurochemistry*, Pergamon Press, Oxford. p. 65
Trevor, A. J. and R. Rodnight (1965) *Biochem. J.*, **95**, 889
Tsukada, Y., Y. Nagata, S. Hirano and T. Matsutani (1963) *J. Neurochem.*, **10**, 241
Ussing, H. H. (1952) *Adv. Enzymol.*, **13**, 21
Van Harreveld, A., J. Crowell and S. K. Malhrota (1965) *J. Cell Biol.*, **25**, 117
Van Harreveld, A. and S. Ochs (1956) *Amer. J. Physiol.*, **187**, 180
Varon, S. and H. McIlwain (1961) *J. Neurochem.*, **8**, 262
Whittaker, V. P. (1959) *Biochem. J.*, **72**, 694
Whittaker, V. P. (1965) *Progr. Biophys. Molec. Biol.*, **15**, 39
Whittam, R. (1962a) *Biochem. J.*, **82**, 205
Whittam, R. (1962b) *Biochem. J.*, **84**, 110
Whittam, R. and M. E. Ager (1964) *Biochem. J.*, **93**, 337
Whittam, R. and D. M. Blond (1964) *Biochem. J.*, **92**, 147
Wollenberger, A. (1947) *J. Pharmacol. Exptl. Therap.*, **91**, 39
Woodbury, D. M. (1968) In A. Lajtha and D. H. Ford (Eds.), *Progress in Brain Research, Vol.* 29, Elsevier, Amsterdam, p. 297
Yamamoto, C. and H. McIlwain (1966) *J. Neurochem.*, **13**, 1333
Yoshida, H., K. Kadota and H. Fujisawa (1966) *Nature*, **212**, 291
Yoshida, H., K. Kaniike and J. Namba (1963) *Nature*, **198**, 191
Yoshida H., T. Nukada and H. Fujisawa (1961) *Biochim. Biophys. Acta.*, **48**, 614

CHAPTER 6

Ion movements in red blood cells

P. J. Garrahan*
Departamento de Quimica Biologica,
Facultad de Farmacia y Bioquimica,
Universidad do Buenos Aires, Argentina

*Established Investigator, Consejo Nacional de Investigaciones Científicas y Técnicas, Argentina

I. SOME BIOPHYSICAL PROPERTIES OF RED BLOOD CELLS

It is generally taken for granted that the activity coefficients of the intracellular ions in red cells are similar to those of an aqueous solution of the same ionic strength. This implies that differences between the concentration of ions inside and outside the cell are the result of special properties of the cell membrane, and not due to the binding of ions in the cell interior. The following is an account of the experimental evidence for this view.

1. *The State of Intracellular Water*

Freezing-point measurements show that the water activity inside the red cell is equal to that of the suspending medium (Brodsky and coworkers, 1953; Appelboom and coworkers, 1958) even when the medium is not isotonic with the plasma (Williams and coworkers, 1959). The volume changes of red cells suspended in anisotonic solutions have been extensively studied (see Ponder, 1948; Dick and Lowenstein, 1958; Whittam, 1964, for reviews of techniques and literature). From the available results it is clear that the small deviations of the osmotic behaviour of the erythrocyte from the van't Hoff law can be accounted for either by the anomalous changes with concentration of the osmotic coefficient of haemoglobin (Dick and Lowenstein, 1958 McConaghey and Maizels, 1961), or by the existence of a fraction of non-solvent water (LeFevre, 1964; Savitz and coworkers, 1964).

Although the changes with concentration in the osmotic coefficient of haemoglobin are a well documented fact, the existence and amount of non-solvent water in red cells is still a matter of dispute. If crystalline conditions held for haemoglobin in the erythrocyte 15 to 20 per cent of the cell water would be non-solvent (Drabkin, 1950). However, direct measurements of the fraction of intracellular water that is accessible to polar solutes have yielded lower values that range from complete absence (Hutchinson, 1952; Miller, 1964; Cook, 1967) to at most 12 per cent of non-accessible water (Gary Bobo, 1967).

As the intracellular solute concentration is around 150 mM, it seems reasonable to conclude that in order to account for the equality in water activity between the cell and its surroundings and for the fact that almost all the intracellular water participates in osmotic volume changes, the main intracellular electrolytes must be in free solution.

2. *The Mobility of Intracellular Ions*

The electrical resistance of the red cell interior can be estimated by passing very high frequency currents through a red cell suspension (Fricke, 1925; Fricke and Morse, 1925; Pauly and Schwan, 1966). This technique yields results which suggest that the ion mobilities in the intracellular fluid are reduced by a factor of about two with respect to a water solution having the same ionic composition (Pauly and Schwan, 1966). The reduction in mobility persists after treating the cells with toluene or saponin in concentrations that abolish the asymmetric ion distribution between the cell and the medium (Pauly and Schwan, 1966), suggesting that it is unrelated to the mechanisms that govern this distribution.

The restriction in intracellular ion mobility probably arises from the fact that the red cell interior is a very concentrated solution of haemoglobin which presumably imposes hydrodynamic hindrances to the free movements of ions (Pauly and Schwan, 1966).

3. *Membrane Potential and Anion Activities*

The main anion in red cells, namely chloride and bicarbonate, seem to be in thermodynamic equilibrium with those in the suspending medium (see below). On this assumption the erythrocyte membrane potential can be calculated by inserting into the Nernst equation the concentration ratio of any of them, provided that their activity coefficients in the intra- and extracellular fluid are equal. Direct determinations of the red cell membrane potential (Lassen and Sten Knudsen, 1968) have yielded values with a large scatter, but none of them exceeding the figure that can be calculated on the equal activity coefficient hypothesis (10–14 mV, inside negative). Any significant decrease of the intracellular activity coefficient would result in a larger than theoretical membrane potential.

4. *Ion Binding by Haemoglobin*

Haemoglobin comprises about 90 per cent of the solid matter of red cells and about 97 per cent of its total protein (see Whittam, 1964); thus any specific ion binding must be attributed mainly to it. However, solutions of haemoglobin in the physiological pH range do not show any specific binding of either sodium or potassium (Morris and Wright, 1954; Carr, 1956; Tanford and Nozaki, 1964).

5. *Reversible Haemolysis and Reconstituted Ghosts*

Red cells submitted to hypotonic haemolysis can lose about 90 to 95 per cent of their contents. After haemolysis a fraction of the membranes regains its low permeability to sodium spontaneously in the haemolysate (Teorell, 1952; Hoffman, 1962b). Almost normal sodium and potassium permeability is restored in a much larger fraction of the membranes if the haemolysate is brought back to isotonicity and incubated at 37 °C (Hoffman and coworkers, 1960).

Reconstituted ghosts, apart from providing an extremely useful technique for studying membrane phenomena in red cells, give one of the best experimental proofs that the mechanisms that govern ion distribution between the red cell and the external medium are located in the cell membrane.

II. ANION PERMEABILITY

A. Chloride and Bicarbonate

Distribution: Since the work of Van Slyke and coworkers (1923) it is accepted that the distribution of chloride, bicarbonate, hydrogen and hydroxyl ions between the intra- and extracellular media obeys the Gibbs–Donnan equilibrium. Assuming equal activity coefficients, one can thus write

$$\frac{[Cl^-]_i}{[Cl^-]_e} = \frac{[OH^-]_i}{[OH^-]_e} = \frac{[CO_3H^-]_i}{[CO_3H^-]_e} = \frac{[H^+]_e}{[H^+]_i} = 0{\cdot}7 \qquad (1)$$

(where i means intracellular and e means extracellular).

From the values of the concentration ratios in equation (1) one can calculate that about 60 per cent of the intracellular cations are electrically balanced by chloride and bicarbonate, the remainder being accounted for mainly by haemoglobin and to a lesser extent by inorganic phosphate and phosphoric esters.

The main evidence in favour of passive chloride and bicarbonate distribution is first, that the ratios in equation (1) change according to theory if the pH of the suspending medium is altered (Harris and Maizels, 1952; Fitzsimons and Sendroy, 1961; Bromberg and coworkers, 1965; Funder and Weith, 1966). Secondly, that the distribution ratios are independent of metabolism (Harris and Maizels, 1952; Fitzsimons and Sendroy, 1961). Thirdly, that direct determination of the red cell membrane potential (Lassen and Sten-Knudsen, 1968) yields values that are consistent with those that can be calculated on the basis of the Nernst equation.

Penetration: If the bicarbonate concentration in a suspension of red cells is suddenly altered, the ratios in equation (1) will then be upset and the ions will exchange across the membrane until equilibrium is again established. The time course of this exchange makes it possible to estimate permeability of the membrane to the relevant ions, although the technique does not provide information about which penetrating ion is rate-limiting. Chloride–bicarbonate exchanges have been measured with the Hartridge and Roughton (1923) rapid flow technique (Driken and Mook, 1931) or monitoring the changes in chloride activity with a Ag/AgCl electrode (Luckner, 1939). It is a first order process with a half-time of 0·1 to 0·2 seconds. The Q_{10} for the exchange lies between 1·2 and 1·4 (Luckner, 1949).

In beef and human red cells in equilibrium conditions at room temperature, tracer exchange of chloride follows a simple exponential time-course with a rate constant of 3·1 sec^{-1}. Under the same conditions the rate constants for the exchange of the other halides are as follows: Br, 0·6; F, 0·3 and I, 0·06 (sec^{-1}). This series is not related to the ionic radii nor to the ionic conductances of the intervening species (Tosteson, 1959).

The high permeability to chloride and bicarbonate ions is the main cause of the very low electrical resistance of the red cell membrane. Tosteson (1959), using his data on chloride fluxes, calculated that the membrane resistance for chloride current is 19 $ohms/cm^2$. This value is of the same order of magnitude as those obtained by direct measurement of red cell membrane resistance (Johnson and Woodbury, 1964; Lassen and Sten-Knudsen, 1968).

As the membrane permeability to chloride and bicarbonate is about 10^6 times greater than the permeability to alkali metal cations, the red cell membrane potential will always be determined by the anion distribution, and will be practically independent of changes in cation distribution.

The high selectivity for anions is one of the most striking properties of the red cell membrane. A satisfactory explanation for it is still lacking but several facts are suggestive. First, the high permeability to chloride as compared to cations does not seem to be related to the lipid composition of the cell membrane: bimolecular films of total membrane lipids from sheep red cells placed in 0·1 N KCl or NaCl solutions have an electrical resistance of $1\text{–}2 \times 10^8$ $ohms/cm^2$ and a transport number for chloride of 0·2 (Andreoli and coworkers, 1967a). Secondly, the permeability of the red cell membrane to chloride is much less—about 10^6 times—than that of a water layer of the same thickness (Tosteson, 1959). This suggests that either there is a severe restriction in the mobility of chloride across the membrane and/or that only a very small fraction of it is directly involved in chloride passage. The latter possibility seems to have some experimental support from the work of Edelberg (1952) who showed that an amount of tannic acid enough to cover only 0·66 per cent of the red cell surface retards haemolysis in ammonium chloride, the rate of which is limited by the rate of Cl–OH exchange by 17 per cent.

The effects of tannic acid suggest that the cell membrane contains a small number of specialized regions across which chloride and other small anions can move with a mobility that is several orders of magnitude larger than that it can attain in a bimolecular lipid layer. It is tempting to visualize these regions as water-filled pores (see Passow, 1964 and Solomon, 1968). Although water-filled pores may explain the high permeability to anions, they fail to account for the discrimination of the red cell membrane between anions and cations of similar hydrated ionic radii. To explain this it is usually assumed that the pores are lined with positive charges. Further experimental evidence for positive charges as the basis for anion selectivity will be provided below. A good review of the development of this concept has been given by Passow (1964, 1965).

B. Inorganic Phosphate

Distribution: the study of the distribution of inorganic phosphate between the red cell and the suspending medium is complicated by the fact that, due to its participation in glycolysis, phosphate is the only anion that suffers transformations inside the red cell. The question of whether or not inorganic phosphate is distributed according to a Gibbs–Donnan equilibrium is therefore still in dispute. The phosphate concentration in fresh human cells shows large random deviations from that expected from the chloride ratio (Mueller and Hastings, 1951) though it is always lower than that of the plasma (Whittam, 1964). However, inorganic phosphate concentration in cells suspended in high-phosphate solutions follow a Gibbs–Donnan ratio (Vestergaard-Bogind and Hesselbo, 1960).

Penetration: Both net phosphate uptake in exchange for internal chloride (Holton, 1952) and tracer exchange in steady-state conditions (Hahn and Hevesy, 1942; Gourley and Matchiner, 1953; Prankerd and Altman, 1955; Zipursky and Israels, 1961; Chedru and Cartier, 1966) are first-order processes with a half-time of 2 to 3 hours at 37 °C. Radioactive phosphate influx is a linear function of external phosphate concentration showing no evidence of saturation (Hahn and Hevesy, 1942; Zipursky and Israels, 1961; Chedru and Cartier, 1966). Phosphate uptake has a large temperature dependence; its apparent energy of activation lies between 15,000 and 21,700 cal/mole (Hahn and Hevesy, 1942; Gourley and Gemill, 1950; Prankerd and Altman, 1955; Chedru and Cartier, 1966).

The half-time for phosphate uptake can be reversibly reduced to 5 minutes at 37 °C if external chloride is replaced by a *non-penetrating anion* like citrate. (Mollison and coworkers, 1958). The effect is not seen if chloride is replaced by a penetrating anion like sulphate. This result seems to suggest that penetrating anions compete among themselves, perhaps for intramembrane positive charges, as they cross the red cell membrane.

Phosphate *efflux* in human red cells is linear with internal phosphate concentration (in a 0–5 mM range) and has a high apparent energy of activation (Glader and Omachi, 1968).

Phosphate exchange and glycolysis: Phosphate uptake (Gourley and Gemmill, 1950; Gourley, 1951; Prankerd and Altman, 1955; Zipursky and Israels, 1961; Chedru and Cartier, 1966) but not phosphate loss (Glader and

Omachi, 1968) is depressed if glycolysis is stopped by glucose deprivation or by addition of inhibitors. Under certain conditions, however, phosphate uptake can be uncoupled from glycolysis; for example, in iodoacetate-poisoned cells, phosphate influx can be restored to normal values by the addition of adenosine which lowers intracellular inorganic phosphate without restoring lactic acid production (Zipursky and Israels, 1961). Iodoacetate also loses its inhibitory action if cells are previously equilibrated in a high-phosphate medium (Lohman and Passow, cited by Passow, 1964). These results suggest that the role of glycolysis in phosphate uptake is indirect and that the process is not energy-coupled.

C. Sulphate

Sulphate distributes itself between the intra- and extracellular fluids in accordance with the Gibbs–Donnan equilibrium. (Richmond and Hastings, 1960; Passow, 1964, 1965).

In equilibrium conditions isotopic sulphate fluxes follow simple exponential kinetics and their rates are inversely related to the external chloride concentration and to pH. Thus, the plot of sulphate flux against external pH resembles a titration curve with an inflexion point at pH 7·2, and the increase in sulphate flux that can be induced by acidifying the medium can be counteracted by an adequate rise in chloride concentration (Passow, 1964, 1965).

Passow has interpreted the above results to be suggestive of the existence within the red cell membrane of positive charges for which penetrating anions compete and whose ionization would be in part governed by the pH of the suspending media. On this hypothesis a linear relationship between sulphate flux and intramembrane sulphate concentration can be predicted—at any pH or chloride concentration—if it is assumed that positive charges are provided by groups with a pK near nine with which the mobile ions are in Donnan equilibrium (Passow, 1965). However the concentration of positive charges that is required to give the above-mentioned linear relationship between flux and concentration (about 3 moles/l, Passow, 1965) falls short by a factor of about thirty to account for the anion selectivity of the red cell membrane (see Solomon, 1960). Moreover, the apparent energy of activation of sulphate fluxes (29,000 cal/mole) is much larger than the largest reported for ion diffusion in ion-exchange resins (Passow, 1964, 1965).

III. CATION PERMEABILITY

A. Sodium and Potassium

The intracellular concentration of sodium and potassium varies in different species. The values in general fall into two groups: primates, including man, rodents and some ungulates have red cells with a low internal sodium concentration (10–20 mM/l cell water) and a high internal potassium concentration (130–150 mM/l cell water). Some ungulates and most carnivores present the opposite picture; a low internal potassium (46–8 mM/l cell water) and a high internal sodium concentration (98–135 mM/l cell water) (data taken from Ponder, 1948 and Whittam, 1964). A particularly interesting case is that of sheep red cells (see Tosteson and Hoffman, 1960; Tosteson, 1966) which can show either a low or a high potassium content depending on the strain of sheep. This property is genetically determined and apparently segregates as a single gene (see Tosteson, 1963).

1. *Intracellular Distribution*

Red cell *potassium* is completely exchangeable. The uptake of ^{42}K by human red cells at 37 °C is a first-order process with a half-time of 35 hours (Racker, and coworkers, 1950). Solomon and Gold (1955) found that in order to account for the kinetics of potassium efflux it was necessary to assume the existence of a quickly exchanging intracellular compartment, including 5 per cent of the total cell potassium. As pointed out by Glynn (1957b) a similar result could arise from a non-homogeneous population that included a small fraction of leaky cells.

Radioactive *sodium* exchange in human red cells is a first-order process with a half-time of 2 to 3 hours at 37 °C (Harris and Prankerd, 1953). The claim that 15 to 20 per cent of the intracellular sodium exchanges more slowly than the rest (Solomon, 1952; Solomon and Gold, 1955) has not been confirmed by Garrahan and Glynn (1967a) who showed that the specific activity of the sodium lost from cells preincubated in a ^{24}Na-containing solution, is not significantly different from the mean intracellular specific activity.

There seems to be evidence for a small (1–2 mM/l cells) unspecific compartment that can take up or lose quickly any monovalent cation (Clarkson and Maizels, 1955; Maizels and Remington, 1959a, 1959b; Funder and Weith, 1967).

2. *The Mechanism of Sodium and Potassium Distribution*

Both in low and high potassium red cells, the intracellular concentration of sodium is lower and that of potassium higher than the values predicted by the Gibbs–Donnan equilibrium. The deviation from equilibrium distribution is thought to result from a steady-state in which the movements of cations towards their equilibrium position (passive movements) are balanced by opposite energy-coupled movements (active transport). The balance between passive movements and active transport not only determines the intracellular concentration of sodium and potassium but also is indispensable for the constancy of cell volume (Hoffman and Tosteson, 1960). If passive movements exceed the rate of active transport then cations will tend towards a Gibbs–Donnan equilibrium with an associated increase in internal osmolality, cell swelling and haemolysis (colloid-osmotic haemolysis, Willbrandt, 1941).

The distinction between passive movements and active transport: The study of sodium and potassium permeation through the cell membrane requires separate assessment of the active and passive components of their unidirectional fluxes. For this to be possible the pump and leak fluxes must be intrinsically independent in the sense that abolition of one of them leaves the other unaltered. If this independence be the case, then active and passive fluxes can be separated by selective inhibition of the active component.

The independence of pump and leak is the basis of most models of cation distribution in red cells (Harris, 1954; Shaw, 1955; Glynn, 1956; Tosteson and Hoffman, 1960); and cardiac glycosides act as selective inhibitors of active transport (Schatzman, 1953; Joyce and Weatherall, 1955; Kahn and Acheson, 1955; Solomon and coworkers, 1956; Glynn, 1957a; Gill and Solomon, 1959). The validity of the independence of pump and leak and of the selective action of cardiac glycosides have been experimentally reexamined by Post and coworkers (1967). Their findings supersede previous work by many authors and can be summarized as follows:

i) Cardiac glycosides have no effect on net sodium or potassium movements when active transport is paralysed by depletion of ATP or by removal of intracellular sodium or extracellular potassium.

ii) In the presence of glycosides, net sodium and potassium movements become linear with their concentration gradients tending to an equilibrium distribution which suggests that the intracellular medium is slightly more selective to potassium than to sodium (about 1·13 times). This apparent selectivity can be explained by a residual pump activity of less than 2 per cent of its normal value.

iii) The total net movements of sodium or potassium can always be shown to be the sum of the rates observed when active and passive pathways are allowed to work in isolation.

It must, however, be stressed that the interpretation of passive fluxes as synonymous with glycoside-insensitive fluxes has become somewhat uncertain since the discovery that ethacrynic acid inhibits a fraction of the glycoside-resistant sodium efflux and potassium influx (Hoffman, 1966; Hoffman and Kregenow, 1966). The exact interpretation of this effect, which will be analyzed in more detail below, is still not completely clear, but undoubtedly it may lead in the near future to a reappraisal of the present concept of pump and leak fluxes.

3. *Kinetics of Passive Sodium and Potassium Fluxes*

Passive permeability of the red cell membrane to sodium and potassium is low and shows little selectivity; it can be represented by an electrical resistance of the order of 10^7 ohms/cm^2. This value is in marked contrast with the already mentioned resistance to chloride and bicarbonate ions, and it is about ten times less than the electrical resistance of bimolecular films prepared from red cell membrane lipids (Andreoli and coworkers, 1967a). All passive fluxes have large apparent energies of activation (around 15,000 cal/mole; see Glynn, 1957b).

Sodium fluxes: The ouabain-resistant sodium *efflux* seems to be a linear function of internal sodium concentration (Hoffman 1962b; Maizels, 1968 but compare Hoffman and Kregenow, 1966; Garrahan and Glynn, 1967c). It represents about one-third of the total sodium efflux in steady-state conditions in fresh human red cells (Glynn, 1957a).

Sodium *influx* in human red cells is not linear with external sodium concentration; its rate constant falls as external sodium is raised (Garrahan and Glynn, 1967a). It is somewhat reduced upon removal of external potassium (Garrahan and Glynn, 1967c).

Under physiological conditions the ratio of glycoside insensitive sodium influx to glycoside insensitive sodium efflux is about three, a figure which is much nearer to unity than that predicted from Ussing's independence principle (1950). The reason for this is not clear, as there is no evidence for exchange diffusion of sodium in the presence of glycosides (Garrahan and Glynn, 1965, 1967a; Hoffman and Kregenow, 1966). The flux ratio approaches the value predicted from the independence principle if the ethacrynic acid-sensitive component of the sodium efflux is taken as representative of a different pathway from that of passive diffusion.

Potassium fluxes: No data are available concerning the effect of internal potassium concentration on the rate of potassium *efflux*. Glycoside-insensitive potassium efflux is somewhat reduced on removal of external potassium (Glynn and Lüthi, 1968) or sodium (Garrahan and Glynn, 1967b); the reason for both effects is unknown.

Glycoside-insensitive potassium *influx* appears as the sum of a linear and a saturable component, when plotted against external potassium concentration (Glynn, 1957a). The saturable component disappears and the flux becomes linear with external potassium if cells are depleted of substrates by incubation in glucose-free media (Sachs and Welt, 1967). In this condition the flux ratio obeys the independence principle. The relationship between the saturable component of the glycoside-resistant potassium influx and the ethacrynic acid-sensitive potassium influx has not yet been elucidated.

4. *Factors affecting Passive Permeability to Sodium and Potassium*

It is beyond the scope of this chapter to present a comprehensive survey of the literature on this aspect of the subject. Excellent reviews dealing with this subject are available (Passow, 1964; Whittam, 1964).

Ionic strength: If the electrolytes of the suspending medium are isoosmotically replaced by a non-penetrant non-electrolyte, e.g. sucrose, the rate of salt loss from red cells is considerably increased and cells shrink, thereby attaining a high osmotic resistance (Joel, 1915; Mond, 1927; Maizels, 1935; Davson, 1939; Wilbrandt and Schatzmann, 1960). The main features of this phenomenon are:

i) The rate of loss is comparatively independent of the nature of the intracellular cation. This phenomenon has been observed in cells loaded with sodium (Willbrandt and Schatzmann, 1960), rubidium, caesium or lithium (LaCelle and Rothstein, 1966) by cold storage.

ii) The rate of leakage is dependent on the external salt concentration but independent of the nature of the extracellular monovalent cations or anions (LaCelle and Rothstein, 1966).

iii) If the rate of leakage is plotted as a function of the logarithm of the external salt concentration, two straight lines with negative slopes that intersect at about 0·18 mM are obtained (figure 1). The slope at lower salt concentrations is about thirty times larger than that at higher salt concentrations (LaCelle and Rothstein, 1966).

Two factors play a role in the effect of low electrolyte solutions. On the one hand, reduction in extracellular salt concentration increases the driving force for outward cation movements. This not only results from the creation of a concentration gradient but also arises from the changes in membrane

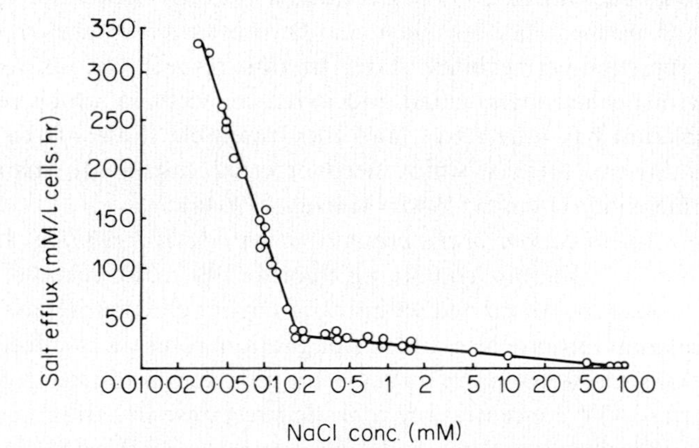

Figure 1. Effect of external NaCl concentration on the rate of salt loss into isotonic sucrose. The ionic strength of the suspending media was maintained constant by continuous addition of isotonic sucrose. (From LaCelle and Rothstein, 1966)

potential that result from the drop in extracellular chloride concentration (Davson, 1939). On the other hand, low extracellular ionic strength may increase the membrane permeability to cations. A possible mechanism for such an increase has been suggested by Passow (1964, 1965). If mobile ions were in equilibrium with intramembrane positive charges, a reduction in extracellular chloride concentration would *raise* the intramembrane pH and thus decrease the concentration of intramembrane positive charges with the concomitant rise in cation permeability. In agreement with his hypothesis Passow (1965) found that the rate of salt loss in low ionic strength solutions increases as the pH of the medium is raised, despite a reduction in the driving force due to the increment with pH of the negative charge of haemoglobin. LaCelle and Rothstein (1966) found that, provided the constant field equation (Goldman, 1942) is applicable, the red cell membrane behaves as if it has a constant permeability to cations over a range as low as 0·18 mM external salt concentration. When this value is exceeded a sudden change in permeability takes place. This now reaches a new constant value forty-five times as large as before. The abrupt change in permeability cannot be explained on any simple hypothesis and suggests an alteration in the tertiary structure of the cell membrane. Confirming Passow's findings, LaCelle and Rothstein (1966) showed that increases in pH raise the cell membrane permeability but this effect is only manifest when the *external* pH is altered.

The low salt effect is reversible, and the rate of cation loss returns to control values if the external salt concentration is brought back to its physiological level (Davson, 1939; LaCelle and Rothstein, 1966). If, however, red cells are incubated for three hours in electrolyte-free lactose solutions they will only regain their normal cation permeability if ionic calcium is present (Bolingbrooke and Maizels, 1959). Calcium is not only needed for restoration of low cation permeability but must also be present afterwards to avoid the return to the high-permeability state. In this respect lactose-treated cells contrast with normal mammalian and avian red cells in which removal of external calcium has little effect, and they resemble those of the snapping turtle and African tortoise which need external calcium to maintain low cation permeability (Lyman, 1945; Maizels, 1954a).

Effect of divalent cations in the presence of metabolic inhibition: If red cells are suspended in solutions containing fluoride (40 mM), metabolism stops and active transport is inhibited without significant changes in passive cation fluxes. If calcium ions are also added (magnesium ions are as effective but at concentrations ten times larger) after a lag that can be correlated with the fall in internal ATP concentration, a graded response in cation permeability takes place. Specifically, at low metal concentration (0·05 mM Ca, 0·5 mM Mg) there is a selective increase in potassium permeability (which can reach twenty times its normal value); if calcium or magnesium concentrations are sufficiently raised, sodium permeability also increases and the cells undergo colloid-osmotic haemolysis (Passow, 1964; Lepke and Passow, 1968). The specific effect on potassium permeability in intact cells can be prevented by substrates like adenosine or pyruvate (Passow, 1964) and it does not take place in reconstituted ghosts if ATP or a chelating agent like EDTA is incorporated into them (Lepke and Passow, 1968).

Effects which are essentially similar to those of fluoride have been observed with iodoacetate but with this compound only calcium is effective and in fresh cells or in ATP-containing reconstituted ghosts, adenosine or inosine plus a suitable source of ammonia must also be present (Gardos, 1959; Passow, 1964; Hoffman, 1966).

The actions of fluoride (Passow, 1964; Lepke and Passow, 1968) and iodoacetate (Hoffman, 1966) are enhanced and take place without a lag period if cells are previously depleted of substrates by incubation in a glucose-free medium. Under these conditions calcium can induce by itself, without the need of inhibitors, a large increase in potassium permeability (Hoffman, 1966). In contrast to what happens in fresh cells adenosine *prevents* the effect of calcium in depleted cells.

Metabolic depletion plus suitable concentrations of calcium ions seem therefore to be the two prerequisites for eliciting selective increases in potassium permeability. Whittam (1968) has incorporated these minimal requirements into an unifying hypothesis which assumes that metabolic inhibition, through the blockage of active calcium extrusion for which there is experimental evidence (see below), results in an increase in intracellular calcium concentration which in turn leads to an increase in cation permeability There is experimental evidence that intracellular calcium *increases* potassium permeability, i.e. if it is incorporated into reconstituted ghosts at a concentration of 0·2 mM it prevents the reappearance of low potassium permeability (Hoffman, 1962a).

Sulphydryl groups: The study of the *direct* role of membrane sulphydryl groups in the control of passive cation permeability is hindered by the indirect effects that can result from the blockage of glycolysis or from the production of irreversible alterations in membrane structure. Ideally, a sulphydryl-blocking reagent should be highly specific, and able to act on the membrane without significantly affecting the intracellular metabolic processes. Its effect should also be reversible. Some organic mercurials, for example, *para*-chloromercuribenzenesulphonate (PCMBS), fulfill these requirements. PCMBS crosses the red cell membrane slowly so that an appreciable build-up can take place within the membrane before there is a significant intracellular accumulation. The slow penetration allows the detection of different pools of sulphydryl groups on the basis of their rate of saturation with the agent.

The kinetics of PCMBS uptake and release indicate that at least two populations of functionally important sulphydryl groups exist within the cell membrane:

i) A small population seems to be located at the external surface of the cell. The blockage of this leads to inhibition of sugar transport (Van Steveninck and coworkers, 1965).

ii) A larger group is only slowly accessible through what seems to be a diffusion-limited process. Saturation of this results in an increase in passive sodium and potassium permeability (Sutherland and coworkers, 1967). A slowly accessible sulphydryl group pool is also necessary for active transport (Rega and coworkers, 1967).

The effects of PCMBS are completely reversible, either spontaneously or after the addition of a sulphydryl group-containing agent. The reversible increase in cation permeability induced by PCMBS provides a useful procedure for preparing red cells with different cation composition (Garrahan and Rega, 1967).

Amino groups: As already mentioned the low permeability of the red cell to cations as compared to anions has been attributed to the existence of positive charges within the cell membrane. The most likely source of such charges are the terminal or lateral amino groups of cell membrane proteins (Passow, 1964, 1965). Some evidence concerning the role of these groups in the control of cation permeability has been obtained by treating red cells with 1-fluoro-2,4-dinitrobenzene, or with 1,5-difluoro-2,4-dinitrobenzene. With the first compound, which forms stable combinations with single amino groups, cation permeability increases at least thirty-five times and cells undergo colloid-osmotic haemolysis. With the second compound, which reacts with two amino groups if they are 5 Å apart, a much larger increase in cation permeability takes place (at least 10^3 times) and the cells become extremely resistant to lytic agents, presumably because the cell membrane becomes rigid (Berg and coworkers, 1965). N-ε-dinitrophenyl lysine has been identified in red cell membranes after treatment with 1-flouro-2-4-dinitrobenzene. In agreement with the predictions of the positive fixed charge hypothesis, blockage of cell membrane amino groups is associated with a decrease in the permeability to anions (Passow, and Schnell, 1969).

Permeability inducing antibiotics: A group of permeability inducing antibiotics have recently been described (see for instance Lardy and coworkers, 1967). They include lipid-soluble uncharged compounds like valinomycin, gramicidin D and nonactin and carboxyl group-containing substances like nigericin and dianemycin (Pressman and coworkers, 1967). They have the ability of forming lipid-soluble complexes with monovalent cations (see Kilbourn and coworkers, 1967) which in some cases, for example, for valinomycin and nonactin, can show a remarkable specificity towards potassium. Probably due to their complexing properties they are able to carry cations across lipid phases (see Stein, 1968) inducing in that way large increases in the cation permeability of intact red cells and in the electrical conductance of lipid bilayers prepared from red cell membrane lipids. According to the experimental conditions and to the cation specificity of the agent employed, the effects of antibiotics on red cell permeability can result in an exchange of sodium for potassium, in unbalanced potassium movements or in cation–hydrogen ion exchanges (Chappell and Crofts, 1965; Andreoli and coworkers, 1967b; Harris and Pressman, 1967; Tosteson and coworkers, 1967, 1968; Bielawski, 1968).

IV. ACTIVE TRANSPORT OF SODIUM AND POTASSIUM

A. Active Transport and Metabolism

As the term implies active transport requires metabolic energy which is provided by glycolysis in mammalian red cells (Danowski, 1941; Harris, 1941; Maizels, 1951) and by respiration in nucleated erythrocytes from birds and reptiles (Maizels, 1954b).

Although glucose is the physiological substrate in red cells active transport can also be maintained by purine nucleosides (Whittam, 1960). These compounds enter the glycolytic pathway without the need of ATP and are therefore effective even when cell ATP stores are depleted. The rate of lactic acid production can be increased up to five times in red cells if adequate amounts of a purine nucleoside and inorganic phosphate are added to the incubation medium. In these conditions there is no change in the rate of active transport, suggesting that glycolysis is *not* rate-limiting for the pump mechanism in fresh red cells (Whittam and Wiley, 1967).

It is now well established that glycolysis energizes active transport because it supplies the system with ATP. The main evidence for the role of ATP as the immediate energy source comes from the demonstration that active transport reappears in reconstituted ghosts if ATP is trapped during haemolysis (Gardos, 1954; Hoffman, 1962a). Other nucleotides are not effective (Hoffman, 1962a) with the exception of ADP. (Garrahan and Glynn, 1967c) probably because it is transformed into ATP by the very active adenylate kinase of red cell membranes (Sen and Post, 1964). In order to energize active transport ATP must have access to the internal surface of the cell membrane (Whittam, 1962), and its hydrolysis products are also released at the internal surface of the membrane (Schatzmann, 1964; Marchesi and Palade, 1967).

Control of metabolism by active transport: In contrast with previous views (Schatzmann, 1953) it is now accepted that the rate of lactic acid production depends in part on the rate of active transport (Murphy, 1963; Whittam and Ager, 1965; Eckel and coworkers, 1966), so that removal of external potassium or addition of cardiac glycosides causes a fall in lactic acid production. The magnitude of this drop, that is a measure of the control of metabolism by the pump system, depends on the rate of transport, varying from 15 to 70 per cent (Whittam and Ager, 1965). The lactic acid production in whole haemolysates is also inhibited by cardiac glycosides, if ATP and sodium are present. This effect disappears if the cell membranes are removed and is only manifest if substrates above the phosphoglycerate kinase step are used. These results suggest that the fraction of glycolysis that is controlled by the pump system is that which takes place through the membrane-bound phosphoglycerate kinase, whose supply of ADP would come only from the active transport system (Parker and Hoffman, 1967).

B. Properties of the Sodium Pump

Coupling: Active movements of sodium and potassium in red cells are linked in the sense that sodium is transported outwards only if there is potassium in the external solution to be transported inwards and vice versa. Such coupling was first suggested by Harris and Maizels (1951), and since then evidence has accumulated which substantiates their view. The main lines of evidence are:

i) In the absence of potassium, sodium efflux is unaffected by glucose deprivation and the cells cannot achieve a net loss of sodium against even small concentration gradients (Glynn, 1956).

ii) External potassium activates sodium efflux and the curve obtained is similar to the potassium influx versus external potassium concentration curve (Glynn, 1956; Sachs and Welt, 1967).

iii) Internal sodium activates potassium influx and the curve is similar to the sodium efflux versus internal sodium concentration curve (Whittam and Ager, 1965).

The ratio of active sodium efflux to active potassium influx measured on the basis of net movements (Post and Jolly, 1957; Post and coworkers, 1967) or with isotopes (Garrahan and Glynn, 1966a, 1967d), is significantly greater than one. The coupling ratio seems to be fixed and independent of the ionic gradients (Post and Jolly, 1957; Whittam and Ager, 1965).

Number of transport sites: Glynn (1957a) has calculated on the assumption that one glycoside molecule blocks one transport site, that the number of these sites in each red cell is of the order of 1,000. Although the size of the transport system is unknown, the smallness of this figure makes it very difficult to avoid the conclusion that only a minute fraction of the cell surface can be directly concerned with active transport.

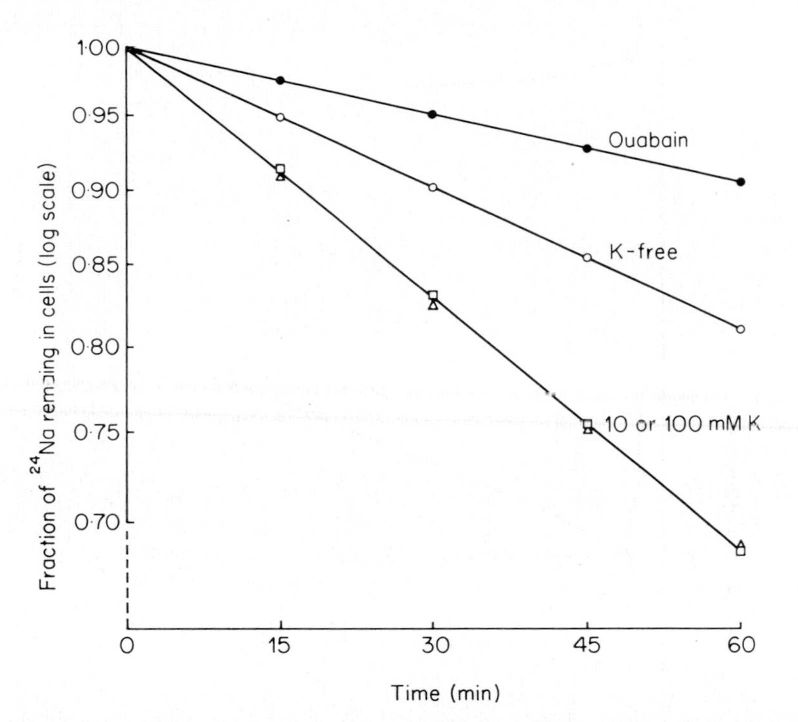

Figure 2. The effects of external potassium and of ouabain on the efflux of sodium from fresh human red cells suspended in a balanced salt solution. K replaced an equivalent quantity of Na in the 10 mM K and 100 mM K media. □ = 10 mM K; Δ = 100 mM K; ○ = K-free; ● = K-free with ouabain. (From Garrahan and Glynn, 1967a.)

C. Kinetics of the Active Fluxes

1. *Sodium Efflux*

Under physiological conditions the glycoside-sensitive sodium efflux in fresh human red cells represents about two-thirds of the total sodium efflux (Glynn, 1957a: figure 2). Its rate is a function of both intra- and extra-cellular cation concentrations.

Internal cations: Glycoside-sensitive sodium efflux needs intracellular magnesium and is inhibited by intracellular calcium (Hoffman, 1962a). In the presence of extracellular potassium it exhibits saturation kinetics when measured as a function of internal sodium concentration (Post and Jolly, 1957; Hoffman, 1962b; Whittam and Ager, 1965; Maizels, 1968). The half-maximal value for internal sodium lies around 20 mM and very little change in sodium efflux is seen above 50 mM internal sodium. The cation pump is highly selective for sodium; in fact no other cation can replace it for

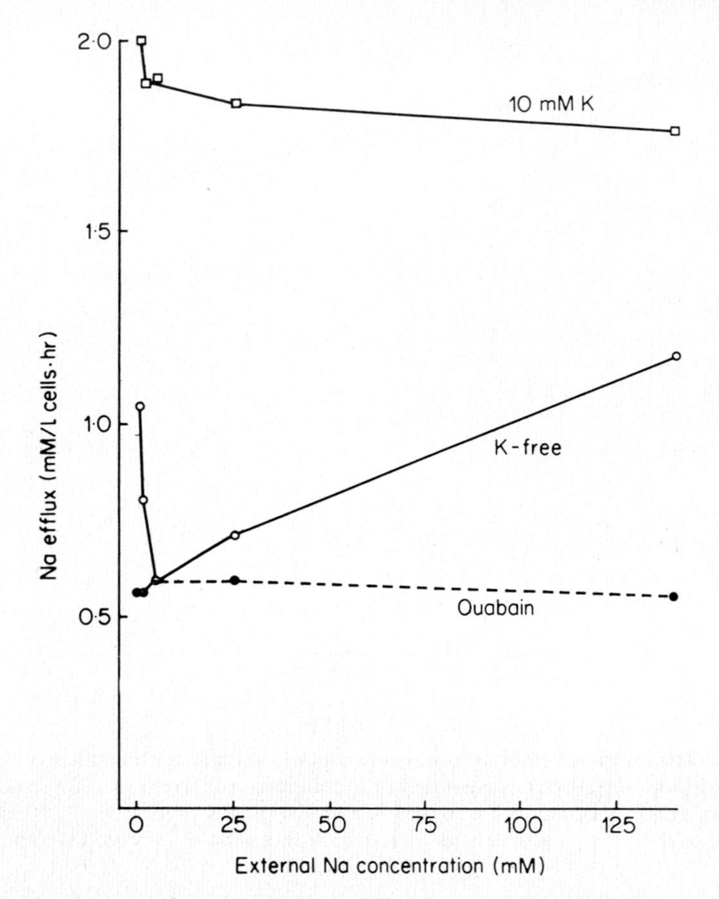

Figure 3. Effect of external sodium on the efflux of sodium from fresh human red cells in the presence and absence of potassium or ouabain. Sodium was replaced by equivalent amounts of choline chloride. (From Garrahan and Glynn, 1967a.)

outward transport. Potassium (Hoffman, 1962b), lithium, rubidium and caesium (Maizels, 1968) seem to compete with sodium at the intracellular pump site.

External cations: sodium–potassium and sodium–sodium exchanges: Removal of extracellular potassium, in the presence of physiological concentrations of external sodium, reduces sodium efflux by about one-third. In these conditions addition of cardiac glycoside results in a further 50 per cent drop in the rate of efflux (Glynn, 1957a: figure 2). The existence of a glycoside-sensitive sodium efflux in the *absence* of external potassium seems at first hand to contradict the experimental evidence that active transport stops when extracellular potassium is removed. This paradox can be solved by the suggestion that when external potassium is removed the pump system changes its mode of behavior and instead of exchanging sodium for potassium it exchanges sodium for *sodium*. A one to one sodium exchange across the cell membrane would only appear as an isotopic flux and would be incapable of achieving net movements (Garrahan and Glynn, 1965; 1966a; 1967a). The following experimental facts support this point of view:

i) Progressive replacement of extracellular sodium with choline has very little effect on sodium efflux in the presence of potassium or cardiac glycosides, but the glycoside-sensitive sodium efflux in the *absence* of external potassium is reduced until with only 5 mM of sodium externally it practically disappears (Garrahan and Glynn, 1965, 1967a: figure 3).

ii) In the absence of external potassium the glycoside-sensitive sodium *efflux* is accompanied by a glycoside-sensitive sodium *influx*. Both fluxes are equal at any external sodium concentration (Garrahan and Glynn, 1966a, 1967a: figure 4).

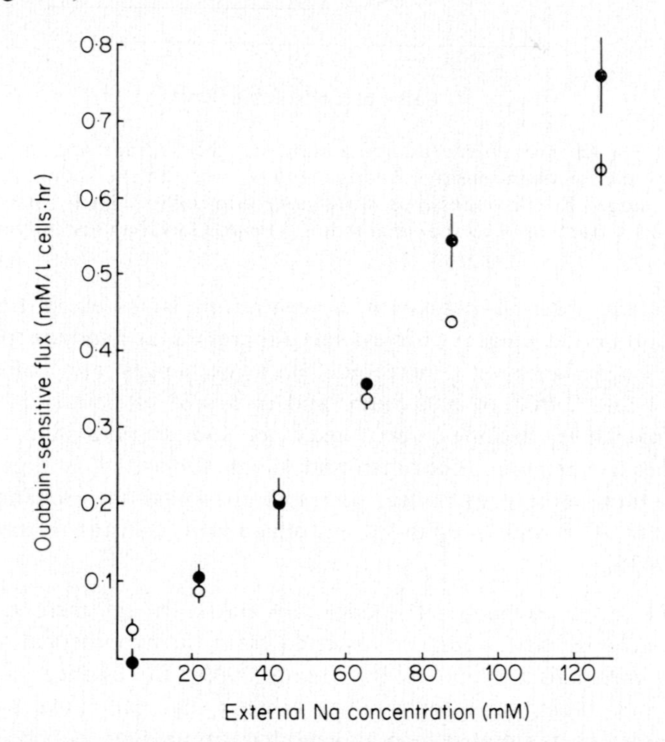

Figure 4. Ouabain-sensitive sodium efflux and ouabain-sensitive sodium influx in the absence of external potassium and at six different levels of external sodium. The experiment was performed on fresh human red cells. o = efflux; ● = influx. The vertical lines show ±1 S.E. (From Garrahan and Glynn, 1967a.)

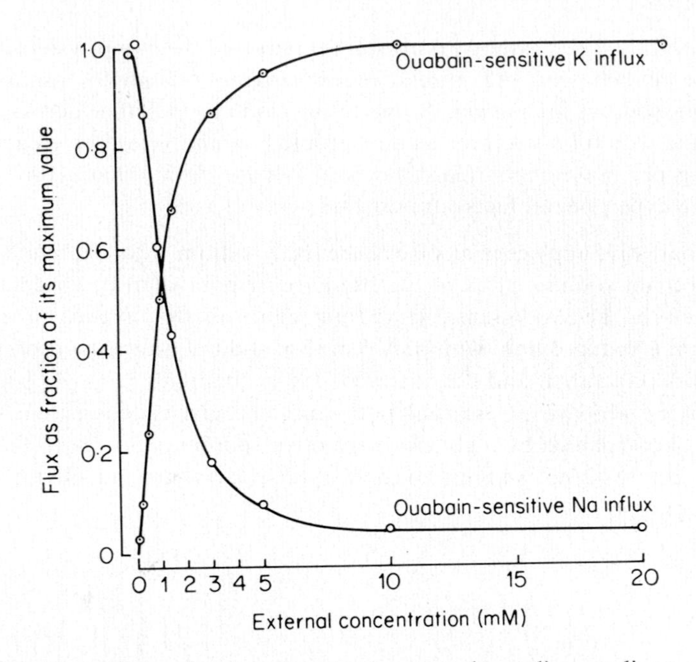

Figure 5. The effects of external potassium on the sodium–sodium exchange, judged by the ouabain-sensitive sodium influx, and on the sodium–potassium exchange, judged by the ouabain-sensitive potassium influx. Each curve has been expressed as a fraction of its maximal value. (From Garrahan and Glynn, 1967c.)

iii) As the external potassium concentration is increased from zero, sodium–sodium exchange is progressively suppressed, and sodium–potassium exchange is progressively increased. Both exchanges are half-maximal between 1 and 2 mM of potassium, and at 5 mM of potassium sodium–sodium exchange becomes very small as sodium–potassium exchange approaches a maximum. (Garrahan and Glynn, 1966a, 1967c: figure 5).

iv) Sodium–sodium exchange, like sodium–potassium exchange, needs intracellular ATP and is inhibited by oligomycin. (Garrahan and Glynn, 1967c, 1967d).

The one to one exchange of sodium ions across the cell membrane is the kind of exchange that would be expected from the hypothetical 'exchange diffusion' mechanism proposed by Ussing (1949), the essence of which is that the ions cross the membrane by carriers that can move only when carrying ions. In its simplest form this mechanism predicts that the exchange should be independent of external sodium concentration when this is much greater than internal sodium concentration (Garrahan and Glynn, 1967a). However, sodium–sodium exchange in red cells is roughly linear with external sodium even at values ten times as large as the internal sodium concentration (figures 3 and 4), suggesting that the apparent affinity for sodium is much lower at the external surface of the cell membrane (see Garrahan and Glynn, 1967a).

ATP hydrolysis during sodium–sodium exchange: Sodium–sodium exchange needs intracellular ATP but in contrast with sodium–potassium exchange, it is associated with no ouabain-sensitive ATP hydrolysis (Garrahan and Glynn, 1965, 1967d). A possible interpretation of the need of ATP during sodium–sodium exchange is that the nucleotide is used to form some phosphorylated intermediate that is not broken down due to the absence of extracellular potassium. This interpretation fits with current views on the intermediate steps of ATP hydrolysis by the cation transport system but in its simplest form it does not account for the oligomycin and glycoside sensitivity of the exchange, suggesting therefore that other steps apart from phosphorylation are operative.

Effects of intracellular composition on sodium–sodium exchange: The relative magnitude of sodium–sodium exchange, as compared to sodium–potassium exchange, varies in a complicated way with internal sodium, ATP and inorganic phosphate. These effects can be summarized as follows (Garrahan and Glynn, 1965, 1967c):

i) If intracellular sodium concentration is increased, sodium–sodium exchange is progressively reduced until it disappears when sodium becomes the only internal monovalent cation (figure 6). High sodium reconstituted ghosts *recover* their ability to carry out sodium–sodium exchange if the intracellular ATP/P_i ratio is adequately decreased.

ii) In low sodium ghosts, reduction of intracellular ATP concentration increases the relative magnitude of the sodium–sodium exchange which approaches values similar to those of the sodium–potassium exchange.

iii) If low sodium, low ATP ghosts are supplied with enough inorganic phosphate, sodium–sodium exchange becomes *greater* than sodium–potassium exchange, so that addition of external potassium *inhibits* the glycoside sensitive sodium efflux (figure 7).

There is no clear-cut interpretation of the mechanism of the above-mentioned effects, although several models have been proposed (Baker and Conelly, 1966; Garrahan and Glynn, 1967e; Stone, 1968). It is clear, however, that the effects of inorganic phosphate are not due to competitive inhibition of ATP utilization (Garrahan and Glynn, 1967c).

Glycoside-sensitive sodium efflux in sodium and potassium-free solutions: When red cells are incubated in a potassium-free solution the glycoside-sensitive sodium efflux is nearly absent at 5 mM sodium externally, but it increases as the external sodium concentration is reduced from 5 mM to zero (Garrahan and Glynn, 1967b: figure 3). One of the most striking

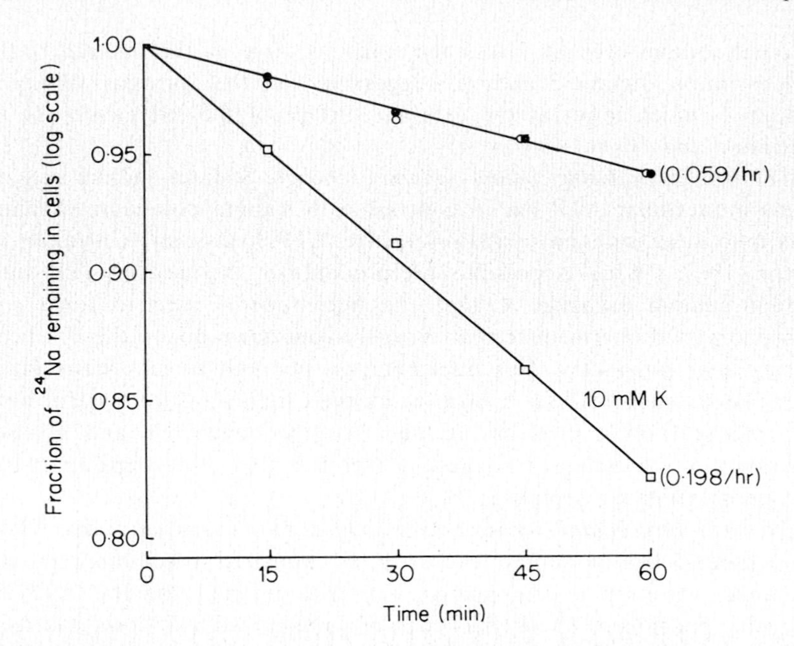

Figure 6. The efflux of sodium from ghosts of human red cells prepared to contain 5 mM ATP and sodium as the only univalent cation. The ghosts were suspended in high-sodium media. □ = 10 mM K; ○ = K-free; ● = K-free with ouabain. The figures in brackets are the rate constants for the efflux. (From Garrahan and Glynn, 1967c.)

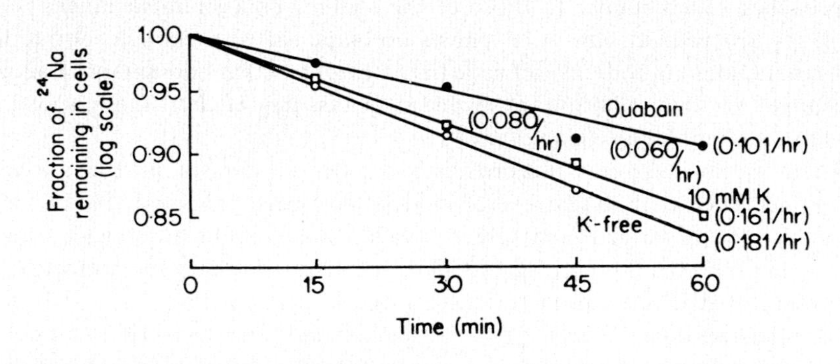

Figure 7. Efflux of sodium from ghosts of human red cells containing (in mM): ATP, 0·92; P_i, 4·6; Na, 6·4; K, 140. □ = 10 mM K; ○ = K-free; ● = K-free with ouabain. The figures in brackets are the rate constants for the efflux. (From Garrahan and Glynn, 1967c.)

properties of this efflux is its very high sensitivity to external sodium: 1 mM sodium in the external medium reduces it by 50 per cent.

Sodium efflux in potassium and sodium-free solutions could conceivably represent an exchange of sodium for traces of potassium leaking from the cells. This explanation, however, probably does not hold as the phenomenon persists even at very low haematocrits at which potassium accumulation is negligible. Furthermore glycoside-sensitive potassium *efflux* into nominally sodium and potassium-free solutions is unaffected by adding 5 mM sodium.

The glycoside-sensitive sodium efflux into sodium and potassium-free solutions resembles the sodium loss observed by Baker (1964) in crab nerves immersed in isotonic dextrose and probably is the expression of a third mode of behavior of the sodium pump distinct both from the normal sodium–potassium exchange and from the abnormal sodium–sodium exchange.

Ethacrynic acid and 'Pump II': About two-thirds of the glycoside-resistant sodium efflux is abolished by 10^{-3} M ethacrynic acid (Hoffman, 1966; Hoffman and Kregenow, 1966). According to Hoffman the ethacrynic acid-sensitive sodium efflux is the expression of the working in parallel of a sodium pump ('Pump II') that is different and independent from the glycoside-sensitive transport system ('Pump I'), the main difference being that 'Pump II' saturates at a lower internal sodium concentration and needs a much longer substrate-free incubation period to stop working than 'Pump I'. In Hoffman's experiments a fraction of the glycoside-sensitive sodium efflux in the presence of extracellular potassium, and *all* the ethacrynic acid-sensitive sodium efflux disappear when external sodium is replaced by magnesium, though in neither case can the effect be attributed to sodium exchange diffusion. These results contrast with those obtained by Garrahan and Glynn (1965; 1967a) who could not detect any effect on sodium efflux in the presence of potassium or ouabain of replacing external sodium with choline (see figure 3), which suggests that the effects observed by Hoffman may have been due to magnesium rather than to the removal of external sodium.

2. *Potassium Influx*

In the presence of physiological concentrations of extracellular sodium the potassium influx versus external potassium concentration curve is S-shaped with a half-maximal value at about 1·5 mM (Garrahan and Glynn, 1967b; Sachs and Welt, 1967). The sigmoid-shaped part of the influx curve lies below 1 mM $[K]_e$, this being probably the reason why the influx curve was considered for a rather long time to be adequately represented by a Michaelis-like curve (Glynn, 1956; Post and Jolly, 1957). As external sodium concentration is progressively reduced, both the half-maximal potassium concentration and the S-shaped region are progressively shifted towards zero.

When sodium is virtually absent the potassium influx is half-maximal at 0·14 mM and the influx curve is a rectangular hyperbola down to 0·020 mM external potassium, though there seems to be a trace of an inflexion at 0·015 mM external potassium (Garrahan and Glynn, 1967b: figure 8).

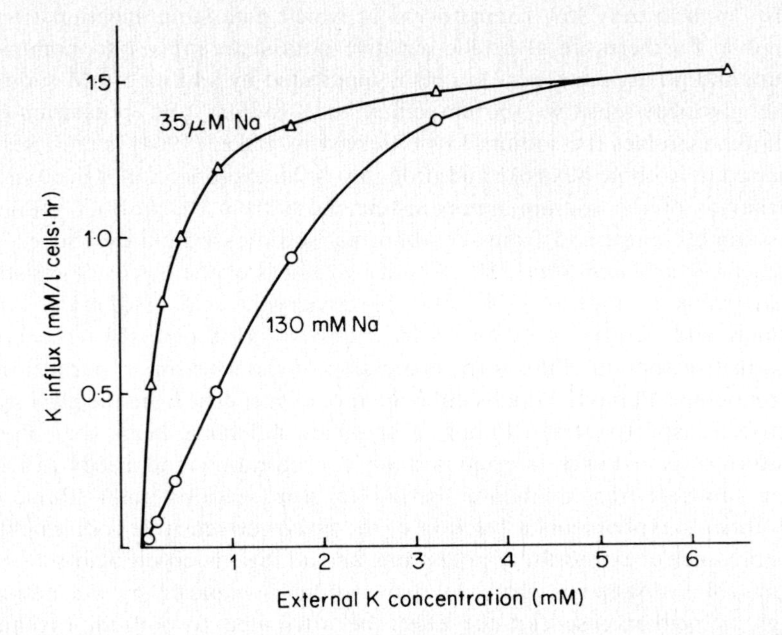

Figure 8. The effect of external sodium on potassium influx. Isotonicity was maintained with choline. (From Garrahan and Glynn, 1967b.)

A possible interpretation of the sigmoid-shape of the influx curve is that in order to induce active transport *two* potassium ions have to combine with the transport system. (Garrahan and Glynn, 1967b; Sachs and Welt, 1967)* This assumption seems to fit with observations on the stoicheiometry of the sodium pump (see below). On the two site-carrier hypothesis the effect occurring on removing external sodium would be due to an increase in the apparent affinity of the system for potassium. When external sodium is

*It may be worthwhile to point out here that if the system has to be saturated with more than one ion to produce observable changes, there is no need to postulate indirect 'cooperative' interactions between separate sites to account for the sigmoid shape of the influx curve. Such interactions are thought to occur in enzymes that exhibit S-shaped responses to substrate concentration because there is no reason to suppose that more than one site has to be occupied in order for the reaction to proceed (see, for instance, Monod and coworkers, 1965).

absent the apparent affinity for potassium would be such that one site would be completely saturated except at very low potassium concentrations. In these circumstances the rate of influx would be limited by the saturation of only *one* site giving, as in fact happens experimentally, an almost Michaelis-like curve.

The most immediate explanation of the decrease in the apparent affinity for potassium induced by sodium is direct competition between sodium and potassium at the potassium site (Post and coworkers, 1960; Whittam and Ager, 1964; Garrahan and Glynn, 1967b; Sachs and Welt, 1967). However, explanations involving indirect allosteric effects like those discussed by Monod and coworkers (1965) can not be dismissed.

The potassium site of the cation transport system is less selective than the intracellular sodium-sensitive site; potassium can be replaced for inward transport in exchange for sodium by rubidium, caesium (Love and Burch, 1953), ammonium (Post and Jolly, 1957) and lithium (McConaghey and Maizels, 1962).

Potassium influx, at low concentrations of external potassium, is *activated* by low concentrations of rubidium, lithium, caesium (Sachs and Welt, 1967) or ammonium (Sachs, 1967). Moreover, in the presence of physiological concentrations of external sodium, addition of adequate concentrations of any of the potassium-like ions changes the shape of the potassium influx curve, making it closer to a rectangular hyperbola. Both phenomena can be quantitatively accounted for on the two site-carrier hypothesis, if it is assumed that potassium is still transported when one of the sites is occupied by any of the other monovalent cations (Sachs and Welt, 1967).

Cardiac glycosides and potassium selective sites: Cardiac glycosides inhibit active transport by acting on the external surface of the red cell (Hoffman, 1966). The effect on active transport of low glycoside concentrations can be abolished by raising the extracellular potassium concentration (Glynn, 1957a). This is apparently due to a decrease in the amount of bound glycoside (Hoffman, 1966).

Both the kinetics of the potassium effect and the lack of action of potassium when optimal glycoside concentrations are present (Glynn, 1957a) make it very unlikely that the action of potassium results from simple competition with glycosides; this view is further strengthened by the fact that of the potassium-like ions rubidium but *not* caesium can abolish the effect of low glycoside concentrations (Hoffman, 1966).

The interactions of potassium and cardiac glycosides are also affected by external sodium ions. At low potassium concentrations the inhibitory effect of ouabain is enhanced by sodium and conversely the glycoside potentiates the inhibitory effect of sodium on the activation by potassium of the cation pump (Schatzmann, 1965).

3. *Ouabain-sensitive Potassium Efflux*

Cardiac glycosides reduce by about 25 per cent the potassium *efflux* in human red cells (Glynn, 1957a). The glycoside-sensitive potassium efflux is dependent on the presence of *intracellular* inorganic phosphate and disappears if external sodium and potassium are replaced by choline. These requirements suggest that the glycoside-sensitive potassium efflux represents an exchange of potassium ions across the cell membrane through a reversal of the last stages of cation transport (Glynn and Lüthi, 1968). In partially starved red cells suspended in K-free media the glycoside sensitive K efflux is a linear function of external Na concentration and is accompanied by an ouabain-sensitive incorporation of inorganic phosphate into ATP. The K-efflux under this condition therefore appears to be associated with a reversal of the entire pump cycle (Glynn and Lew, 1969).

D. Biochemistry of Active Sodium and Potassium Transport

The study of the biochemical events associated with active sodium and potassium transport is at present mainly concerned with the detailed analysis of the Na^+–K^+-activated ATPase system of cell membranes ('transport ATPase'). A detailed treatment of this subject is found in Volume I. Here we will only consider those aspects of the 'transport ATPase' activity that are directly related to the problem of active transport in red cells.

Stoicheiometry: The number of sodium and potassium ions that move across the red cell membrane for each molecule of ATP that is split by the glycoside-sensitive ATPase system, has been measured by a number of authors using different techniques (Gárdos, 1964; Glynn, 1962; Sen and Post, 1964; Whittam and Ager, 1965). The values obtained are not significantly different from three sodium ions and two potassium ions moved per molecule of ATP split. This ratio seems to be fixed and independent of the ionic gradients imposed across the cell membrane (Whittam and Ager, 1965).

It must however be remembered that the red cell membrane is able to hydrolyse ATP at about half its maximal rate when sodium and potassium are omitted or cardiac glycosides are present (Post and coworkers, 1960; Dunham and Glynn, 1961). The figure for stoicheiometry mentioned above is based on the assumption that the glycoside-insensitive hydrolysis is completely unrelated to the transport ATPase system. No unequivocal demonstration of this independence is yet available and until such proof is forthcoming the definite physical meaning of the measured stoicheiometry will remain uncertain (see Garrahan and coworkers, 1969).

Reversibility: If we take for granted that the cation pump in red cells expels three sodium ions and takes up about two potassium ions for each molecule of ATP it hydrolyses, the free energy available to drive the transport reaction forwards must be quite small. Under physiological conditions this is less than 4,000 cal (Garrahan and Glynn, 1967e). Thus by making the concentration gradients even more adverse than under physiological conditions, it should be thermodynamically possible to drive the pump backwards and hence synthesize ATP, though theory cannot predict the rate of this reaction.

The reversibility of the transport reaction has been measured by means of incorporation of $^{32}P_i$ into ATP and ADP in reconstituted ghosts containing much potassium, very little sodium and a low $[ATP]/[ADP] \times [P_i]$ ratio. (Garrahan and Glynn, 1966b, 1967e). Under these conditions they found that there is always an extra incorporation of inorganic phosphate into the nucleotides when ghosts are incubated in a high-sodium, potassium-free medium; this extra incorporation is largely abolished by cardiac glycosides or by oligomycin. These findings have been confirmed and extended by Lant and Whittam, 1969).

The rate of the reverse reaction under the conditions tested is about 1·4 per cent of the maximum forward rate. The low rate of inorganic phosphate incorporation when the ionic gradients favour the running backwards of the system, together with the lack of ATP–P_i exchange when the cation pump is exchanging sodium for potassium or sodium for sodium suggests rather strongly that it is not until sodium and potassium are moved that inorganic phosphate is released by the transport ATPase system (see Garrahan and Glynn, 1967e).

Partial reactions: Current experimental evidence seems to indicate that the active transport system splits ATP into ADP and inorganic phosphate in at least two steps. First there is a sodium-dependent phosphorylation of a membrane protein followed by a potassium-activated dephosphorylation (see Chapter 8 in Volume 1).

Neither Heinz and Hoffman (1965) nor Blake and coworkers, (1967) could detect any specific incorporation of ^{32}P into red cell membrane preparations exposed to (^{32}P-ATP) under conditions similar to those successfully employed in other tissues. This failure was probably due to the existence of a large unspecific phosphate incorporation. Sodium-dependent phosphate incorporation has afterwards been demonstrated in red blood cell membranes employing conditions chosen to minimize unspecific phosphate incorporation, such as the use of very low ATP concentrations (Blonstein, 1968) or electrophoretic separation of labelled membrane material after partial enzymatic hydrolysis (Bader and coworkers, 1968).

The potassium-activated dephosphorylation step has been related to a $Mg^{2+}+K^{+}$-activated and glycoside-sensitive phosphatase activity which is present in most cell membrane preparations (see for instance, Bader and coworkers, 1968). Such activity has been shown to be present in red cell membranes by Judah and coworkers (1962). Like the transport-ATPase system, the K^+-activated phosphatase in red cells shows spatial asymmetry with respect to the membrane, i.e. the substrate has to be at the *internal* surface of the cell membrane in order to be hydrolysed (Garrahan and coworkers, 1969) and potassium activates only when it is at the *external* surface of the cell membrane (Rega and coworkers, 1970).

V. DIVALENT CATIONS

A. Magnesium

The magnesium concentration in red blood cells is about 3·5 mM/l cells and after hypotonic haemolysis it remains in the soluble fraction (Harrison and coworkers, 1968). Intracellular magnesium seems to be an essential cofactor in active sodium and potassium transport and in the associated ATP hydrolysis. Magnesium is also an essential component of red cell membranes. If red cells are haemolysed in media containing magnesium-chelating agents, e.g. EDTA, the membranes fail to regain their low permeability to sodium and potassium (Hoffman 1962a).

Radioactive magnesium exchanges with intracellular magnesium with a half-time of 24 hours (Rogers, 1961).

B. Calcium

The calcium content of red cells is very low—about 0·015 mM/l cell and it is mostly associated with the cell membrane (Harrison and coworkers, 1968). The low intracellular calcium concentration seems to be the result of an active ATP-dependent mechanism that extrudes calcium out of the cell (Schatzman, 1966). The calcium transport system may be related to the activation by calcium of the glycoside-insensitive ATPase activity of red cell membranes (Dunham and Glynn, 1961; Wins and Schoffeniels, 1966).

In the presence of ATP, calcium ions also increase the rate of hydrolysis of *p*-nitrophenylphosphate by the potassium-activated phosphatase of red cell membranes (Pouchan and coworkers, 1969). How this is related to active calcium transport is still unknown.

Acknowledgement

The work reported here was supported in part by the Consejo Nacional de Investigaciones Científicas y Técnicas, Argentina.

REFERENCES

Andreoli, T. E., J. A. Bangham and D. C. Tosteson (1967a) *J. Gen. Physiol.*, **50**, 1729
Andreoli, T. E., M. Tieffenberg and D. C. Tosteson (1967b) *J. Gen. Physiol.*, **50**, 2527
Appelboom, J. W. T., W. A. Brodsky, W. S. Tuttle and J. Diamond (1958) *J. Gen. Physiol.*, **41**, 1153
Bader, H., R. L. Post and G. H. Bond (1968) *Biochim. Biophys. Acta.*, **150**, 41
Baker, P. F. (1964) *Biochim. Biophys. Acta.*, **88**, 458

Baker, P. F. and C. M. Connelly (1966) *J. Physiol.* (*London*), **185**, 270
Berg, H. C., J. M. Diamond and P. S. Marfey (1965) *Science*, **150**, 64
Bielawski, J. (1968) *Eur. J. Biochem.*, **4**, 181
Blake, A., D. P. Leader and R. Whittam (1967) *J. Physiol.* (*London*), **193**, 467
Blostein, R. (1968) *J. Biol. Chem.*, **243**, 1957
Bolingbroke, V. and M. Maizels (1959) *J. Physiol.* (*London*), **149**, 563
Brodsky, W. M., W. S. Rehm and B. J. McIntosh (1953) *J. Clin. Invest.*, **32**, 556
Bromberg, P. A., J. Theodore, E. D. Robin and W. N. Jensen (1965) *J. Lab. Clin. Med.*, **66**, 464
Carr, C. W. (1956) *Arch. Biochem. Biophys.*, **62**, 476
Chappell, B. and A. Crofts (1965) In J. M. Tager, S. Papa, E. Quagliariello and E. C. Slater (Eds.), *Regulation of Metabolic Processes in Mitochondria.* Elsevier, Amsterdam
Chedru, J. and P. Cartier (1966) *Biochim. Biophys. Acta.*, **126**, 500
Clarkson, E. M. and M. Maizels (1955) *J. Physiol.* (*London*), **129**, 476
Cook, J. S. (1967) *J. Gen. Physiol.*, **50**, 1311
Danowski, T. S. (1941) *J. Biol. Chem.*, **139**, 693
Davson, H. (1939) *Biochem. J.*, **33**, 389
Dick, D. A. T. and L. M. Lowenstein (1958) *Proc. Roy. Soc.* (*London*) *Ser. B.*, **148**, 241
Dirken, M. N. J. and H. W. Mook (1931) *J. Physiol.* (*London*), **73**, 349
Drabkin, D. L. (1950) *J. Biol. Chem.*, **185**, 231
Dunham, E. T. and I. M. Glynn (1961) *J. Physiol.* (*London*), **156**, 274
Eckel, R., S. C. Rizzio, A. Lodish and A. B. Bergren (1966) *Amer. J. Physiol.*, **210**, 737
Edelberg, R. (1952) *J. Cellular Comp. Physiol.*, **40**, 529
Fitzsimons, E. J. and J. Sendroy (1961) *J. Biol. Chem.*, **236**, 1595
Fricke, H. (1925) *J. Gen. Physiol.*, **9**, 137
Fricke, H. and S. Morse (1925) *J. Gen. Physiol.*, **9**, 153
Funder, J. and J. O. Wieth (1966) *Acta Physiol. Scand.*, **68**, 234
Funder, J. and J. O. Wieth (1967) *Acta Physiol. Scand.*, **71**, 105
Gárdos, G. (1954) *Acta Physiol. Hung.*, **6**, 191
Gárdos, G. (1959) *Acta Physiol. Hung.*, **15**, 121
Gárdos, G. (1964) *Experientia*, **20**, 387
Garrahan, P. J. and I. M. Glynn (1965) *Nature*, **207**, 1098
Garrahan, P. J. and I. M. Glynn (1966a) *J. Physiol.* (*London*), **185**, 31P
Garrahan, P. J. and I. M. Glynn (1966b) *Nature*, **211**, 1414
Garrahan, P. J. and I. M. Glynn (1967a) *J. Physiol.* (*London*), **192**, 159
Garrahan, P. J. and I. M. Glynn (1967b) *J. Physiol.* (*London*), **192**, 175
Garrahan, P. J. and I. M. Glynn (1967c) *J. Physiol.* (*London*), **192**, 189
Garrahan P. J. and I. M. Glynn (1967d) *J. Physiol.* (*London*), **192**, 217
Garrahan, P. J. and I. M. Glynn (1967e) *J. Physiol.* (*London*), **192**, 237
Garrahan, P. J. and A. F. Rega (1967) *J. Physiol.* (*London*), **193**, 459
Garrahan, P. J., M. I. Pouchan and A. F. Rega (1969) *J. Physiol.* (*London*), **202**, 305
Gary Bobo, C. M. (1967) *J. Gen. Physiol.*, **50**, 2547
Gill, T. J. and A. K. Solomon (1959) *Nature*, **183**, 1127
Glader, B. E. and A. Omachi (1968) *Biochim. Biophys. Acta.*, **163**, 30
Glynn, I. M. (1956) *J. Physiol.* (*London*), **134**, 278
Glynn, I. M. (1957a) *J. Physiol.* (*London*), **136**, 148
Glynn, I. M. (1957b) *Progr. Biophys.*, **8**, 241
Glynn, I. M. (1962) *J. Physiol.* (*London*), **160**, 18P
Glynn, I. M. and L. Lüthi (1968) *J. Gen. Physiol.*, **51**, 385s
Glynn, I. M. and V. L. Lew (1969) *J. Physiol.* (*London*), **202**, 89P
Goldman, D. E. (1943) *J. Gen. Physiol.*, **27**, 37
Gourley, D. H. R. (1951) *Amer. J. Physiol.*, **164**, 213
Gourley, D. R. H. and C. L. Gemill (1950) *J. Cellular Comp. Physiol.*, **35**, 341
Gourley, D. R. H. and J. T. Matschiner (1953) *J. Cellular Comp. Physiol.*, **41**, 225
Hahn, L. A. and C. C. H. Hevesy (1942) *Acta Physiol. Scand.*, **3**, 193
Harris, E. J. (1954) *Symp. Soc. Exp. Biol.*, **8**, 228

Harris, E. J. and M. Maizels (1951) *J. Physiol.* (*London*), **113**, 506
Harris, E. J. and M. Maizels (1952) *J. Physiol.* (*London*), **118**, 40
Harris, E. J. and T. A. J. Prankerd (1953) *J. Physiol.* (*London*), **121**, 470
Harris, E. J. and B. C. Pressman (1967) *Nature*, **216**, 918
Harris, J. E. (1941) *J. Biol. Chem.*, **141**, 579
Harrison, D. G., C. Long and A. B. Sidle (1968) *Biochem. J.*, **108**, 40P
Hartridge, H. and F. J. W. Roughton (1923) *Proc. Roy. Soc.* (*London*) *Ser. B.*, **94**, 336
Heinz, E. and J. F. Hoffman (1965) *J. Cellular Comp. Physiol.*, **65**, 31
Hoffman, J. F. (1962a) *Circulation*, **26**, 1201
Hoffman, J. F. (1962b) *J. Gen. Physiol.*, **45**, 837
Hoffman, J. F. (1966) *Amer. J. Med.*, **41**, 666
Hoffman, J. F and F. M. Kregenow (1966) *Ann. N.Y. Acad. Sci.*, **137**, 566
Hoffman, J. F., D. C. Tosteson and R. Whittam (1960) *Nature*, **185**, 186
Holton F. A. (1952) *Biochem. J.*, **52**, 506
Hutchinson, E. (1952) *Arch. Biochem. Biophys.*, **38**, 35
Joel, A. (1915) *Arch. Ges. Physiol.*, **161**, 5
Johnson, S. L. and J. W. Woodbury (1964) *J. Gen. Physiol.*, **47**, 827
Joyce, C. R. B. and M. Weatherall (1955) *J. Physiol.* (*London*), **127**, 33P
Judah, J. D., K. Ahmed and A. E. M. McLean (1962) *Biochim. Biophys. Acta.*, **65**, 472
Kahn, J. B. and G. H. Acheson (1955) *J. Pharmacol.*, **115**, 305
Kilbourn, B. T., J. D. Dunitz, L. A. R. Pioda and W. Simon (1967) *J. Mol. Biol.*, **30**, 559
LaCelle, P. L. and A. Rothstein (1966) *J. Gen. Physiol.*, **50**, 171
Lardy, H. A., S. N. Graven and S. Estrada (1967) *Federation Proc.*, **26**, 1355
Lassen, U. V. and O. Sten-Knudsen (1968) *J. Physiol.* (*London*), **195**, 681
Lant, A. F. and R. Whittam (1969) *J. Physiol.* (*London*), **199**, 457
LeFevre, P. G. (1964) *J. Gen. Physiol.*, **47**, 585
Lepke, S. and H. Passow (1968) *J. Gen. Physiol.*, **51**, 365s
Love, W. D. and E. Burch (1953) *J. Lab. Clin. Med.*, **41**, 351
Luckner, H. (1939) *Arch. Ges. Physiol.*, **241**, 753
Luckner, H. (1949) *Arch. Ges. Physiol.*, **250**, 303
Lyman, R. A. (1945) *J. Cellular Comp. Physiol.*, **25**, 65
McConaghey, P. D. and M. Maizels (1961) *J. Physiol.* (*London*), **155**, 28
McConaghey, P. D. and M. Maizels (1962) *J. Physiol.* (*London*), **162**, 485
Maizels, M. (1935) *Biochem. J.*, **29**, 1970
Maizels, M. (1951) *J. Physiol.* (*London*), **112**, 59
Maizels, M. (1954a) *Synp. Soc. Exp. Biol.*, **8**, 202
Maizels, M. (1954b) *J. Physiol.* (*London*), **125**, 263
Maizels, M. (1968) *J. Physiol.* (*London*), **195**, 657
Maizels, M. and M. Remington (1959a) *J. Physiol.* (*London*), **145**, 641
Maizels, M. and M. Remington (1959b) *J. Physiol.* (*London*), **145**, 658
Marchesi, V. T. and G. E. Palade (1967) *J. Cell. Biol.*, **35**, 385
Miller, D. M. (1964) *J. Physiol.* (*London*), **170**, 219
Mollison, P. L., M. A. Robinson and D. A. Hunter (1958) *Lancet*, **1**, 766
Mond, R. (1927) *Arch. Ges. Physiol.*, **217**, 618
Monod, J., J. Wyman and J. P. Changeux (1965) *J. Mol. Biol.*, **12**, 88
Morris, R. and R. D. Wright (1954) *Australian J. Exp. Biol. Med. Sci.*, **32**, 669
Mueller, C. B. and A. B. Hastings (1951) *J. Biol. Chem.*, **189**, 869
Murphy, J. R. (1963) *J. Lab. Clin. Med.*, **61**, 567
Parker, A. C. and J. F. Hoffman (1967) *J. Gen. Physiol.*, **50**, 893
Passow, H. (1964) In C. Bishop (Ed.), *The Red Blood Cell*. Academic Press, New York, p. 71
Passow, H. (1965) *Proc. XXIII Intern. Congr. Physiol. Sci.*, Tokyo, p. 555
Passow, H. and K. F. Schnell (1969), *Experientia*, **25**, 460
Pauly, H. and H. P. Schwan (1966) *Biophys. J.*, **6**, 621
Ponder, E. (1948) *Hemolysis and Related Phenomena*. Grune and Stratton, New York
Post, R. L. and P. C. Jolly (1957) *Biochim. Biophys. Acta.*, **25**, 118

Post, R. L., C. R. Merrit, C. R. Kinsolving and C. D. Albright (1960) *J. Biol. Chem.*, **235**, 1796
Post, R. L., C. D. Albright and K. Dayani (1967) *J. Gen. Physiol.*, **50**, 1201
Pouchan, M. I., P. J. Garrahan and A. F. Rega (1969) *Biochim. Biophys. Acta.*, **173**, 153
Prankerd, T. A. J. and K. I. Altman (1955) *Biochem. J.*, **58**, 622
Pressman, P. C., E. J. Harris and W. S. Jaeger (1967) *Proc. Natl. Acad. Sci. U.S.*, **58**, 1949
Raker, J. W., I. H. Taylor, J. M. Weller and A. B. Hastings (1950) *J. Gen. Physiol.*, **33**, 691
Rega, A. F., A. Rothstein and R. I. Weed (1967) *J. Cellular Comp. Physiol.*, **70**, 45
Rega, A. F., M. I. Pouchan and P. J. Garrahan (1970) *Science*, **167**, 55
Richmond, J. E. and A. B. Hastings (1960) *Amer. J. Physiol.*, **199**, 821
Rogers, T. A. (1961) *J. Cellular Comp. Physiol.*, **57**, 119
Sachs, J. R. (1967) *J. Clin. Invest.*, **46**, 1433
Sachs, J. R. and L. G. Welt (1967) *J. Clin. Invest.*, **46**, 65
Savitz, D., V. W. Sidel and A. K. Solomon (1964) *J. Gen. Physiol.*, **48**, 79
Sen, A. K. and R. L. Post (1964) *J. Biol. Chem.*, **239**, 345
Schatzmann, H. J. (1953) *Helv. Physiol. Pharmacol. Acta.*, **11**, 346
Schatzmann, H. J. (1964) *Experientia*, **20**, 551
Schatzmann, H. J. (1965) *Biochim. Biophys. Acta.*, **94**, 89
Schatzmann, H. J. (1966) *Experientia*, **22**, 364
Shaw, T. I. (1955) *J. Physiol. (London)*, **129**, 464
Solomon, A. K. (1952) *J. Gen. Physiol.*, **36**, 57
Solomon, A. K. (1960) *J. Gen. Physiol.*, **43**, Suppl., 1
Solomon, A. K. (1968) *J. Gen. Physiol.*, **51**, 335s
Solomon, A. K. and G. L. Gold (1955) *J. Gen. Physiol.*, **38**, 371
Solomon, A. K., T. J. Gill and G. L. Gold (1956) *J. Gen. Physiol.*, **40**, 327
Stein, W. D. (1968) *Nature*, **218**, 570
Stone, A. J. (1968) *Biochim. Biophys. Acta.*, **150**, 578
Sutherland, R., A. Rothstein and R. I. Weed (1967) *J. Cell Physiol.*, **69**, 185
Tanford, C. and Y. Nozaki (1966) *J. Biol. Chem.*, **241**, 2832
Teorell, T. (1952) *J. Gen. Physiol.*, **35**, 669
Tosteson, D. C. (1959) *Acta Physiol. Scand.*, **46**, 19
Tosteson, D. C. (1963) *Federation Proc.*, **22**, 19
Tosteson, D. C. (1966) *Ann. N.Y. Acad. Sci.*, **137**, 577
Tosteson, D. C. and J. F. Hoffman (1960) *J. Gen. Physiol.*, **44**, 169
Tosteson, D. C., P. Cook, T. Andreoli and M. Tieffemberg (1967) *J. Gen. Physiol.*, **50**, 2513
Tosteson, D. C., T. E. Andreoli, M. Tieffemberg and P. Cook (1968) *Gen. J. Physiol.*, **51**, 373s
Ussing, H. H. (1949) *Physiol. Rev.*, **29**, 127
Ussing, H. H. (1950) *Acta Physiol. Scand.*, **19**, 43
Van Slyke, D. D., H. Wu and F. C. McLean (1923) *J. Biol. Chem.*, **56**, 765
Van Steveninck, J., R. I. Weed and A. Rothstein (1965) *J. Gen. Physiol.*, **48**, 617
Vestergaard-Bogind, B. and T. Hesselbo (1960) *Biochim. Biophys. Acta.*, **44**, 117
Whittam, R. (1960) *J. Physiol. (London)*, **154**, 608
Whittam, R. (1962) *Biochem. J.*, **84**, 110
Whittam, R. (1964) *Transport and Diffusion in Red Blood Cells.* Williams and Wilkins, Baltimore
Whittam, R. (1968) *Nature*, **219**, 610
Whittam, R. and M. E. Ager (1964) *Biochem. J.*, **93**, 337
Whittam, R. and M. E. Ager (1965) *Biochem. J.*, **97**, 214
Whittam, R. and J. S. Wiley (1967) *J. Physiol. (London)*, **191**, 633
Wilbrandt, W. (1941) *Arch. Ges. Physiol.*, **245**, 22
Wilbrandt, W. and H. J. Shatzmann (1960) *Ciba Foundation Study Group*, **5**, 33
Williams, T. F., C. C. Fordham, W. Hollander and L. G. Welt (1959) *J. Clin. Invest.*, **38**, 1587
Wins, P. and E. Schoffeniels (1966) *Biochim. Biophys. Acta*, **120**, 341
Zipursky, A. and L. G. Israels (1961) *Nature*, **189**, 1013

CHAPTER 7

Ion movements in liver tissue

Irving Seidman

Department of Pathology,
New York University Medical Center,
New York, N.Y., 10016, U.S.A.

I. INTRODUCTION

In order for cells to perform their essential metabolic functions, the proper intracellular environment must be maintained. Since the chemical composition of intra- and extracellular fluid is different, it is necessary that a mechanism(s) exists by which cells are able to transport selectively ions and molecules against concentration gradients. The importance of this process is readily apparent and the mechanism by which this active transport occurs is one of the major fields of investigation in modern biology.

The active transport of ions, Na^+ and K^+ in particular, has been extensively studied in a variety of tissues and certain generalities have emerged from these investigations. It has been determined that ion transport against a concentration gradient is an endergonic process, requiring ATP as the energy source. Inhibition of ATP production results in the cessation of active ion movements. An ATPase (whether this represents a single molecule or multiple molecular species is as yet unclear) located at the cell membrane is now considered an essential component in active transport; inhibitors of this enzyme also inhibit active transport.

The general aspects of active transport have already been considered in detail in this volume together with special properties peculiar to individual tissues. This chapter will be limited to the description and analysis of studies concerned with liver which relate to the general problem of active ion transport.

II. ION TRANSPORT IN LIVER SLICES

One of the earliest studies on ion movements in liver slices was that of Flink and coworkers (1950) who incubated rat liver slices in a Ringer's–bicarbonate medium under an atmosphere of 95 per cent O_2–5 per cent CO_2. They observed a 60 per cent loss in tissue K^+ during the first fifteen minutes of incubation with a return to 80 per cent of control values by forty-five minutes. Tissue Na^+, on the other hand, increased 300 per cent during the first fifteen minutes and declined only slightly during subsequent incubation. When these experiments were performed under an anaerobic atmosphere, tissue K^+ decreased rapidly during the first fifteen minutes and then more slowly, reaching a plateau of about 25 per cent of the control value in ninety minutes. Similar results were obtained under aerobic conditions when sodium fluoride (16 mM) was added to the incubation fluid. These observations supported the view that aerobic metabolism was essential for the maintenance of proper intracellular Na^+ and K^+ concentrations. It was also shown that anaerobic incubation for fifteen minutes interfered significantly with the ability of the liver slices to reaccumulate K^+ when they were subsequently transferred to an aerobic atmosphere.

The initial loss of tissue K^+ and gain of Na^+, even under aerobic conditions, was thought to be due to tissue anoxia which occurred during the preparation of the tissue slices for study. Since the time required for the preparation of liver slices might be expected to vary in different laboratories or experimental situations, McLean (1963) developed a procedure for the study of ion transport in rat liver slices which was intended to yield more reproducible results. The procedure was based on the observation (Mudge, 1951; Leaf, 1956) that cooling of tissue slices led to the entry of Na^+ and water and the loss of K^+; subsequent warming of the slices in a proper environment resulted in a reaccumulation of K^+ and a variable extrusion of Na^+. McLean incubated rat liver slices in cold (2 °C) Ringer's–phosphate under an O_2 atmosphere and, after thirty-five minutes, he found that the tissue K^+ fell about 65 per cent while Na^+ and water increased about 500 per cent and 40 per cent, respectively. Within ten minutes of being transferred to a warm (38 °C) incubation solution, the liver slices began to lose Na^+ and water.

Potassium reaccumulation exhibited a lag period of about ten minutes. After thirty-five minutes of warm incubation, the liver slice had regained about 90 per cent of its initial K^+ and the water content had fallen to control levels.

While tissue Na^+ decreased during the warm incubation, the tissue levels after sixty minutes were still 2–3 times higher than those of controls. Parsons and van Rossum (1962) found that incubation of liver slices in warm Ringer's solution resulted in an increase of extracellular water. Since the incubation medium has a very high Na^+ concentration relative to the tissue, it is probable that a loss in intracellular Na^+ might not be recognized if measurements were made on the slices; an increase in the amount of Na^+-rich fluid in the extracellular space would mask a loss of intracellular Na^+. The concentration of K^+ in the incubation fluid, on the other hand, is low relative to that of the liver slice. An increase in the extracellular fluid volume would therefore not produce a large error in tissue K^+ determinations. It is believed that changes of tissue K^+ are a more accurate reflection of ion transport than Na^+ measurements because of these considerations.

A detailed kinetic analysis of Na^+ and K^+ movements in rat liver slices during cooling and rewarming was performed by Elshove and van Rossum (1963). During incubation at 1 °C for ninety minutes, tissue K^+ fell from 303 to 95 mmole/kg fat-free dry weight; Na^+ content increased from 134 to 677 mmole/kg fat-free dry weight and water increased from 2·43 to 3·94 mmole/kg water. When the slices were transferred to a medium warmed to 38 °C and oxygenated, the changes in tissue electrolyte and water content were largely reversed. As noted above (McLean, 1963) there was an initial rapid loss of Na^+ with little change in K^+ during the first fifteen minutes. Between fifteen and thirty minutes, the uptake of K^+ equalled the extrusion of Na^+ on a molar basis. The uptake of oxygen by the liver slice was constant throughout the period of warm incubation.

A. The Effect of Changes in the Cation Content of the Medium

The experiments described above were performed using an incubation medium similar to plasma in its electrolyte composition. It was found that variation in the concentration of several cations resulted in alterations in ion movements and respiration (Elshove and van Rossum, 1963).

1. *Sodium*

The liver slices were incubated in cold and warm solutions of the same ionic composition. The reaccumulation of K^+ during the incubation at 38 °C was dependent on the Na^+ concentration of the medium. When the Na^+ concentration was 25 mM or less, there was a slight loss of intracellular K^+, and at 50 mM Na^+, a small influx was observed. Further increases in the Na^+

concentration to 145 mM were accompanied by an almost linear increase in the accumulation of K^+. The loss of Na^+ from the slices at 38 °C was likewise dependent on the Na^+ concentration of the medium.

2. *Potassium*

Similar experiments were performed in which the K^+ concentration of the medium was varied. In a K^+-free medium there was no uptake of K^+ at 38 °C. Increasing the K^+ concentration led to an increasing accumulation of tissue K^+. At a K^+ concentration of 4 mM, the reaccumulation of K^+ by the slices had reached a maximum. Na^+ extrusion was half-maximum in a K^+-free medium and achieved maximum values at 10 mM K^+.

3. *Magnesium*

Magnesium (8 mM) was required in the incubation medium for maximum K^+ uptake at 38 °C. Higher concentrations inhibited K^+ accumulation, with a 68 per cent inhibition occurring at 50 mM Mg^{2+}.

4. *Calcium*

In a Ca^{2+}-free medium, K^+ uptake at 38 °C was markedly inhibited and was not sustained for more than twenty minutes (McLean, 1963). At a Ca^{2+} concentration of 0·1 mM, K^+ uptake was essentially normal.

B. Effect of Cation Content of the Medium on Respiration

Changes in the cation composition of the medium had effects on the respiratory rate of liver slices which closely paralleled the effects on Na^+ and K^+ movement. Increasing the Na^+ concentration from 25 to 147 mM resulted in an almost linear increase in the rate of respiration. Similarly, the respiration of slices was maximum in a medium containing K^+ and Mg^{2+} concentrations producing maximum stimulation of cation movements. Respiration was inhibited by about 35% in either a Na^+ or K^+-free medium and only about 15 per cent when Mg^{2+} was omitted.

C. Physiologic and Pharmacologic Agents Affecting Ion Transport

The cardiac glycosides were found to be specific inhibitors of active ion transport (Schatzmann, 1953). In rat liver slices, it was determined that strophanthin-k at a concentration of 10^{-3} g/ml inhibited net Na^+ and K^+ transport completely (Elshove and van Rossum, 1963). At this concentration of strophanthin-k, respiration was inhibited by 36 per cent. Lower concentrations produced less inhibition of ion transport, and at 10^{-5} g/ml, no effect on Na^+ and K^+ movements was observed. It is worthy of note that all of the studies reported here have been carried out on liver slices prepared from rats whose tissues are known to be relatively resistant to the actions of the cardiac glycosides. This explains why the concentrations required for complete inhibition of ion transport are high when compared with results of experiments on other species (Skou, 1965).

A variety of metabolic inhibitors have been shown to depress or completely abolish active ion transport in liver slices, presumably by interfering with the production of ATP. The inhibition of glycolysis by sodium fluoride (16 mM) led to complete cessation of K^+ uptake in liver slices incubated at 38 °C (Flink and coworkers, 1950). The interruption of aerobic metabolism by 1 mM cyanide or a nitrogen atmosphere (Elshove and van Rossum, 1963) resulted in a marked inhibition of K^+ uptake. Concentrations of cyanide producing a maximum inhibition of respiration still permitted about 10 per cent K^+ uptake, suggesting that energy provided by anaerobic metabolism is responsible for this small uptake in slices prepared from adult rats.

The uncoupling of oxidative phosphorylation also inhibits K^+ movements. This is supporting evidence for the view that the major source of energy for active transport in liver slices comes from oxidative metabolism. The uncoupler 2,4-dinitrophenol almost completely inhibits K^+ uptake when used in a concentration of 2×10^{-4}M. Oligomycin, a fungicide which is believed to inhibit an energy-rich intermediate, also inhibits active K^+ uptake in liver slices (van Rossum, 1964). Oligomycin, at a concentration of 10 μg/ml, produces a maximum inhibition of K^+ uptake by 40–50 per cent. This result implies that a portion of the energy required for active transport is provided by an energy-rich intermediate and, therefore, that ATP is not the immediate energy source. However this interpretation is not certain as it has been suggested that oligomycin might act directly on the transport mechanism (Glynn, 1962).

Furthermore there are indications that several drugs with antihistamine-like properties inhibit active cation transport (Judah and Ahmed, 1964).

D. Anaerobic Stimulation of Active Cation Transport in Rat Liver Slices

It is now generally accepted that ATP is the principal, if not the only, direct energy source for active ion transport. In cells incapable of respiration, such as the mature mammalian erythrocyte, ATP is generated by the reactions of the glycolytic pathway. Although there exists a considerable capacity for anaerobic glycolysis in the adult rat liver, it has been demonstrated that the incubation of adult liver slices in an anaerobic atmosphere or in the presence of cyanide results in a striking reduction in active cation transport.

Liver slices prepared from foetal rats have a greater capacity for anaerobic glycolysis than those prepared from adults (van Rossum, 1963). Correspondingly, cyanide inhibits potassium uptake by 60 per cent in liver slices prepared from foetal rats of twenty-two days gestation and 90 per cent in slices prepared from adult rats.

Experiments were undertaken in the author's laboratory to determine if anaerobic energy production could be augmented sufficiently to lead to significant active cation transport in adult rat liver. These experiments were motivated by the finding that mitochondria can generate ATP under anaerobic conditions provided the proper substrates are available to act as electron acceptors (Hunter, 1949; Seidman and Entner, 1961). It is possible that these electron acceptors are able to regenerate cytoplasmic NAD and so lead to stimulation of glycolysis.

The experimental design was similar to that described above, i.e. rat liver slices were cooled in iced Krebs–Ringer bicarbonate solution for ninety minutes followed by incubation at 37 °C for sixty minutes under an atmosphere of either 95 per cent O_2–5 per cent CO_2 or 95 per cent N_2–5 per cent CO_2. Metabolites or inhibitors were added to the medium ten minutes prior to warm incubation. An isoosmotic volume of metabolite or inhibitor was substituted for 0·9 per cent sodium chloride to give the desired concentration and to maintain isotonicity. At the end of the warm incubation, the liver slices were dried to constant weight. Sodium and potassium were extracted and assayed by flame photometry (Cascarano and Seidman, 1965; Seidman and Cascarano, 1966).

As can be seen in table 1, cooling the liver slices resulted in a loss of tissue

Table 1. Stimulation of anaerobic cation transport in liver slices by oxalacetate and pyruvate

Incubation Conditions	Na^+ (mequiv./kg dry wt)	K^+ (mequiv./kg dry wt)
Controls	249 ± 15·5[2]	179 ± 7·9
Precooled for 90 min[1]	467 ± 11·6	79 ± 5·5
Incubation at 37 °C for 60 min after precooling		
Oxygen[1]	153 ± 9·3	277 ± 6·5
Nitrogen[3]	510 ± 146·0	114 ± 12·7
Nitrogen[3] + 23 mM oxalacetate	222 ± 8·4	261 ± 5·7
Nitrogen[3] + 23 mM pyruvate	256 ± 11·1	216 ± 5·6

From Seidman and Cascarano (1966)

[1]Gas phase 95 per cent O_2, 5 per cent CO_2
[2]Values in the table represent averages of ten determinations ±SE
[3]Gas phase 95 per cent N_2, 5 per cent CO_2

K^+ and an increase of Na^+. Since the 'control' slices were first rinsed briefly in iced saline prior to electrolyte determination, these values do not represent the true content of these cations in the liver under physiologic conditions.

Following the period of cooling, incubation of the liver slices at 37 °C under aerobic conditions resulted in a significant gain of tissue K^+ and loss of Na^+. Only a slight increase in K^+ was observed when the warm incubation was performed under an anaerobic atmosphere. In neither instance did the addition of 55 mM glucose to the incubation medium enhance K^+ accumulation.

The idea of using oxalacetate as an electron acceptor was due to the work of Hunter (1949) who demonstrated that when used in conjunction with α-ketoglutarate it stimulated anaerobic ATP production in isolated rat liver mitochondria. The addition of 23 mM oxalacetate to the incubation medium ten minutes before the warm incubation resulted in an accumulation of K^+ and loss of Na^+ equivalent to that seen in the oxygenated slices. Other citric acid cycle intermediates, alone or in various combinations, were ineffective in stimulating the anaerobic uptake of K^+ or extrusion of Na^+. Pyruvate, like oxalacetate, contains a carbonyl group capable of accepting electrons, and this metabolic intermediate was found to be as effective as oxalacetate in stimulating anaerobic cation movement.

It thus became necessary to determine whether this phenomenon was, in fact, a stimulation of active transport. Ouabain is known to inhibit active cation transport and to have little or no influence on passive ion diffusion, glycolysis or respiration. Addition of ouabain (1·0 mg/ml) to the incubation medium caused an inhibition of anaerobic cation transport. A similar effect was obtained, as would be expected, in the oxygenated slices.

The mechanism of this stimulation by oxalacetate and pyruvate was then considered. These intermediates could have acted either by stimulating glycolysis or by increasing mitochondrial ATP production. Experiments were therefore performed with the metabolic inhibitors iodoacetate and 2,4-dinitrophenol to test these possibilities. Iodoacetate inhibits a number of enzymes requiring sulphhydryl groups for their activity and is generally regarded as an inhibitor of glycolysis because of its potent action on phosphoglyceraldehyde dehydrogenase. Iodoacetate (1 mM) caused a marked reduction in the effect of oxalacetate or pyruvate on anaerobic cation transport. Iodoacetate also inhibited K^+ uptake in oxygenated slices even in the presence of a variety of citric acid cycle intermediates. These particular experiments were inconclusive, since iodoacetate appeared to inhibit metabolic reactions other than those related to glycolysis.

Dinitrophenol is a known uncoupler of oxidative phosphorylation and therefore inhibits aerobic mitochondrial ATP production. It was thus reasoned that if oxalacetate and pyruvate act solely by enhancing glycolysis, then the stimulation of anaerobic cation transport supported by these intermediates would not be affected by 2,4-dinitrophenol. It was found that 2,4-dinitrophenol, in concentrations ineffective in inhibiting glycolysis in human erythrocytes, inhibited anaerobic cation transport. Oxygenated slices were likewise inhibited. These results were taken as evidence that oxalacetate and pyruvate have the ability to stimulate anaerobic mitochondrial ATP production.

The effect of oxalacetate and pyruvate on anaerobic cation transport was then tested in liver slices prepared from rats that had been starved for twenty-four hours. Starvation is known to markedly decrease the activities of the glycolytic enzymes glukokinase, phosphofructokinase and pyruvate kinase in rat liver (Weber and coworkers, 1965), thereby diminishing the rate of glycolysis without affecting respiration. In liver slices prepared from rats which had been starved for twenty-four hours, oxalacetate or pyruvate failed to stimulate anaerobic cation exchange (Cascarano and Seidman, unpublished observations). On the other hand, starvation for twenty-four hours failed to diminish the ability of oxygenated slices to reaccumulate K^+ and extrude Na^+. This raised the question whether liver slices prepared from starved rats respond to a direct energy source. Hoffman (1962) showed that exogenous ATP could support active ion transport in red cell ghosts. We attempted similar experiments with liver slices but failed to demonstrate any effect; it is probable that the highly polar ATP is unable to penetrate an intact cell membrane. Phosphoenolpyruvate (23 mM) was found to stimulate active cation movements in liver slices from fed rats but had no effect in slices from twenty-four hour starved rats. Since hepatic pyruvate kinase activity is markedly decreased after a twenty-four hour fast (Krebs and Eggleston, 1965) the generation of ATP from phosphoenolpyruvate would be expected to be diminished. From these observations it was concluded that an active glycolytic pathway is essential for oxalacetate or pyruvate to stimulate active cation transport under anaerobic conditions.

III. ATPases PREPARED FROM MEMBRANE FRACTIONS OF RAT LIVER

Once certain properties of active cation transport had been established—dependence on ATP, the necessity for Na^+ and K^+ in the external medium, inhibition by cardiac glycosides, etc., biologists began to investigate the mechanism by which the transport system operates. One of the prerequisites for such a system is that its activity be affected by activators or inhibitors in a manner similar to active transport. Skou (1957) demonstrated a (Na^+ and K^+)-activated ATP hydrolase in crab nerve which met these requirements for the transport system.

The first report of a similar enzyme in rat liver came from Emmelot and Bos (1962). Employing the method of Neville (1960) for the preparation of membrane fractions from solid tissues, Emmelot and Bos identified a (Na^+ and K^+)-activated ATPase in a membrane fraction which was identified as originating from plasma membrane. The identification of a similar enzyme, located in the microsomal fraction was subsequently reported by Schwartz (1963). The activity of the fresh preparation, although depressed by ouabain in the presence of Na^+ and K^+, was not stimulated by Na^+ and K^+. It was shown that the fresh microsomal preparation had a very active Mg^{2+}-stimulated ATPase which obscured the stimulation by Na^+ and K^+. Storing the microsomal fraction at $-5\,°C$ for periods of five to twelve days resulted in a preferential loss of Mg^{2+}-stimulated activity, yielding a preparation whose activity could be measurably stimulated by Na^+ and K^+.

The preparation of a lipoprotein fraction containing a cation-dependent ATPase was described by Ahmed and Judah (1964). These authors stressed the use of mannitol rather than sucrose in the preparation of the tissue homogenates since the former yielded a more stable preparation. Microsomes prepared from such a homogenate were layered over a solution of dense sucrose and centrifuged. The lipoprotein material floating to the surface contained the enzyme. This fresh preparation also contained an active Mg^{2+}-stimulated ATPase the activity of which could be selectively reduced by freezing for 24–48 hours. A stimulation in the order of 500–700 per cent was observed in the presence of Na^+ and K^+. At a Na^+ concentration of 115 mM, 10 mM K^+ gave maximum enzyme activity. The pH optimum was found to be 7·2–7·4 but the Mg^{2+}-activated enzyme was relatively insensitive to pH alterations over a wide range. Strophanthin-g (10^{-4}M) inhibited enzyme activity by 68 per cent. Raising the K^+ levels tended to overcome the inhibitory effect of the glycoside.

With this lipoprotein fraction, Judah and Ahmed (1964) determined that several quinidine-like drugs inhibited the (Na^+ and K^+)-activated ATPase. As with the cardiac glycosides, the action of these drugs could be overcome by raising the external K^+ levels.

The experiments by Bakkeren and Bonting (1968) with aqueous homogenates which were subsequently lyophilized and frozen showed that the active Mg^{2+}-stimulated enzyme was markedly reduced by treating the homogenate with 1·5 M urea. This treatment resulted in the (Na^+ and K^+)-stimulated activity being about 35 per cent of the total. It was also found that, with a K^+ concentration of 5 mM, maximum activity was obtained with 40 mM Na^+ and half-maximum activity with 6 mM Na^+. When the Na^+ concentration was held at 60 mM maximum activity was reached at 0·9 mM K^+. It was determined that Mg^{2+} was essential for (Na^+ and K^+)-stimulated ATPase activity, maximum activity occurring at 2 mM Mg^{2+} in the presence of

2 mM ATP. The pH and temperature optima were 7·3 and 45 °C, respectively.

Ahmed and Judah (1965) have suggested that the phosphorylation of protein is the first step in the utilization of ATP for active cation transport. These workers showed that the presence of Na^+ in the incubation medium stimulated the incorporation of ^{32}P into the phosphoprotein of liver slices. The addition of K^+ to the medium led to decreased incorporation and ouabain inhibited the effect of K^+. It was also shown that the lipoprotein fraction containing the (Na^+ and K^+)-activated ATPase contained a phosphatase that was stimulated by K^+. Na^+ stimulated the transfer of the terminal phosphate of ATP to the protein in the lipoprotein fraction and this reaction was inhibited by K^+. These studies gave support to the suggestion that Na^+ is required for the phosphorylation of protein and that K^+ stimulates dephosphorylation. These are both features of an ATPase system which may be operative in active cation transport.

IV. CONCLUDING REMARKS

This account has been an attempt to summarize much of the existing information on cation transport in liver tissue. Admittedly, many fundamental questions relating to the precise mechanism of this transport remain unanswered.

Virtually all of the work to date has been limited to liver slices. With the exception of the mitochondrion, the study of the compartmentalization of ions and metabolites in various subcellular organelles has barely begun. The pioneer experiments of Okazaki and coworkers (1968) indicate that the feeding of the hepatocarcinogen ethionine to rats results in an alteration of the ionic composition of the liver cell nucleus. The rate and type of ribonucleic acid synthesized by the nucleus appears to be related to the ionic environment of the polymerase (Pogo and coworkers, 1967). Thus alterations in intranuclear ion composition may be regulating factors in cell growth and differentiation.

REFERENCES

Ahmed, K. and J. D. Judah (1964) *Biochim. Biophys. Acta*, **93**, 603
Ahmed, K. and J. D. Judah (1965) *Biochim. Biophys. Acta*, **104**, 112
Bakkeren, J. A. J. M. and S. L. Bonting (1968) *Biochim. Biophys. Acta*, **150**, 460
Cascarano, J. and I. Seidman (1965) *Biochim. Biophys. Acta*, **100**, 301
Elshove, A. and G. D. V. van Rossum (1963) *J. Physiol.* (*London*), **168**, 531
Emmelot, P. and C. J. Bos (1962) *Biochim. Biophys. Acta*, **58**, 374
Flink, E. B., A. B. Hastings and J. K. Lowry (1950) *Amer. J. Physiol.*, **163**, 598
Glynn, I. M. (1962) *Biochem. J.*, **84**, 75P
Hoffman, J. F. (1962) *Circulation*, **26**, 1201
Hunter, F. E. (1949) *J. Biol. Chem.*, **177**, 361

Judah, J. D., K. Ahmed and A. E. M. McLean (1962) *Nature*, **196**, 484
Judah, J. D. and K. Ahmed (1963) *Biochim. Biophys. Acta*, **71**, 34
Judah, J. D. and K. Ahmed (1964) *J. Cellular Comp. Physiol.*, **64**, 355
Krebs, H. A. and L. V. Eggleston (1965) *Biochem. J.*, **94**, 3c
Leaf, A. (1956) *Biochem. J.*, **62**, 241
McLean, A. E. M. (1963) *Biochem. J.*, **87**, 161
Mudge, G. H. (1951) *Amer. J. Physiol.*, **167**, 206
Neville, Jr., D. M. (1960) *J. Biophys. Biochem. Cytol.*, **8**, 413
Okazaki, K., K. H. Shull and E. Farber (1968) *J. Biol. Chem.*, **243**, 4661
Parsons, D. S. and G. D. V. van Rossum (1962) *J. Physiol. (London)*, **164**, 116
Pogo, A. O., V. C. Littau, V. G. Allfrey and A. E. Mirsky (1967) *Proc. Natl. Acad. Sci. U.S.*, **57**, 743
Schatzmann, H. J. (1953) *Helv. Physiol. Pharmacol. Acta*, **11**, 346
Schwartz, A. (1963) *Biochim. Biophys. Acta*, **67**, 329
Seidman, I. and N. Entner (1961) *J. Biol. Chem.*, **236**, 915
Seidman, I. and J. Cascarano (1966) *Amer. J. Physiol.*, **211**, 1165
Skou, J. C. (1957) *Biochim. Biophys. Acta*, **23**, 394
Skou, J. C. (1965) *Physiol. Rev.*, **45**, 596
van Rossum, G. D. V. (1963) *Biochim. Biophys. Acta*, **74**, 1 and 15
van Rossum, G. D. V. (1964) *Biochim. Biophys. Acta*, **82**, 556
Weber, G., R. L. Singhal, N. B. Stamm and S. K. Srivastava (1965) *Federation Proc.*, **24**, 745

II

Ion Movements in Sub-cellular Organelles

CHAPTER 8

Ion Movements in Mitochondria

Dr. R. Cereijo-Santaló

The New Mount Sinai Hospital
Toronto, Ontario
Canada

I. INTRODUCTION

The mitochondrion could better be described as an enzyme-rich membrane complex rather than as an enzymatic system surrounded by a membrane. It is true that the isolated mitochondrion behaves as an osmometer. However, the mitochondrial outer membrane, which resembles the smooth endoplasmic reticulum (Parsons and coworkers, 1967; Parsons and Yano, 1967) encloses a sucrose-accessible space (Werkheiser and Bartley, 1957) and seems to be permeable to solutes up to a molecular weight of about 10,000 (Brierley and Green, 1965; Pfaff and coworkers, 1968). The inner membrane, on the other hand, is relatively impermeable to many small ions, and is considered to be the mitochondrial osmotic barrier which encloses the sucrose-inaccessible matrix space. The significance of this 'barrier' in relation to the function of mitochondria *in vivo* is, however, obscured by the observation that, in contrast to the cell membrane, the mitochondrial membrane separates two media of similar electrolyte composition (Gamble and Hess, 1966). Thus it 'would seem to minimize the need for a mechanism of active transport' (Tarr and Gamble, 1966). Furthermore, transient periods of increased permeability of mitochondrial membranes can be induced by the metabolic activity of the extramitochondrial cytoplasm (Slautterback, 1965; Cereijo-Santaló, 1966a,b,c). In this respect it should be noted that the work to be reviewed here deals with isolated mitochondria incubated in a structureless, metabolically inactive medium which, with the doubtful exception of tonicity, has no similarity to the cytoplasmic environment. Indeed it can safely be said that results obtained under these conditions reflect the functional potentiality of mitochondria free from cellular regulatory controls. Thus, for example, isolated liver mitochondria are capable of accumulating nearly 3 μmoles of Ca^{2+}, Sr^{2+} or Mn^{2+} per mg of protein, when only traces of these cations are present in any compartment of the liver cell *in vivo*.

With these reservations in mind, the movements of ions between incubation media and isolated mitochondria will be described. This will be followed by an analysis of mitochondrial reactions to the entry of ions: the tendency to maintain electrical neutrality and osmotic balance, and the changes in respiratory and ATPase activities.

Owing to the enormous amount of work published in this field, a comprehensive treatment of this subject cannot be attempted here and, regretfully, some important contributions will have to be omitted.

II. THE NATURE OF ION MOVEMENTS

If by 'active' we mean the movement of an ion against an electrochemical gradient (Ussing, 1949) then, strictly speaking, mitochondrial ion movements cannot be classified as 'active' or 'passive'. In many cases the existence of a chemical concentration difference is doubtful, inasmuch as ions could be osmotically inactive either because they are bound to mitochondrial fixed sites or because they form insoluble salts. Furthermore the electrical potential across the membrane remains unknown. Nonetheless it is known that, whereas in some cases ion movements can be observed in the absence of an energy supply, in other cases the movement is associated or linked to energy expenditure. In agreement with this, mitochondrial ion movements are described as 'energy linked' or 'non-energy linked' processes.

A. Non-Energy Linked Ion Movements

1. *Monovalent Cations*

Early studies showed that, in the absence of respiration, there is a rapid exchange between mitochondrial K^+ and the K^+ of the incubation medium (Stanbury and Mudge, 1953). External K^+ was also found to exchange with mitochondrial H^+. Thus the addition of KCl to a suspension of mitochondria in sucrose results in acidification of the medium in which K^+ exchanges for H^+ (Amoore, 1960). This type of exchange has been studied in great detail (Gear and Lehninger, 1967, 1968) and it appears that several cations can support H^+ ejection. The rapid Na^+–H^+ exchange increases in magnitude with pH and with NaCl concentration and is not inhibited by antimycin. At pH 8·5, with 80 mM NaCl, mitochondria bind about 50 nmoles of Na^+/mg protein. Titration of mitochondrial suspensions in the presence of increasing concentrations of NaCl shows a shift towards the acid side, indicating an increased dissociation of mitochondrial anionic groups. These findings suggest that mitochondria behave like cation exchange resins, and there is some evidence that the protonated anionic groups are located in the inner membrane, presumably in the phospholipids (Gear and Lehninger, 1968). Similar ion exchange properties have been found in microsomes. The suggestion has been made that the effect of ATP on the binding of Na^+ by these particles is due to its ability to complex divalent cations, thus decreasing their competition with Na^+ for binding sites. This view is strengthened by the observation that EDTA can replace ATP (Sanui and Pace, 1959, 1965, 1967).

More puzzling is the observation that the binding of Na^+ by mitochondria is associated with the release of K^+; the Na^+–K^+ exchange is very rapid. Since it involves a large amount of K^+ (about 80 per cent of that in control mitochondria) it has to be assumed that this exchange takes place across the inner membrane. This ion translocation in the absence of permeability-inducing agents poses a serious problem and it has been suggested that an exchange-diffusion carrier could account for this phenomenon (Gear and Lehninger, 1968).

Mitochondria and submitochondrial particles bind Rb^+ in the absence of respiration and there is evidence that the binding sites are mainly phospholipids. The sonic particles bind up to 50 nmoles of Rb^+/mg protein (Azzi and Azzone, 1967; Scarpa and Azzi, 1968).

Respiratory independent exchanges of mitochondrial K^+ for external H^+

have also been described (Moore and Pressman, 1964; Ogata and Rasmussen, 1966).

2. *Divalent Cations*

Heart mitochondria bind large amounts of Ca^{2+} by a process which may be independent of respiration. Phosphate compounds, mainly phospholipids, have been suggested as the binding sites (Slater and Cleland, 1953). Liver mitochondria also bind Ca^{2+} in the absence of respiration. This binding, which increases with the Ca^{2+} concentration, follows a saturation curve and also increases with pH. At pH 7, mitochondria bind 25–30 nmoles of Ca^{2+}/mg protein. At low pH, Ca^{2+} exchanges mainly for K^+, whereas at high pH it exchanges mainly for H^+ (Rossi and coworkers, 1967a). The binding of Ca^{2+} is inhibited by monovalent cations (Scarpa and Azzone, 1968; Wenner and Hackney, 1967). Studies with submitochondrial particles show that the inhibition is competitive in nature and suggest that mitochondria have a common binding site for cations. These binding sites are phospholipids (Scarpa and Azzi, 1968).

It has been suggested that the surface binding of Ca^{2+} is the first step in the mechanism of its translocation across the membrane (Scarpa and Azzi, 1968; Scarpa and Azzone, 1968). There are, however, experiments indicating that the binding of Ca^{2+} involves sites other than those of the membrane surface and that Ca^{2+} is a permeant cation the distribution of which on both sides of the membrane, but not its transfer across the membrane, is affected by mitochondrial metabolism (Rossi and coworkers, 1967a).

Binding of 30 nmoles of Mn^{2+}/mg protein in liver mitochondria (Chappell and coworkers, 1963) and of 60 nmoles of Mg^{2+}/mg protein in heart mitochondria (O'Brien and Brierley, 1965) have been reported to occur in the absence of respiration.

The observation that liver mitochondria accumulate about 500 nmoles of Sr^{2+}/mg protein in the presence of ATP, and the absence of respiration, raises an intriguing problem. Since this ATP-supported rapid uptake of Sr^{2+} is not inhibited by oligomycin, a mechanism other than the reversal of the energy-transfer of oxidative phosphorylation must be involved (Carafoli and coworkers, 1965c), as was suggested for the ATP-supported Ca^{2+} uptake by non-phosphorylating water-washed mitochondria (Vasington and Greenawalt, 1964). However, the postulated pathway of ATP utilization in Ca^{2+} uptake is oligomycin-sensitive; furthermore, the hydrolysis of ATP is stimulated by Ca^{2+}. In contrast, Sr^{2+} strongly inhibits even the spontaneous low rate of ATPase activity observed in intact mitochondria (Caplan and Carafoli, 1965; Vasington, 1966). Therefore it seems that in this case the role of ATP cannot be related to its hydrolysis but rather to the presence of ATP as such.

3. *Anions*

Inorganic phosphate accumulates in mitochondria during the energy linked uptake of divalent cations (Brierley and coworkers, 1962; Lehninger and coworkers, 1963) and it has been shown that the anion follows passively the translocation of the cation (Chappell and coworkers, 1963). During Ca^{2+} uptake P_i can be substituted for acetate, but not for chloride which appears to be impermeant (Rasmussen and coworkers, 1965). However, respiratory independent swelling of mitochondria observed in a KCl medium in the presence of valinomycin indicates that chloride can penetrate the membrane when the pH is raised (Azzi and Azzone, 1967). In contrast with this is the observation that a high pH favours the uptake of cations, whereas low pH favours the uptake of anions (Gamble, 1963a). In any case, it should be noted that a respiratory independent swelling in isotonic KCl occurs also in the absence of valinomycin when the pH is increased or decreased, which suggests that, under these conditions, mitochondria behave as a protein gel (Cereijo-Santaló, 1966c).

Accumulation of citrate and malate against large concentration gradients has been described in rabbit liver mitochondria as a respiratory-independent process. The anion uptake was associated with release of nearly equal amounts of phosphate (Gamble, 1965). Accumulation of citrate accompanied by P_i ejection has also been observed in rat liver mitochondria. Since in this case the P_i released came from the hydrolysis of added ATP, the citrate uptake was considered to be driven by an ATP derived non-phosphorylated high-energy intermediate (Max and Purvis, 1965). In contrast, it has been reported that rat liver mitochondria accumulate in the matrix space substrate anions at concentrations higher than that of the medium under conditions where no energy was available. The anions exchange for the endogenous substrates and phosphate, presumably bound to cations immobilized due to membrane impermeability (van Dam and Tsou, 1968). This is in agreement with the observation that deterioration of mitochondrial structure during aging leads to a parallel release of K^+ and P_i (Gamble and Hess, 1966). It is pointed out that although the anion exchange process does not require energy, the formation and maintenance of the original pool of cations and anions does require energy expenditure (van Dam and Tsou, 1968). Actually the uptake of anions after incubation of mitochondria for ten minutes with an uncoupler to deplete endogenous reserves, does show an energy requirement (Harris, 1968).

An elegant method of studying anion translocation is to measure the permeability of the membrane to solutes by following the light scattering changes of mitochondria (Chappell and Crofts, 1966). The anions, in the form of ammonium salts, are added to mitochondria incubated with respiratory inhibitors. Since NH_3 is assumed to rapidly penetrate the membrane, the rate of swelling of mitochondria will depend upon the rate of anion translocation. The anion enters in exchange for OH^- ions produced as the result of intramitochondrial protonation of NH_3. Weak acids could also enter in the unionized form and dissociate inside, thereby carrying protons from the external NH_4^+ to the internal NH_3. Therefore, an OH^-/anion antiport, or an H^+–anion symport type of translocation (Mitchell, 1967a) will occur, depending upon the pK of the acid and the pH of the medium. With this method it was found that mitochondria are impermeable to chloride, bromide, sulphate, fumarate and malate, whereas they are permeable to formate, propionate, butyrate, acetate, phosphate and arsenate. The translocation of succinate, malate, malonate and mesotartrate requires the presence of P_i, and that of citrate, *cis*-aconitate, tartrate and oxoglutarate requires L-malate in addition to P_i. The translocation of these anions would be facilitated by a number of specific exchange-diffusion carriers (Chappell and Crofts, 1966; Chappell and Haarhoff, 1967). It seems possible that the activation by malate involves as a first step the entry of malate into mitochondria, and then a tricarboxylic acid–malate exchange diffusion, since 2-n-butymalonate, which inhibits competitively the entry of malate, inhibits also the effect of malate on the entry of the tricarboxylic acids (Chappell and Robinson, 1968).

Permeability to substrate anions has also been studied by following their rates of utilization (Chappell, 1964; Ferguson and Williams, 1966; Meijer and Tager, 1966; Azzi and coworkers, 1967; Klingenberg, 1967; Meijer and coworkers, 1967). The possible significance of the anion-transporting system in cellular metabolism has been discussed by Chappell (1968). It should be noted that the need for a mechanism of substrate translocation is based on the finding that most of the citric acid cycle enzymes are located inside the osmotic barrier (Parsons and coworkers, 1967; Schnaitman and Greenawalt, 1968). Were these enzymes located in the outer membrane (Allmann and Bachmann, 1965, 1967; Green and coworkers, 1966) such a mechanism would be unnecessary.

Many papers, which cannot be reviewed here, are concerned with the atractyloside-sensitive carrier which catalyses the adenine nucleotide exchange across the inner membrane. The interested reader is referred to the recent publications of Klingenberg and Pfaff (1968), Winkler and coworkers, (1968), and Winkler and Lehninger (1968).

B. Energy-Linked Ion Movements

1. *Monovalent Cations*

The early work on the respiration or ATP-supported K^+ uptake by intact mitochondria usually involved long periods of incubation and rather low rates of K^+ uptake (Ulrich, 1960; Gamble, 1963b; Christie and coworkers, 1965; Lynn and Brown, 1965b; Rottenberg and Solomon, 1965). However, the situation in this field changed dramatically with the introduction of K^+-specific glass electrodes and following the discovery of agents which can increase mitochondrial permeability to cations. These agents include a number of antibiotics, mercurial reagents, heavy metals and chelators.

Up to 800 nmoles of K^+/mg protein can be taken up by mitochondria in the presence of valinomycin with an initial rate of K^+ uptake of 1 μmole/mg protein . min. (Harris and coworkers, 1966). The uptake of K^+ is accompanied by H^+ ejection. Rb^+ or Cs^+ can substitute for K^+ (Moore and Pressman, 1964). Rather surprising is the finding that the $K^+:\sim$ accumulation ratios for the ATP- and respiration-supported uptakes are 7 and 3·2 respectively (Cockrell and coworkers, 1966). Of great interest also are those observations which suggest that valinomycin could have a dual action, to increase both membrane permeability and mitochondrial affinity for K^+. Thus valinomycin increases the turnover of K^+. Furthermore, if mitochondria are incubated with a low K^+ concentration (0·23 mM), some K^+ leaks out. The addition of valinomycin *reverses* this movement causing a K^+ influx (Harris and coworkers, 1966).

Gramicidin also causes accumulation of K^+ and ejection of H^+. Any alkali metal cation can substitute for K^+ (Chappell and Crofts, 1965b). The nonactin analogues constitute another group of antibiotics which induce mitochondrial uptake of alkali metal cations except Li^+ (Graven and coworkers, 1966b; Graven and coworkers, 1967). In contrast, the carboxylic antibiotics nigericin and dianemycin inhibit the cation uptake induced by valinomycin, gramicidin and nonactin (Graven and coworkers, 1966a).

Changes in the energized state or in the affinity for cations of a membrane carrier have been invoked to explain these ion movements (Cockrell and coworkers, 1966; Höfer and Pressman, 1966; Lardy and coworkers, 1967; Pressman, 1968). However, antibiotic-induced ion movements have been observed in erythrocytes (Chappell and Crofts, 1966), phospholipid micelles (Chappell and Haarhoff, 1967), lipid bilayer membranes (Mueller and Rudin, 1967) and even across a two-phase water-mock lipid system (Pressman and coworkers, 1967). Observations of this kind support the view that the antibiotics merely increase membrane permeability to cations (Chappell and Crofts, 1956b), possibly by formation of lipid-soluble complexes (Pressman and coworkers, 1967). The interaction of this complex with an ion pump has been postulated to explain the energy requirement of mitochondrial systems (Pressman, 1968).

Increased permeability to monovalent cations has been observed in the presence of heavy metal ions (Brierley and Settlemire, 1967) as well as in the presence of EDTA which presumably acts by removing Mg^{2+} (Azzone and

Azzi, 1966). An interesting and incisive analysis of the mechanism of action of EDTA has been made recently (Settlemire and coworkers, 1968). Mercurial reagents which were reported to induce exchange of external K^+ with tightly bound mitochondrial K^+ (Gamble, 1957; Scott and Gamble, 1961) also cause an energy linked cation uptake in heart mitochondria (Brierley and coworkers, 1967; Brierley and coworkers, 1968b).

2. *Divalent Cations*

As in the case of monovalent cations, respiration or ATP can support the uptake of divalent cations. However, the penetration of Ca^{2+}, Sr^{2+} or Mn^{2+} does not require the presence of permeability-inducing agents.

In the absence of permeant anions. Ca^{2+} can be added to a mitochondrial suspension without altering the rate of state 4 respiration as long as the addition is made by continuous infusion at a rate similar to that of its disappearance from the medium. Under these conditions the $Ca^{2+}:\sim$ accumulation ratio is only 0·25 (Drahota and coworkers, 1965). Taking this experiment at its face value it would appear that Ca^{2+} uptake can occur without energy expenditure. However, if resting respiration represents the basal energetic requirement to maintain mitochondrial structure, the possibility exists that other endergonic reactions can compete for this energy.

Under similar experimental conditions ADP can be phosphorylated without alteration of the rate of resting respiration. The efficiency of respiration is again very low and only 0·12 ADP molecules are phosphorylated per energy conserving site (Carafoli and coworkers, 1965b).

In contrast, when 50–100 nmoles of Ca^{2+}/mg protein are added in a single pulse, a sharp increase of respiration is produced which then returns to its initial resting state (Chance, 1965). During the respiratory jump Ca^{2+} is taken up by mitochondria. There is a stoicheiometric relationship between the degree of respiratory stimulation and the amount of Ca^{2+} accumulated. The $Ca^{2+}:\sim$ ratio is about 1·8 (Rossi and Lehninger, 1964).

A respiratory dependent binding of 200–300 nmoles of Mn^{2+}/mg protein, which is prevented and reversed by DNP and respiratory inhibitors, has been observed (Chappell and coworkers, 1963).

In the presence of permeant anions. The mitochondrial capacity for binding Ca^{2+} is rather limited. This capacity could be increased if Ca^{2+} were sequestered in the form of insoluble calcium phosphate. Thus, in the presence of P_i, ATP and Mg^{2+}, mitochondria accumulate Ca^{2+} far beyond their binding possibilities (Rossi and Lehninger, 1964). Early observations show that, under these conditions, mitochondria accumulate up to 2·7 μmoles of Ca^{2+}/mg protein. The interesting point was made that Ca^{2+}, at the concentrations used in these experiments, inhibits respiration as much as 60 per cent and completely uncouples oxidative phosphorylation. However, DNP inhibits

Ca^{2+} binding even though the system is already fully uncoupled by the presence of Ca^{2+} (Vasington and Murphy, 1962). P_i accumulates with Ca^{2+} with a Ca^{2+}:P_i ratio of 1·7–1·8, which is similar to that in hydroxyapatite (Lehninger and coworkers, 1963; Rossi and Lehninger, 1963b).

Mitochondrial accumulation of 2·5 μmoles of Sr^{2+}/mg protein (Carafoli and coworkers, 1965c) and of more than 2 μmoles of Mn^{2+}/mg protein (Chappell and coworkers, 1963) has been reported.

The deposit of divalent cations in the matrix as dense granules has been extensively studied (Brierley and Slautterback, 1964; Greenawalt and coworkers, 1964; Peachey, 1964; Greenawalt and Carafoli, 1966). The granules of Ca^{2+}-loaded mitochondria contain calcium phosphate, adenine nucleotides and a large amount of organic material (Weinbach and von Brand, 1965). The calcium phosphate may be a colloidal precursor of hydroxyapatite with a Ca^{2+}:P_i ratio of about 1·5 (Thomas and Greenawalt, 1968).

III. ELECTROCHEMICAL BALANCE

A. Monovalent Cations

The accumulation of K^+ by mitochondria appears to be largely compensated by the extrusion of an equivalent amount of H^+. However, it is possible to find in the literature values for the H^+:K^+ ratio varying from 0·1 to 1·0 (Moore and Pressman, 1964; Christie and coworkers, 1965; Azzi and Azzone, 1966; Harris and coworkers, 1966). Simultaneous uptake of anions could be responsible for these differences. Thus, if fresh mitochondria are incubated in the absence of permeant anions the ratio H^+:K^+ is about 1. However, with aged mitochondria values as low as 0·5 can be observed, presumably because P_i is released during storage and re-enters mitochondria accompanying K^+. If aged mitochondria are deprived of the released P_i by washing, the H^+:K^+ ratio remains close to 1 (Rossi and coworkers, 1967c). Similarly, the presence of acetate decreases the amount of H^+ ejected (Ogata and Rasmussen, 1966) and also increases K^+ uptake (Harris and coworkers, 1966), thus causing a considerable decrease of the H^+:K^+ ratio. Succinate uptake also decreases the ratio H^+:K^+. It should be noted that the uptake of both K^+ and succinate, as well as the ejection of H^+, are influenced by the pH of the medium. An increase of pH from 6·6 to 8·5 increases the H^+:K^+ ratio from 0·1 to 0·9 as a consequence of decreased uptake of K^+ and succinate and increased H^+ ejection (Rossi and coworkers, 1967c).

B. Divalent Cations

When a divalent cation such as Ca^{2+} enters mitochondria, electrochemical balance could be obtained if two monovalent cations were ejected or if Ca^{2+}

were accompanied by anions. Early workers pointed out that during Ca^{2+} uptake there is an acidification of the medium (Chappell and coworkers, 1963; Saris, 1963; Brierley and others, 1964; Engström and DeLuca, 1964; Chance, 1965; Chappell and Crofts, 1965a; Judah and coworkers, 1965a). However, the observed ratio of H^+ ejected to Ca^{2+} taken up was not 2 but close to 1. Therefore, either other ions are translocated during this process or a large electrochemical potential across the membrane results. It has been calculated that the translocation of 100 nmoles of Ca^{2+}/mg protein with a $H^+:Ca^{2+}$ ratio of 1 would lead to a potential of about 20 V (Azzone and coworkers, 1967a). However, later work demonstrated that the $H^+:Ca^{2+}$ ratio depends largely upon experimental conditions and values from 0·3 to 2 can be observed. It should be noted that in the calculation of these ratios there is some disagreement with regard to the contribution of the respiratory independent binding of Ca^{2+} (Wenner and Hackney, 1967; Scarpa and Azzone, 1968).

1. *Low $H^+:Ca^{2+}$ Ratios*

It might be expected that when Ca^{2+} uptake is accompanied by anion uptake, the $H^+:Ca^{2+}$ ratio will decrease. In fact, the $H^+:Ca^{2+}$ ratio of 1 observed in a chloride medium decreases to about 0·3 in the presence of acetate (Rasmussen and coworkers, 1965). In the presence of P_i, the $H^+:Ca^{2+}$ value decreases only slightly (Chappell and coworkers, 1963; Saris, 1963; Chance, 1965). It can be assumed that P_i would lower the ratio to about the same level as acetate were it not for the acid evolved in the formation of calcium phosphate. Uptake of substrate anions can also contribute to the decrease of the $H^+:Ca^{2+}$ ratio (Rossi and coworkers, 1967b).

Low $H^+:Ca^{2+}$ values would be expected if Ca^{2+} exchanges not only for H^+ but also for other mitochondrial cations. K^+ is the most likely candidate because of its relatively high concentration. However, it has been stated that the Ca^{2+}–H^+ exchange is not accompanied by K^+ translocation in intact mitochondria (Drahota and Lehninger, 1965; Ogata and Rasmussen, 1966; Carafoli and Rossi, 1967). Some Ca^{2+}-induced movements of K^+ have been described with aged mitochondria, or with high concentrations of Ca^{2+}, but the rate of K^+ ejection was too slow compared with that of H^+ (Carafoli and Rossi, 1967; Rossi and coworkers, 1967). On the other hand, an interesting calculation has been made which shows that $K^+ + H^+:Ca^{2+} = 1{\cdot}9$ (Wenner and Hackney, 1967).

In the presence of valinomycin the $H^+:Ca^{2+}$ ratio decreases and a rapid ejection of K^+ is observed, the $K^+ + H^+:Ca^{2+}$ ratio being 1·3 (Ogata and Rasmussen, 1966). This exchange is limited by the mitochondrial content of K^+. Addition of K^+ to the medium increases the mitochondrial K^+ (in the presence of valinomycin) and the contribution of this cation to the exchange for Ca^{2+} also increases. Under these conditions the H^+:Ca^{2+} ratio drops to 0·5–0·2, whereas the $H^+ + K^+$:Ca^{2+} rises to 1·6–1·9 (Carafoli and Rossi, 1967; Rossi and coworkers, 1967b). Similar observations were made on respiratory independent uptake of Ca^{2+} (Rossi and coworkers, 1967a).

2. *High H^+:Ca^{2+} Ratios*

H^+:Ca^{2+} values approaching 2 are obtained by increasing Ca^{2+} concentration. This effect depends on protein concentration (Rossi and coworkers, 1967b; Wenner and Hackney, 1967).

The absence of buffer also results in an H^+:Ca^{2+} ratio of 2 (Rossi and coworkers, 1966d). A similar observation has been made with Sr^{2+}. It appears that when Tris buffer is omitted or when it is substituted by Tricine buffer, values of 2 can be obtained for the H^+:Sr^{2+} ratio (Wenner, 1966).

H^+:Ca^{2+} values approaching 2 can be observed during Ca^{2+} uptake supported by succinate oxidation by increasing pH (Wenner and Hackney, 1967), which might be due to the inhibition of succinate uptake (Rossi and coworkers, 1967b). However, an increase of the H^+:Ca^{2+} ratio can be induced by high pH in the absence of substrate (Rossi and coworkers, 1967a).

Of great interest is the observation that high H^+:Ca^{2+} ratios can be obtained by suspending mitochondria in a medium containing Ca^{2+} (Rossi and coworkers, 1967b). Usually the experiment is performed by adding Ca^{2+} to a medium in which mitochondria have already been suspended. It can be assumed that in the equilibration period preceding the addition of Ca^{2+}, monovalent cations could exchange for mitochondrial H^+ (Gear and Lehninger, 1968) and some of the Ca^{2+} enters in exchange for these cations. Therefore, the $Na^+ + K^+ + H^+$:Ca^{2+} ratio should be measured in this case.

These observations clearly indicate the existence of mechanisms by which electrochemical balance could be maintained during cation uptake. Whether a complete balance is achieved or not, cannot be decided at present.

C. Mitochondrial Alkalinization

When cation uptake is balanced mainly by H^+ ejection, intramitochondrial alkalinization might be expected to occur. This has been confirmed experimentally by three elegant methods: titration of 'opened' mitochondria (Rossi and coworkers, 1966a; Gear and coworkers, 1967), measuring the changes in absorbance of bromthymol blue which is tightly bound to the osmotically inactive intracristal space (Chance and Mela, 1966a,b,c,d, 1967), and calculation of intramitochondrial pH by the distribution of the weak acid 5,5-dimethyl-2,4-oxazolidinedione (Addanki and coworkers, 1968). In spite of some discrepancies, the results obtained with these three methods indicate that cation uptake, in the absence of permeant anions, causes an intramitochondrial alkalinization equivalent to the external acidification.

IV. ISOSMOTIC EQUILIBRATION

Mitochondrial uptake of cations in exchange for H^+ ions is not associated with movement of water, presumably because the cations become bound to fixed anionic sites and are osmotically inactive (Azzone and coworkers, 1967b; Rossi and coworkers, 1967c). However, if an anion is also translocated, movement of water and swelling will take place, provided that a soluble salt is formed (Chappell and Crofts, 1965b; Ogata and Rasmussen, 1966; Rossi and coworkers, 1967c). This implies permeability to both anion and cation, as well as to water.

It is clear that natural and artificial lipid membranes are permeable to water although the mechanism of this permeability remains obscure and controversial (Hanai and coworkers, 1966; Thompson and Huang, 1966; Stein, 1967). Indeed, the observation that water permeates lipid membranes at a higher rate than could be predicted, indicates how little is known about the structure of either the membrane or liquid water (Stein, 1967). Whatever the mechanism involved, the fact that the mitochondrion can behave as an 'osmometer' (Dianzani, 1953; Raaflaub, 1953; Tedeschi and Harris, 1955), indicates that water moves rapidly across its membranes.

The swelling technique described above (see section II.A.3) is based on the fact that penetration of mitochondria by a soluble salt increases the internal osmotic pressure, causing entry of water. Obviously both anion and cation need to be permeant. Thus a respiratory independent swelling is observed with the ammonium salts of acetate, formate, propionate, butyrate, phosphate and arsenate, but not of chloride, bromide or sulphate. Sodium and potassium cannot substitute for ammonium, and it is suggested that ammonium enters mitochondria as uncharged ammonia (Chappell and Crofts, 1966).

The pattern of mitochondrial permeability to anions which emerges from these studies appears quite sound. It is not so clear, however, that permeability to cations occurs as described, since a respiratory independent swelling of liver mitochondria has been observed in the presence of 0·1 M sodium acetate (Lynn and Brown, 1965a). A similar swelling of heart mitochondria has been reported with sodium salts of acetate, formate and phosphate (Brierley and coworkers, 1968a). Permeability to Na^+ is also implicated in the observed Na^+–K^+ exchange across the inner membrane (Gear and Lehninger, 1968). These results suggest that a mechanism exists for the penetration of the membrane by Na^+ which perhaps could be applied to NH_4^+. Certainly lipid membranes might be expected to be permeable to the lipid soluble unionized NH_3 rather than to the charged NH_4^+. However, the high pK' (about 9·4) of the dissociation reaction: $NH_4^+ \longleftrightarrow NH_3 + H^+$, could pose some problems. Indeed, in the kidney, the existing pH gradient between tubular cell and tubular lumen (maintained by an outward-H^+

pump), together with the continuous intracellular production of ammonia, produces favourable conditions for the extrusion of NH_3 and its trapping in the lumen as NH_4^+. The equilibrium of the above reaction will be shifted towards the right in the cell and towards the left in the lumen. The opposite situation occurs in chloroplasts in which the inward-H^+ pump is able to trap ammonia as NH_4^+ in the cell (Deamer and Packer, 1967). It can be questioned whether such conditions exist in the experiments with mitochondria where the concentration of NH_3 depends solely on the dissociation of NH_4^+ in a medium at pH 7·2–7·4, let alone the fact that the intramitochondrial pH is unknown. Even so, it would appear that some findings support the view that uncharged ammonia penetrates mitochondria (Chappell and Crofts, 1966).

The relation between ion translocation and the movement of water can also be studied during the energy linked uptake of Ca^{2+}. When Ca^{2+} is added in the form of $CaCl_2$ the cation enters mitochondria in exchange for H^+ and no swelling occurs. However, in the presence of acetate the uptake of Ca^{2+} is followed by mitochondrial swelling in proportion to the concentration of the anion. The movement of water appears to be related to the osmotic activity of the soluble calcium acetate. This view is strengthened by the observation that if P_i is now added, an insoluble salt of calcium is formed, acetate is extruded and an increase in light scattering, proportional to P_i concentration, is observed. This P_i-induced contraction of mitochondria occurs also in the absence of respiration (Rasmussen and coworkers, 1965).

Similar observations have been made during the energy linked uptake of K^+ (Chappell and Crofts, 1965b; Rossi and coworkers, 1967c). In the presence of valinomycin KCl produces only a slight swelling of mitochondria which can be increased by the addition of P_i. When KCl is substituted by potassium acetate a marked swelling takes place. Antimycin A causes ejection of both potassium and acetate and mitochondrial contraction. The slight swelling observed in the presence of KCl could be attributed to the uptake of substrate anions. The movement of water is considered to be a consequence of osmotic equilibration (Ogata and Rasmussen, 1966). Of special interest is the reciprocal relationship found between the H^+:K^+ ratio and mitochondrial swelling, the uptake of water increasing with the decreasing H^+:K^+ ratio (Rossi and coworkers, 1967c). It should be noted, however, that there is some disagreement with regard to the osmotic nature of the swelling during the energy-linked K^+ uptake. It has been pointed out that the amount of K^+ taken up cannot account for the observed movement of water, and furthermore there is not enough energy available in the ATP-supported K^+ uptake for an isosmotic concentration of this cation (Pressman, 1965; Harris and coworkers, 1967a; see also Blondin and Green, 1967; Graven and coworkers, 1967).

V. ION MOVEMENTS AND RESPIRATORY ACTIVITY

A. Monovalent Cations

Antibiotic-induced cation uptake and H^+ ejection occurs without change in the rate of respiration when it takes place in a chloride medium. However, if a permeant anion is present, the cation–H^+ exchange is accompanied by a sharp increase in respiratory activity (Moore and Pressman, 1964; Chappell and Crofts, 1965b). Two explanations have been offered for these phenomena.

i) The K^+–H^+ exchange does not require energy but is limited by the number of negative charges in mitochondria. A permeant anion which either enters mitochondria in exchange for OH^- or dissociates inside, provides more H^+ and thus activates the energy-linked H^+ pump and increases the K^+–H^+ exchange (Chappell and Crofts, 1965b). Indeed, K^+ uptake increases in the presence of P_i or acetate (Azzi and Azzone, 1966; Chappell and Crofts, 1966; Harris and coworkers, 1966).

ii) Mitochondrial saturation with K^+ can be achieved with a small energy requirement. However, in the presence of P_i, which facilitates the passive efflux of K^+, a cyclic transport of this cation takes place, the active transport into mitochondria being followed by a passive efflux resulting in considerable increase of the total amount of K^+ transported through the membrane and, therefore, of the energy requirement (Moore and Pressman, 1964).

In apparent contrast with this, it has been shown that K^+ plus valinomycin stimulates respiration and pyridine nucleotide oxidation in a chloride medium (Azzi and Azzone, 1966). It has also been noted that the P_i requirement for the respiratory stimulation induced by gramicidin plus cation is not absolute (Chappell and Crofts, 1965b). In fact, in the presence of gramicidin considerable increases of oxygen uptake (200–300 per cent) have been observed with 5 mM NaCl or KCl in a chloride medium (see figure 1, series A, B and C of Harris and coworkers, 1967b). Perhaps the permeation of substrate anions should be considered in these cases. Actually the relation between the uptake of substrate and that of cation constitutes the basis of a new concept of respiratory regulation. It is postulated that respiratory activity can be limited by the availability of substrate (Harris and coworkers, 1967c,d), and that the permeation of substrates is facilitated by the simultaneous uptake of K^+ (Lynn and Brown, 1966; Harris and coworkers, 1967b). However, this cannot account for the stimulation of oxidation of endogenous substrates (Cereijo-Santaló, 1968a); furthermore, the respiratory stimulation induced by ADP does not appear to be related to an increased uptake of K^+. In fact ADP phosphorylation completely inhibits the valinomycin-induced K^+ uptake, presumably by diverting energy from the cation pump to the synthesis of ATP (Harris and coworkers, 1967b). However, the ADP-activated respiration implies the permeation of three anions: ADP_i, P and substrate.

B. Divalent Cations

It has been known for some years that Ca^{2+} uncouples oxidative phosphorylation (Lehninger, 1949) and stimulates respiration (Potter and coworkers, 1953; Siekevitz and Potter, 1953; Lindberg and Ernster, 1954). In this early work, Ca^{2+} was added in concentrations which cause structural damage and the reaction of Ca^{2+} with tightly coupled mitochondria could not be observed. This deleterious effect of Ca^{2+} can be avoided by using low Ca^{2+}: protein ratios (Chance, 1956, 1965). Thus, addition of less than 100 nmoles of Ca^{2+}/mg protein induces a change in the oxidation–reduction levels of the respiratory carriers (which corresponds to the state 4 to 3 transition), H^+ ejection and a respiratory stimulation proportional to the amount of Ca^{2+} added, the respiration then returning to the resting state. Since the reaction rate of Ca^{2+} with the electron carriers is extremely high, showing half-times of 70 milliseconds at 26°, it is suggested that 'any intervening processes between the arrival of Ca^{2+} at the outer membrane of the mitochondria and its subsequent reaction with respiratory enzymes of the cristae are either non-existing or non-rate limiting' (Chance, 1965).

The similarities between the response of mitochondria to Ca^{2+} and to $ADP + P_i$ suggest that the chemical mechanism of Ca^{2+} accumulation is the same as that of ADP phosphorylation (Chance, 1965). However, there are some important differences:

i) The respiratory stimulation induced by Ca^{2+} is 50 per cent higher than that induced by ADP (Chance, 1956, 1965) and it does not require the presence of P_i (Rossi and Lehninger, 1964; Chance, 1965).

ii) In the reaction of mitochondria with Ca^{2+}, hydrogen ions are ejected (Chappell and coworkers, 1963; Saris, 1963; Engström and DeLuca, 1964; Chance, 1965; Chappell and Crofts, 1965a), whereas in the reaction with ADP, hydrogen ions are taken up (Swanson, 1957; Nishimura and coworkers, 1962; Chance, 1965; Wenner, 1966).

iii) To activate each energy-conserving site only 1 ADP molecule is required, compared with 2 calcium atoms (Chance, 1963, 1965; Rossi and Lehninger, 1964).

iv) The reaction of ADP with mitochondria is almost completely inhibited at 7°, whereas that with Ca^{2+} proceeds effectively even at 0° (Chance, 1965).

v) When both ADP and Ca^{2+} are present, the phosphorylation of ADP does not start until all Ca^{2+} has been taken up, either because the affinity for Ca^{2+} is higher than that for ADP (Rossi and Lehninger, 1964) or because Ca^{2+} reacts with a site which is closer to the chain than that of ADP (Chance, 1965).

vi) Oligomycin does not inhibit the respiratory response to Ca^{2+} (Rossi and Lehninger, 1964; Chance, 1965).

When Ca^{2+} is added in the presence of P_i, respiration continues indefinitely in the activated state. Under these conditions less than 10 per cent of the Ca^{2+} is accumulated. It seems that P_i either prevents the binding of Ca^{2+} or causes the discharge of bound Ca^{2+}. This effect of P_i is not observed in the presence of either oligomycin or ATP (Rossi and Lehninger, 1964). In this respect the reaction of mitochondria with Ca^{2+} differs from that with Sr^{2+}, which otherwise is quite similar. The respiratory activation induced by Sr^{2+}, in the absence of P_i, returns to the resting state when most of the Sr^{2+} has been accumulated. However, the binding sites for Sr^{2+} are saturated at about 150 nmoles of Sr^{2+}/mg protein, and when larger amounts of Sr^{2+} are present respiratory stimulation continues indefinitely. Addition of P_i increases the accumulative capacity of mitochondria by trapping Sr^{2+} as an insoluble salt and respiration eventually returns to its initial state 4 (Carafoli, 1965). The differences observed between Ca^{2+} and Sr^{2+} may be related to the fact that Ca^{2+} is a powerful swelling agent, whereas Sr^{2+} inhibits swelling (Caplan and Carafoli, 1965). Thus, P_i is able to trap the cation inside the mitochondrion as long as the membranes are not 'leaky', as is the case with $Sr^{2+}+P_i$, but not with $Ca^{2+}+P_i$.

The respiratory response of mitochondria to Ca^{2+} varies with the existing experimental conditions. Addition of Ca^{2+} to P_i-depleted mitochondria in a chloride medium leads to an inhibited state (state 6) in which the respiratory carriers are in the oxidized state with the exception of cytochrome *b* which is reduced. This inhibition is released by permeant anions (Chance, 1964, 1965, 1967; Chance and Yoshioka, 1965; Chance and Schoener, 1966).

There are experimental conditions under which the extra oxygen uptake evoked by Ca^{2+} decreases *without* decreasing the amount of Ca^{2+} accumulated (Carafoli and coworkers, 1965a; Rossi and Azzone, 1965). The extra oxygen uptake induced by Ca^{2+} diminishes, for example, when the concentration of NaCl is increased from 40 to 240 mM, the $Ca^{2+}:\sim$ accumulation ratio increasing from 2 to 4 or 5. NaCl can be substituted by KCl, LiCl, KBr or KCNS. The action of these salts is prevented by permeant anions. The effect of Cl^-, Br^- and CNS^- can be mimicked by OH^-. Thus, keeping the salt concentration constant but increasing the pH produces an analogous increase of the $Ca^{2+}:\sim$ ratio (Carafoli and coworkers, 1967).

Since in the resting state which follows the Ca^{2+}-induced respiratory jump there is a continuous movement of Ca^{2+} in and out of mitochondria (Drahota and coworkers, 1965; Rossi and coworkers, 1966c), alterations in the equilibrium of this dynamic balance can influence the amount of retained Ca^{2+}. Strong inhibition of the efflux rate has been observed by increasing salt concentration or pH (Rossi and coworkers, 1966b). Therefore, it has been suggested that the increased $Ca^{2+}:\sim$ values could be due, at least in part, to a diminished Ca^{2+} efflux. High concentrations of impermeant anions could also favour the uptake of Ca^{2+} by altering the membrane potential (Carafoli and coworkers, 1967).

It should be noted, however, that these increased $Ca^{2+}:\sim$ values are due to a decrease in the oxygen consumed rather than to an increased Ca^{2+} uptake. It has been pointed out that the charging of mitochondria with K^+ by the respiratory independent K^+–H^+ exchange, which is favoured by high pH or high salt concentration (Gear and Lehninger, 1968), 'may result in a smaller stimulation of oxygen uptake by the added Ca^{2+} because some of the latter may replace the previously bound K^+' (Lehninger and coworkers, 1967). This is an interesting suggestion since it implies that the uptake of Ca^{2+} in exchange for H^+, but not for K^+, stimulates respiration.

Recently it has been proposed that the amount of extra oxygen uptake evoked by Ca^{2+} depends on anion uptake (Rossi and Azzone, 1968). When Ca^{2+} concentration is increased, both H^+ ejection and Ca^{2+} uptake increase proportionately, whereas the respiratory response tends to level off above 60 nmoles of Ca^{2+}/mg protein. Consequently the $Ca^{2+}:\sim$ ratio increases with Ca^{2+} concentration. It would appear that at high Ca^{2+} concentration, both H^+ ejection and Ca^{2+} uptake lose their quantitative relationship with oxygen consumption. Furthermore, the concentration of Ca^{2+} required to produce a high $Ca^{2+}:\sim$ ratio seems to depend upon the concentration of permeant anions. When respiration is maintained by endogenous substrates, the addition of only 5 nmoles of Ca^{2+}/mg protein is enough to cause high $Ca^{2+}:\sim$ values. With added substrate anions, the concentration of Ca^{2+} required to inhibit respiration increases with the concentration of substrate present. It can be concluded that any condition which impedes the entry of anions will favour the appearance of high $Ca^{2+}:\sim$ ratios (Rossi and Azzone, 1968). An inhibitory effect of high pH on succinate uptake has been reported (Rossi and coworkers, 1967b; Rossi and coworkers, 1967c). The same action has been suggested for increased concentrations of NaCl (Rossi and Azzone, 1968).

Whether or not this kind of respiratory inhibition may be related to that described as state 6, is open to discussion (Carafoli and coworkers, 1967; Rossi and Azzone, 1968). In this respect it is of interest that the characteristics of the bromthymol blue response are similar in both cases (Rossi and Azzone, 1968). The oxidation–reduction levels of the respiratory carriers have not, to my knowledge, been studied.

VI. ION MOVEMENTS AND ATPase ACTIVITY

Mitochondria incubated in an isotonic medium containing KCl show little or no ATPase activity. However, high rates of ATP hydrolysis are observed after addition of valinomycin or gramicidin (Lardy, 1961; Lehninger and coworkers, 1959; Pressman, 1963; Lardy and coworkers, 1964), and this can be attributed to increased permeability of the inner membrane to monovalent cations (Moore and Pressman, 1964; Chappell and Crofts, 1965b; Pressman, 1965; Cockrell and coworkers, 1966). The same mechanism is suggested for the ATPase activity induced by the nonactin homologues (Graven and coworkers, 1966b). The stimulation of ATPase by a series of surface-active agents also shows a monovalent cation dependence (Cereijo-Santaló, 1968a). All these observations suggest that hydrolysis of ATP occurs when an alkali metal cation enters mitochondria. In agreement with this, the antibiotic-induced ATPase activity is partially inhibited by another antibiotic, nigericin (Estrada-O and coworkers, 1967), which has been shown to block cation uptake (Graven and coworkers, 1966a). A Ca^{2+}-induced activation of mitochondrial ATPase has also been described (Potter and others, 1953; Rossi and Lehninger, 1963a; Brierley and coworkers, 1964; Bielawski and Lehninger, 1966).

There are, however, other observations suggesting an apparent lack of relationship between cation translocation and ATP hydrolysis. Thus, the ATPase activity induced by monazomycin (Ferguson and Lardy, 1968), as well as that induced by nigericin at relatively high concentrations (Estrada-O and coworkers, 1967), appear to be unrelated to cation uptake. Further complications arise from studies with the classical activator of ATPase, 2,4-dinitrophenol (DNP). Early observations showed that maximal activation of ATPase by DNP could be observed in the presence of KCl or NaCl (Lardy and Wellman, 1953). If mitochondrial permeability to cations were increased by DNP, then the activation of ATPase by this uncoupler could be explained by a mechanism similar to that for valinomycin (Cereijo-Santaló, 1968a). However, if in the presence of DNP the membrane remains impermeable to alkali metal cations, then the requirement for these ions is difficult to explain. Recently the view has been advanced that the cation requirement may be 'related to the difference in charge between ATP and ADP' (Amons and coworkers, 1968). The alternative, that there is no cation requirement for the activation of ATPase by DNP, has also been considered (Cereijo-Santaló, 1968c). This idea comes from the observation that the salt requirement is completely unspecific with regard to the cation but not with regard to the anion. Indeed the maximal stimulation of DNP-activated ATPase observed with KCl can be obtained also when K^+ is replaced with Cs^+, Rb^+, Na^+, Li^+, Tris (Amons and coworkers, 1968; Cereijo-Santaló, 1968c), NH_4^+, triethylammonium (Amons and coworkers, 1968), monoethanolamine, diethanolamine or choline (Cereijo-Santaló, 1968c). However, variations in the rate of the DNP-induced ATPase are observed when Cl^- is substituted by other anions (Amons and coworkers, 1968; Cereijo-Santaló, 1968c). Since in these substitutions, the weaker the base the higher the rate of ATPase activity observed, the anion may act as a proton conductor and DNP could make the enzyme site accessible to the anion (Cereijo-Santaló, 1968c). It is apparent that this view implies that permeant anions should be able to stimulate ATPase even in the absence of DNP. An alternative could be the involvement of a Gibbs–Donnan equilibrium (Cereijo-Santaló, 1968c) which could explain the observation that the ATPase activity induced by DNP in a chloride medium is inhibited by permeant anions. Such inhibition has been attributed to competition with DNP for entry into mitochondria (Veldsema-Currie and Slater, 1968). However, even if the lipid-soluble DNP should require a carrier to penetrate the membrane, since it is postulated that it enters non-specifically via a number of specific anion carriers including those for acetate, malate, tricarballylate, aspartate and possibly those for ATP, P_i and other amino acids, it seems unlikely that the competition for a single carrier can have any detectable effect on DNP permeation.

It is apparent that the salt requirement for the DNP-activated ATPase is still an unsolved problem. The idea that ion permeability plays a role in this phenomenon should not be disregarded. Although the action of uncouplers on the permeability of phospholipid bilayer membranes seems to be limited to H^+ and OH^- (Hopper and coworkers, 1968), their effect on the more complex membranes of mitochondria could be different. Binding of DNP to mitochondrial proteins and subsequent derangement of the protein–phospholipid organization has been suggested (Weinbach and Garbus, 1968a,b). Changes in mitochondrial permeability to K^+ induced by uncouplers (Judah and coworkers, 1965b; Kimmich and Rasmussen, 1967; Caswell and Pressman, 1968), as well as a two to three-fold increment in the mitochondrial content of Cl^- after incubation with DNP (Gamble, 1963b) have been reported. In this respect, the observation that 10 mM $MgCl_2$ completely inhibits the effect of KCl on DNP-activated ATPase (Amons and coworkers, 1968) might be especially meaningful. Since there is much evidence that Mg^{2+} plays an important role in the control of membrane permeability (Hoffman, 1962; Azzone and Azzi, 1966; Cereijo-Santaló, 1966a; Harris and coworkers, 1967a; Leive, 1968; Packer and coworkers, 1966; Settlemire and coworkers, 1968), the inhibitory action of $MgCl_2$ might indicate that DNP had increased the permeability of mitochondria to K^+ and/or to Cl^-.

VII. MECHANISMS FOR ION TRANSLOCATION

The observation that cation uptake is usually linked to H^+ ejection, and that there is uncertainty with regard to the cause–effect relationship between these processes, makes it possible to explain mitochondrial ion movements in two ways:

i) There is a 'cation pump' which actively translocates cations into mitochondria and, as a consequence, H^+ ions are ejected;

ii) There is a 'H^+ pump' which actively ejects protons from mitochondria and, as a consequence, metal ions are taken up. Since these hypothesis have been reviewed many times, they will be described here very briefly.

A. Cation Pump

Some years ago it was postulated that the transfer of energy from respiratory carriers to ATP involves a series of high-energy intermediates (Slater, 1953; Lehninger, 1955; Chance and Williams, 1956), which may be used not only for ATP synthesis but also for active transport of ions and for mechanochemical changes in the membrane (Lehninger, 1964). On this hypothesis, mitochondrial uptake of Ca^{2+} can be explained by assuming that this cation reacts with the protonated high-energy intermediate, $H_2(X \sim I)$, to form in the outer surface of the inner membrane the compound, $Ca_2(X \sim I)^{2+} + 2\,H^+$. The transport of this compound to the inner surface of the membrane takes place by a rotational translocation which involves energy utilization, resulting in the formation of the low energy complex $(Ca_2{-}X)^{2+} + I$. When all the available X is bound to Ca^{2+}, respiration is inhibited (state 6), and no more Ca^{2+} can be translocated. However, in the presence of a permeant anion, Ca^{2+} can be discharged into the matrix. The release of X will allow respiration to proceed and raise the capacity of mitochondria to accumulate Ca^{2+} (Chance, 1963, 1965; Rasmussen and coworkers, 1965; Chance and Mela, 1966b).

One modification of this hypothesis suggests that the energized form $H_2X \sim I$ is able to carry mono- and divalent cations across the membrane into the matrix, whereas the non-energized form H_2X can shuttle ions in either direction. Since this would imply that the complex $(Ca_2X)^{2+}$ would be able to carry Ca^{2+} out of mitochondria, it is further postulated that in the state 6 calcium is bound to the energized form $X \sim I$ (Rasmussen and Ogata, 1966).

B. H^+ Pump

Attempts to explain gastric secretion of HCl and the accumulation of salts by plants led to the idea that oxidation–reduction reactions constitute the basis of the translocation of ions (see review by Robertson, 1960, 1967). For many years a theory relating this basic idea to the coupling of respiration and phosphorylation could not be found—not until Mitchell suggested the chemiosmotic mechanism (Mitchell, 1961). This hypothesis has been further elaborated (Mitchell, 1963, 1966a,b, 1967a,b; Mitchell and Moyle, 1965a,b, 1967a,b) arousing considerable interest and strong criticism (Chance and

Mela, 1966c,d; Cockrell and coworkers, 1966; Chance and coworkers, 1967; Slater, 1967; Tager and coworkers, 1966).

The chemiosmotic hypothesis can be summarized as follows:

The respiratory carriers form a 'loop' between the two sides of the membrane. The first arm of the loop transfers hydrogen atoms from the inner phase (matrix) to the outer phase (intermembrane compartment), whereas the second arm transfers electrons in the opposite direction. Therefore, the loop works as a H^+ pump. Since three loops are described (i: NAD–Fe,SH; ii: FMN–Cyt.*b*; iii: CoQ–Cytochromes), for each pair of electrons traversing the loops, six H^+ ions are removed from the matrix and released to the outer compartment. Energy is thus conserved as a proton concentration difference and each one of the loops can be considered as an energy-conserving site.

Perhaps the most attractive feature of this hypothesis is the postulation of an anisotropic ATPase which translocates protons in the same direction as the loops. Since this ATPase system is reversible, it could be driven towards ATP synthesis if the pH differential created by respiration reached appropriate values. The values required are unacceptable in practice. Therefore it is proposed that H^+ and OH^- can be substituted by cations and anions and most of the pH differential is converted into a membrane potential, which makes the term 'chemiosmotic' rather ambiguous. It is apparent that this membrane potential would create a tendency for ions to permeate the membrane down the electrical gradient, cations towards the inside and anions towards the outside, resulting in a decrease of the potential. This process could be compensated by two exchange diffusion systems, anion–H^+ symport and cation/H^+ antiport, which restore the membrane potential at the expense of the pH differential.

Although the folding of the respiratory chain in 'loops' remains to be proven (Chance and coworkers, 1967; Slater, 1967), some elegant evidence for the directional translocation of protons has been presented (see Mitchell, 1966b, p. 166). However, even if separation of charges takes place in such a way that H^+ and e^- (or OH^-) appear in the outer and inner surface of the membrane, respectively, the question arises whether protons are ejected into the water phase or remain in the membrane. Experiments with bromthymol blue indicate that no pH differential can be found in state 4 (Chance and Mela, 1966c). However, the interpretation of these experiments has been criticized on the basis that the bromthymol blue, because of its negatively charged groups, is driven out of the membrane by the negative electrical potential of the inner phase in state 4 (Mitchell and coworkers, 1968). Furthermore, it has been reported that addition of oxygen to an anaerobic suspension of mitochondria results in the ejection of protons with a $H^+:\sim$ of 2, and when ATP was added instead of oxygen, a $H^+:P_i$ value of about 2 could be calculated (Mitchell and Moyle, 1965a). However, these experiments were performed with mitochondria preincubated anaerobically for twenty minutes at 25°, conditions under which ions are known to leak out of mitochondria. Therefore, reaccumulation of cations may be responsible for the H^+ ejection observed during respiration (Chance and Mela, 1966d; Tager and coworkers, 1966). In fact, studies on the oxidation of succinate by mitochondria isolated in the presence of EDTA show very low $H^+:\sim$ ratios (0·25–0·75). The value of these ratios can be increased, even above 2, by the addition of Ca^{2+}, Sr^{2+} or Mn^{2+} (Chappell and Haarhoff, 1967). Similarly, the H^+ ejection observed during ATPase activity, and ATP hydrolysis itself, is dependent on the presence of endogenous Ca^{2+}. This cation is released in the presence of respiratory inhibitors and its re-entry into mitochondria when ATP is added is accompanied by ejection of H^+, presumably as a consequence of a Ca^{2+}–H^+ exchange (Rossi and coworkers, 1967d).

These results suggest that H^+ remains in the membrane until it can be exchanged for an external cation. In agreement with this, a cation-dependent H^+ pump has been proposed. The pump, driven by a high energy intermediate, tends to eject H^+ but is unable to do so unless an equivalent positively charged ion takes its place. This exchange would leave an excess of OH^- inside mitochondria which could be discharged by specific anion–OH^- exchange diffusion carriers (Chappell and Crofts, 1965b). An H^+ pump directly coupled to the respiratory chain has also been considered (Chappell and Haarhoff, 1967).

This modification of the H^+ pump hypothesis avoids the long debated question of the existence of a pH gradient during state 4, but it cannot avoid other objections which are common to any H^+ pump theory. Obviously the uptake of external cations in exchange for mitochondrial cations other than protons would be difficult to explain. For example, Ca^{2+} uptake in the presence of valinomycin gives rise to two types of exchanges: Ca^{2+}–H^+ and Ca^{2+}–K^+ (Ogata and Rasmussen, 1966). Even in the absence of an energy supply, mitochondrial K^+ can exchange with external Na^+ (Gear and Lehninger, 1968) or with H^+ (Ogata and Rasmussen, 1966).

Indeed the most crucial problem to be resolved by any energy-linked H^+ or cation pump theory is the observed variability in the ratio of ion translocated to energy expended. Certainly the $K^+:\sim$ accumulation ratio of 7 in ATP-supported systems represents a serious test for any proposed hypothesis (Cockrell and coworkers, 1966). Attempts to relate the uptake of divalent cations to the amount of ATP hydrolysed could prove to be a more difficult task. The uptake of Ca^{2+} and Sr^{2+} can be supported by ATP, but ATP-supported accumulation of Mn^{2+} has not been observed (Carafoli and coworkers, 1965c). Consequently, Ca^{2+} and Sr^{2+}, but not Mn^{2+}, should stimulate the hydrolysis of ATP. Actually the ATPase of intact mitochondria is stimulated by Ca^{2+} and *inhibited* by Sr^{2+} (Caplan and Carafoli, 1965) and by Mn^{2+} (Cereijo-Santaló, 1967). [A reported stimulation by Mg^{2+} and Mn^{2+} at low concentrations (Cereijo-Santaló, 1967) was later shown to be due to an experimental error (Cereijo-Santaló, 1968b)]. Therefore, the ATP-supported uptake of almost equivalent amounts of Ca^{2+} and Sr^{2+} by intact mitochondria (Vasington, 1966) cannot, by any further stretch of stoichiometric mechanisms, be attributed to similar rates of ATP hydrolysis. In addition, ATP cannot support Sr^{2+} accumulation by water-washed mitochondria in spite of the fact that the ATPase activity of these particles is *not* inhibited by Sr^{2+} (Vasington, 1966).

Analogous observations can be made with regard to the uptake of divalent cations supported by respiration. Mitochondria accumulate 1·3–2·5 μmoles of Sr^{2+} (Carafoli and coworkers, 1965c) and more than 2 μmoles of Mn^{2+} (Chappell and coworkers, 1963) per mg of protein. However, the 'extra' oxygen uptake induced by these cations is 101 and 15 natoms oxygen/mg protein . min for Sr^{2+} and Mn^{2+}, respectively (Carafoli, 1965). Most interesting is the finding that the 'extra' oxygen uptake induced by Ca^{2+} can be inhibited up to 70 per cent without any alteration in the amount of Ca^{2+} accumulated (Carafoli and coworkers, 1967). This experiment also reveals another important point, namely that the inhibition of oxygen uptake must be paralleled by an inhibition of substrate utilization unless an 'electron sink' is available. Therefore, the observation that the amount of H^+ ejected is not diminished (Carafoli and coworkers, 1967) indicates that the 'extra' protons come from the dissociation of mitochondrial anionic groups.

C. Mitochondria as Ion Exchangers

The relationship between electron transport and ion translocation would be more flexible if energy were conserved in a mitochondrial energized state such as a conformational change of their proteins. Configurational changes ('swelling–shrinking', 'orthodox–condensed', 'energized–nonenergized') have been studied in both mitochondria and chloroplasts in relation to metabolic states (Packer, 1961; Boyer, 1965; Slautterback, 1965; Hackenbrock, 1966; Packer and coworkers, 1966; Deamer and coworkers, 1967; Utsumi and Packer, 1967; Harris and coworkers, 1968; Penniston and coworkers, 1968). Of special interest is the observation that respiratory energy can be converted into conformational work *without* the mediation of coupling systems (Hackenbrock, 1968). It has also been pointed out that the energized state of mitochondria could be something as simple as localized proton production (Mitchell and coworkers, 1967; Boyer, 1968). This raises the possibility that the conformational changes of proteins are a consequence of proton accumulation. In fact, increased proton concentration in membranes has been related to configurational changes in both chloroplasts (Hind and Jagendorf,

1965; Deamer and coworkers, 1967a; Deamer and Packer, 1967) and mitochondria (Cereijo-Santaló, 1966d). When ADP and P_i are available, the high local concentration of protons in the region of the ATPase could be used to bring about the synthesis of ATP (Williams, 1961, 1962) during which H^+ ions are absorbed, and the 'energized' configuration is discharged (Harris and coworkers, 1968).

The fact that respiring mitochondria take up cations in exchange for H^+ and that this process is reversed when respiration is inhibited, suggests that the energized–non-energized state transition coincides with a change in the pK of mitochondrial anionic groups. Such a change, induced by electron flow, has been proposed in both chloroplasts and mitochondria (Lynn and Brown, 1967). This brings into focus the role of the anionic field strength (Eisenman, 1961) or the parameter '*c*' of the 'association–induction hypothesis' which considers the bulk phase of the cell as a fixed-charge system (Ling, 1952, 1962, 1965, 1966). Whatever the shortcomings of this hypothesis when applied at the cellular level (see Caldwell, 1968) or as a general theory of the living state, the idea of the mitochondrion as a complex of cationic and anionic exchangers deserves serious consideration. The more important and immediate consequence of this view is that the translocation of cations would not require more energy than that of state 4 respiration. [This energy could be used to maintain a certain electronic and configurational state of proteins (Ling, 1965)]. State 3 respiration would be the consequence rather than the cause of cation uptake. This implies that energy is not conserved as an ion gradient but is dissipated as heat, as occurs with DNP. Microcalorimetric techniques could be used to test the validity of this idea.

If cation translocation does not consume energy, the problem arises as to how it stimulates respiratory and ATPase activities. It has been suggested that the latency of ATPase of intact mitochondria may be due to accumulation of protons in the membrane and, therefore, removal of these protons by means of a cation–H^+ exchange activates ATP hydrolysis (Cereijo-Santaló, 1967, 1968a). A similar mechanism has been proposed for the regulation of respiratory activity (Cereijo-Santaló, 1965, 1966d). It is important to realize that only cations which exchange for substrate-derived H^+ can be related to respiratory stimulation, and not those which exchange for H^+ of fixed anionic groups. Even acidic groups of high p*K* could contribute to this exchange, because of the internal alkalinization. Therefore, the values of the H^+:O and cation:O ratios will depend upon the relative contribution of these two sources of protons. An extreme example of this may be found in the respiratory inhibition induced by high salt concentration (Carafoli and coworkers, 1967) in which the main source of H^+ which exchanges for Ca^{2+} seems to be the mitochondrial anionic groups, thus resulting in high cation:O and H^+:O ratios.

Since electron transport produces in the membrane two impermeant ions, H^+ and OH^-, the cation–H^+ exchange would allow respiration to proceed only until the concentration of OH^- builds up to an inhibitory level (state 6), in which case respiration becomes dependent on an anion–OH^- exchange. If a permeant anion is available, the consequent decrease of intramitochondrial pH will favour the exchange of cations for substrate-derived H^+. Under these conditions a 'normal' cation:O value and an apparent low H^+:O ratio (due to the extrusion of OH^-), would be expected.

The respiratory stimulation induced by ADP plus P_i could be explained if P_i enters mitochondria in exchange for OH^- (with the consequent alkalinization of the medium) and protons used in the synthesis of ATP. This view accounts for the observation that the pH gradient created by Ca^{2+} uptake, in which the protons are outside the membrane, cannot be used for ATP synthesis (Reynafarje and coworkers, 1967).

In conclusion it is apparent that many, far too many, questions remain unanswered, regardless of the hypothesis one chooses to favour. Nevertheless the idea of the mitochondrion as an ion exchanger has the distinct advantage of being simple. It is likely that complicated mechanisms exist only in man's imagination, for it is written that 'when the reason dreams, it produces monsters' (Goya-Lucientes, 1797).

Acknowledgement

I am indebted to Dr. Margaret J. Henderson de Cereijo for her invaluable help in the preparation of this manuscript.

REFERENCES

Addanki, S., F. D. Cahill and J. F. Sotos (1968) *J. Biol. Chem.*, **243**, 2337

Allmann, D. W. and E. Bachmann (1965) *Federation Proc.*, **24**, 425

Allmann, D. W. and E. Bachmann (1967) In R. W. Estabrook and M. E. Pullman (Eds.), *Methods in Enzymology*, Vol. X, Academic Press, New York, p. 438

Amons, R., S. G. Van Den Bergh and E. C. Slater (1968) *Biochim. Biophys. Acta.*, **162**, 452

Amoore, J. E. (1960) *Biochem. J.*, **76**, 438

Azzi, A. and G. F. Azzone (1966) *Biochim. Biophys. Acta.*, **113**, 445

Azzi, A. and G. F. Azzone (1967) *Biochim. Biophys. Acta.*, **131**, 468

Azzi, A., J. B. Chappell and B. H. Robinson (1967) *Biochem. Biophys. Res. Commun.*, **29**, 148

Azzone, G. F. and A. Azzi (1966) In J. M. Tager, S. Papa, E. Quagliariello and E. C. Slater (Eds.) *Regulation of metabolic processes in Mitochondria*, Elsevier, Amsterdam, p. 332

Azzone, G. F., A. Azzi and C. Rossi (1967a) In E. Quagliariello, S. Papa, E. C. Slater and T. M. Tager (Eds.), *Mitochondrial Structure and Compartmentation*, Adriatica Editrice, Bari, p. 234

Azzone, G. F., C. Rossi, A. Azzi, E. Rossi and A. Scarpa (1967b) *7th Intern. Congr. Biochem. Tokyo*, p. 311

Bielawski, J. and A. L. Lehninger (1966) *J. Biol. Chem.*, **241**, 4316

Blondin, G. A. and D. E. Green (1967) *Proc. Natl. Acad. Sci. U.S.*, **58**, 612
Boyer, P. D. (1965) In T. E. King, H. S. Mason and M. Morrison (Eds.), *Oxidases and Related Redox Systems*. Vol. II, Wiley, New York, p. 994
Boyer, P. D. (1968) In T. P. Singer (Ed.), *Biological Oxidations*, Interscience, New York, p. 193
Brierley, G. P. and D. P. Slautterback (1964) *Biochim. Biophys. Acta.*, **82**, 183
Brierley, G. P. and D. E. Green (1965) *Proc. Natl. Acad. Sci. U.S.*, **53**, 73
Brierley, G. P. and C. T. Settlemire (1967) *J. Biol. Chem.*, **242**, 4324
Brierley, G. P., E. Bachmann and D. E. Green (1962) *Proc. Natl. Acad. Sci. U.S.*, **48**, 1928
Brierley, G. P., E. Murer and E. Bachmann (1964) *Arch. Biochem. Biophys.*, **105**, 89
Brierley, G. P., C. T. Settlemire and V. A. Knight (1967) *Biochem. Biophys. Res. Commun.*, **28**, 420
Brierley, G. P., C. T. Settlemire and V. A. Knight (1968a) *Arch. Biochem. Biophys.*, **126**, 276
Brierley, G. P., A. Knight and C. T. Settlemire (1968b) *J. Biol. Chem.*, **243**, 5035
Caldwell, P. C. (1968) *Physiol. Rev.*, **48**, 1
Caplan, A. I. and E. Carafoli (1965) *Biochim. Biophys. Acta.*, **104**, 317
Carafoli, E. (1965) *Biochim. Biophys. Acta.*, **97**, 107
Carafoli, E. and C. S. Rossi (1967) *Eur. J. Biochem.*, **2**, 224
Carafoli, E., R. L. Gamble and A. L. Lehninger (1965a) *Biochem. Biophys. Res. Commun.*, **21**, 215
Carafoli, E., C. S. Rossi and A. L. Lehninger (1965b) *Biochem. Biophys. Res. Commun.*, **19**, 609
Carafoli, E., S. Weiland and A. L. Lehninger (1965c) *Biochim. Biophys. Acta.*, **97**, 88
Carafoli, E., R. L. Gamble, C. S. Rossi and A. L. Lehninger (1967) *J. Biol. Chem.*, **242**, 1199
Caswell, A. H. and B. C. Pressman (1968) *Biochem. Biophys. Res. Commun.*, **30**, 637
Cereijo-Santaló, R. (1965) *Can. J. Biochem.*, **43**, 425
Cereijo-Santaló, R. (1966a) *Can. J. Biochem.*, **44**, 67
Cereijo-Santaló, R. (1966b) *Biochem. Biophys. Res. Commun.*, **24**, 650
Cereijo-Santaló, R. (1966c) *Can. J. Biochem.*, **44**, 695
Cereijo-Santaló, R. (1966d) *Can. J. Biochem.*, **44**, 1259
Cereijo-Santaló, R. (1967) *Can. J. Biochem.*, **45**, 897
Cereijo-Santaló, R. (1968a) *Can. J. Biochem.*, **46**, 55
Cereijo-Santaló, R. (1968b) *Can. J. Biochem.*, **46**, 510
Cereijo-Santaló, R. (1968c) *Can. J. Biochem.*, **46**, 1161
Chance, B. (1956) In E. Liébecq (Ed.), *Proc. Third Intern. Congr. Biochem., Brussels*, 1955, Academic Press, New York, p. 300
Chance, B. (1963) In B. Chance (Ed.), *Energy-linked Functions of Mitochondria*, Academic Press, New York, p. 253
Chance, B. (1964) *Federation Proc.*, **23**, 265
Chance, B. (1965) *J. Biol. Chem.*, **240**, 2729
Chance, B. (1967) In E. C. Slater, Z. Kaniuga and L. Wojtczak (Eds.), *Biochemistry of Mitochondria*, Academic Press, London, p. 93
Chance, B. and L. Mela (1966a) *Proc. Natl. Acad. Sci. U.S.*, **55**, 1243
Chance, B. and L. Mela (1966b) *J. Biol. Chem.*, **241**, 4588
Chance, B. and L. Mela (1966c) *Nature*, **212**, 369
Chance, B. and L. Mela (1966d) *Nature*, **212**, 372
Chance, B. and L. Mela (1967) *J. Biol. Chem.*, **242**, 830
Chance, B. and B. Schoener (1966) *J. Biol. Chem.*, **241**, 4577
Chance, B. and G. R. Williams (1956) In F. F. Nord (Ed.), *Advances in Enzymology*, Vol. 17, Interscience, New York, p. 65
Chance, B. and T. Yoshioka (1965) *Federation Proc.*, **24**, 425
Chance, B., C–P. Lee and L. Mela (1967) *Federation Proc.*, **26**, 1341
Chappell, J. B. (1964) *Biochem. J.*, **90**, 225
Chappell, J. B. (1966) *Biochem. J.*, **100**, 43P
Chappell, J. B. (1968) *Brit. Med. Bull.*, **24**, 150
Chappell, J. B. and A. R. Crofts (1965a) *Biochem. J.*, **95**, 378

Chappell, J. B. and A. R. Crofts (1965b) *Biochem. J.*, **95**, 393
Chappell, J. B. and A. R. Crofts (1966) In J. M. Tager, S. Papa, E. Quagliariello and E. C. Slater (Eds.), *Regulation of Metabolic Processes in Mitochondria*, Elsevier, Amsterdam, p. 293
Chappell, J. B. and K. N. Haarhoff (1967) In E. C. Slater, Z. Kaniuga and L. Wojtczak (Eds.), *Biochemistry of Mitochondria*, Academic Press, London, p. 75
Chappell, J. B. and B. H. Robinson (1968) In T. W. Goodwin (Ed.), *The Metabolic Roles of Citrate*, Academic Press, London, p. 123
Chappell, J. B., M. Cohn and G. D. Greville (1963) In B. Chance (Ed.), *Energy-linked Functions of Mitochondria*, Academic Press, New York, p. 219
Christie, G. S., K. Ahmed, A. E. M. McLean and J. D. Judah (1965) *Biochim. Biophys. Acta.*, **94**, 432
Cockrell, R. S., E. J. Harris and B. C. Pressman (1966) *Biochemistry*, **5**, 2326
Deamer, D. W. and L. Packer (1967) *Arch. Biochem. Biophys.*, **119**, 83
Deamer, D. W., A. R. Crofts and L. Packer (1967a) *Biochim. Biophys. Acta.*, **131**, 81
Deamer, D. W., K. Utsumi and L. Packer (1967b) *Arch. Biochem. Biophys.*, **121**, 641
Dianzani, M. U. (1953) *Biochim. Biophys. Acta.*, **11**, 353
Drahota, Z. and A. L. Lehninger (1965) *Biochem. Biophys. Res. Commun.*, **19**, 351
Drahota, Z., E. Carafoli, C. S. Rossi, R. L. Gamble and A. L. Lehninger (1965) *J. Biol. Chem.*, **240**, 2712
Eisenman, G. (1961) In A. Kleinzeller and A. Kotyk (Eds.), *Membrane Transport and Metabolism*, Academic Press, London, p. 163
Engström, G. W. and H. F. DeLuca (1964) *Biochemistry*, **3**, 379
Estrada-O, S., S. N. Graven and H. A. Lardy (1967) *J. Biol. Chem.*, **242**, 2925
Ferguson, S. F. M. and G. R. Williams (1966) *J. Biol. Chem.*, **241**, 3696
Ferguson, S. M. and H. A. Lardy (1968) *Federation Proc.*, **27**, 528
Gamble, J. L., Jr. (1957) *J. Biol. Chem.*, **228**, 955
Gamble, J. L., Jr. (1963a) *Proc. Soc. Exptl. Biol. Med.*, **113**, 375
Gamble, J. L., Jr. (1963b) *Biochim. Biophys. Acta.*, **66**, 158
Gamble, J. L., Jr. (1965) *J. Biol. Chem.*, **240**, 2668
Gamble, J. L., Jr. and R. C. Hess, Jr. (1966) *Amer. J. Physiol.*, **210**, 765
Gear, A. R. L. and A. L. Lehninger (1967) *Biochem. Biophys. Res. Commun.*, **28**, 840
Gear, A. R. L. and A. L. Lehninger (1968) *J. Biol. Chem.*, **243**, 3953
Gear, A. R. L., C. S. Rossi, B. Reynafarje and A. L. Lehninger (1967) *J. Biol. Chem.*, **242**, 3403
Goya-Lucientes, F. (1797) *Los caprichos*, No. 43, Prado Museum, Madrid
Graven, S. N., S. Estrada-O and H. A. Lardy (1966a) *Proc. Natl. Acad. Sci. U.S.*, **56**, 654
Graven, S. N., H. A. Lardy, D. Johnson and A. Rutter (1966b) *Biochemistry*, **5**, 1729
Graven, S. N., H. A. Lardy and S. Estrada-O (1967) *Biochemistry*, **6**, 365
Green, D. E., E. Bachmann, D. W. Allmann and J. F. Perdue (1966) *Arch. Biochem. Biophys.*, **115**, 172
Greenawalt, J. W. and E. Carafoli (1966) *J. Cell Biol.*, **29**, 37
Greenawalt, J. W., C. S. Rossi and A. L. Lehninger (1964) *J. Cell Biol.*, **23**, 21
Hackenbrock, C. R. (1966) *J. Cell Biol.*, **30**, 269
Hackenbrock, C. R. (1968) *J. Cell Biol.*, **37**, 345
Hanai, T., D. A. Haydon and W. R. Redwood (1966) *Ann. N.Y. Acad. Sci.*, **137**, 731
Harris, E. J. (1968) *Biochem. J.*, **109**, 247
Harris, E. J., R. Cockrell and B. C. Pressman (1966) *Biochem. J.*, **99**, 200
Harris, E. J., G. Catlin and B. C. Pressman (1967a) *Biochemistry*, **6**, 1360
Harris, E. J., M. P. Höffer and B. C. Pressman (1967b) *Biochemistry*, **6**, 1348
Harris, E. J., K. Van Dam and B. C. Pressman (1967c) *Federation Proc.*, **26**, 610
Harris, E. J., K. Van Dam and B. C. Pressman (1967d) *Nature*, **213**, 1126
Harris, R. A., J. T. Penniston, J. Asai and D. E. Green (1968) *Proc. Natl. Acad. Sci. U.S.*, **59**, 830
Hind, G. and A. T. Jagendorf (1965) *J. Biol. Chem.*, **240**, 3195
Höfer, M. and B. C. Pressman (1966) *Biochemistry*, **5**, 3919

Hoffman, J. F. (1962) *Circulation*, **26**, 1201
Hopper, U., A. L. Lehninger and T. E. Thompson (1968) *Proc. Natl. Acad. Sci. U.S.*, **59**, 484
Judah, J. D., K. Ahmed, A. E. M. McLean and G. S. Christie (1965a) *Biochim. Biophys. Acta.*, **94**, 452
Judah, J. D., A. E. M. McLean, K. Ahmed and G. S. Christie (1965b) *Biochim. Biophys. Acta.*, **94**, 441
Kimmich, G. A. and H. Rasmussen (1967) *Biochim. Biophys. Acta.*, **131**, 413
Klingenberg, M. (1967) In E. Quagliariello, S. Papa, E. C. Slater and J. M. Tager (Eds.), *Mitochondrial Structure and Compartmentation*, Adriatica Editrice, Bari, p. 216
Klingenberg, M. and E. Pfaff (1968) In T. W. Goodwin (Ed.), *The Metabolic Roles of Citrate*, Academic Press, London, p. 105
Lardy, H. (1961) In T. W. Goodwin and O. Lindberg (Eds.), *Biological Structure and Function*, Vol. II, Academic Press, New York, p. 265
Lardy, H. A. and H. Wellman (1953) *J. Biol. Chem.*, **201**, 357
Lardy, H. A., J. L. Connelly and D. Johnson (1964) *Biochemistry*, **3**, 1961
Lardy, H. A., S. N. Graven and S. Estrada-O (1967) *Federation Proc.*, **26**, 1355
Lehninger, A. L. (1949) *J. Biol. Chem.*, **178**, 625
Lehninger, A. L. (1955) *Harvey Lectures*, Ser. 49 (1953–1954), Academic Press, New York, p. 176
Lehninger, A. L. (1964) *The Mitochondrion*, W. A. Benjamin, New York
Lehninger, A. L., C. I., Wadkins and L. F. Remmert (1959) in G. E. W. Wolstenholme and C. M. O'Connor (Eds.), *Ciba Found. Symp. Regulation Cell Metabolism*, Churchill, London, p. 130
Lehninger, A. L., C. S. Rossi and J. W. Greenawalt (1963) *Biochem. Biophys. Res. Commun.*, **10**, 444
Lehninger, A. L., E. Carafoli and C. S. Rossi (1967) In F. F. Nord (Ed.), *Advances in Enzymology*, Vol. 29, Interscience, New York, p. 259
Leive, L. (1968) *J. Biol. Chem.*, **243**, 2373
Lindberg, O. and L. Ernster (1954) *Nature*, **173**, 1038
Ling, G. N. (1952) In W. D. McElroy and B. Glass (Eds.), *Phosphorus Metabolism*, Vol. II, Johns Hopkins, Baltimore, p. 748
Ling, G. N. (1962) *A Physical Theory of the Living State: the association-induction hypothesis*, Blaisdell, New York
Ling, G. N. (1965) *Federation Proc.*, **24**, Suppl. No. 15, p. 103
Ling, G. N. (1966) *Ann. N.Y. Acad. Sci.*, **137**, 837
Lynn, W. S. and R. H. Brown (1965a) *Biochim. Biophys. Acta.*, **110**, 445
Lynn, W. S. and R. H. Brown (1965b) *Biochim. Biophys. Acta.*, **110**, 459
Lynn, W. S. and R. H. Brown (1966) *Arch. Biochem. Biophys.*, **114**, 260
Lynn, W. S. and R. H. Brown (1967) *J. Biol. Chem.*, **242**, 426
Max, S. R. and J. L. Purvis (1965) *Biochem. Biophys. Res. Commun.*, **21**, 587
Meijer, A. J. and J. M. Tager (1966) *Biochem. J.*, **100**, 79P
Meijer, A. J., E. J. De Haan and J. M. Tager (1967) In E. Quagliariello, S. Papa, E. C. Slater and J. M. Tager (Eds.), *Mitochondrial Structure and Compartmentation*, Adriatica Editrice, Bari, p. 207
Mitchell, P. (1961) *Nature*, **191**, 144
Mitchell, P. (1963) *Biochem. Soc. Symp.* **22**, 142
Mitchell, P. (1966a) In J. M. Tager, S Papa, E. Quagliariello and E. C. Slater (Eds.), *Regulation of Metabolic Processes in Mitochondria*, Elsevier, Amsterdam, p. 65
Mitchell, P. (1966b) *Chemiosmotic Coupling in Oxidative and Photosynthetic Phosphorylation*, Research Report No. 66/1, Glynn Research Ltd., Bodmin, Cornwall
Mitchell, P. (1967a) In F. F. Nord (Ed.), *Advances in Enzymology*, Vol. 29, Intersciences, New York, p. 33
Mitchell, P. (1967b) *Federation Proc.*, **26**, 1370
Mitchell, P. and J. Moyle (1965a) *Nature*, **208**, 147
Mitchell, P. and J. Moyle (1965b) *Nature*, **208**, 1205

Mitchell, P. and J. Moyle (1967a) In E. C. Slater, Z. Kaniuga and L. Wojtczak (Eds.), *Biochemistry of Mitochondria*, Academic Press, London, p. 53
Mitchell, P. and J. Moyle (1967b) *Nature*, **213**, 137
Mitchell, R. A., R. D. Hill and P. D. Boyer (1967) *J. Biol. Chem.* **242**, 1793
Mitchell, P., J. Moyle and L. Smith (1968) *Eur. J. Biochem.*, **4**, 9
Moore, C. and B. C. Pressman (1964) *Biochem. Biophys. Res. Commun.*, **15**, 562
Mueller, P. and D. O. Rudin (1967) *Biochem. Biophys. Res. Commun.*, **26**, 398
Nishimura, M., T. Ito and B. Chance (1962) *Biochim. Biophys. Acta.*, **59**, 177
O'Brien, R. L. and G. Brierley (1965) *J. Biol. Chem.*, **240**, 4527
Ogata, E. and H. Rasmussen (1966) *Biochemistry*, **5**, 57
Packer, L. (1961) In T. W. Goodwin and O. Lindberg (Eds.), *Biological Structure and Function*, Vol. II, Academic Press, London, p. 85
Packer, L., K. Utsumi and M. G. Mustafa (1966) *Arch. Biochem. Biophys.*, **117**, 381
Parsons, D. F. and Y. Yano (1967) *Biochim. Biophys. Acta.*, **135**, 362
Parsons, D. F., G. R. Williams, W. Thompson, D. Wilson and B. Chance (1967) In E. Quagliareillo, S Papa, E. C. Slater and T. M. Tager (Eds.), *Mitochondrial Structure and Compartmentation*, Adriatica Editrice, Bari, p. 29
Peachey, L. D. (1964) *J. Cell Biol.*, **20**, 95
Penniston, J. T., R. A. Harris, J. Asai and D. E. Green (1968) *Proc. Natl. Acad. Sci. U.S.*, **59**, 624
Pfaff, E., M. Klingenberg, E Ritt and W. Vogell (1968) *Eur. J. Biochem.*, **5**, 222
Potter, V. R., P. Siekevitz and H. C. Simonson (1953) *J. Biol. Chem.*, **205**, 893
Pressman, B. C. (1963) In B. Chance (Ed.), *Energy-linked Functions of Mitochondria*, Academic Press, New York, p. 181
Pressman, B. C. (1965) *Proc. Natl. Acad. Sci. U.S.*, **53**, 1076
Pressman, B. C. (1968) *Federation Proc.*, **27**, 1283
Pressman, B. C., E. J. Harris, W. S. Jagger and J. H. Johnson (1967) *Proc. Natl. Acad. Sci. U.S.*, **58**, 1949
Raaflaub, J. (1953) *Helv. Physiol. Acta.*, **11**, 142
Rasmussen, H. and E. Ogata (1966) *Biochemistry*, **5**, 733
Rasmussen, H., B. Chance and E. Ogata (1965) *Proc. Natl. Acad. Sci. U.S.*, **53**, 1069
Reynafarje, B., A. R. L. Gear, C. S. Rossi and A. L. Lehninger (1967) *J. Biol. Chem.*, **242**, 4078
Robertson, R. N. (1960) *Biol. Revs.*, **35**, 231
Robertson, R. N. (1967) *Endeavour*, **26**, 134
Rossi, C. S. and A. L. Lehninger (1963a) *Biochem. Z.*, **338**, 698
Rossi, C. S. and A. L. Lehninger (1963b) *Biochem. Biophys. Res. Commun.*, **11**, 441
Rossi, C. S. and A. L. Lehninger (1964) *J. Biol. Chem.*, **239**, 3971
Rossi, C. S. and G. F. Azzone (1965) *Biochim. Biophys. Acta.*, **110**, 434
Rossi, C. S. and G. F. Azzone (1968) *J. Biol. Chem.*, **243**, 1514
Rossi, C. S., J. Bielawski and A. L. Lehninger (1966a) *J. Biol. Chem.*, **241**, 1919
Rossi, C. S., J. Bielawski, E. Carafoli and A. L. Lehninger (1966b) *Biochem. Biophys. Res. Commun.*, **22**, 206
Rossi, C. S., E. Carafoli, Z. Drahota and A. L. Lehninger (1966c) In J. M. Tager, S. Papa, E. Quagliariello and E. C. Slater (Eds.), *Regulation of Metabolic Processes in Mitochondria*, Elsevier, Amsterdam, p. 317
Rossi, C. S., A. Azzi and G. F. Azzone (1966d) *Biochem. J.*, **100**, 4c
Rossi, C. S., A. Azzi and G. F. Azzone (1967a) *J. Biol. Chem.*, **242**, 951
Rossi, C. S., G. F. Azzone and A. Azzi (1967b) *Eur. J. Biochem.*, **1**, 141
Rossi, C. S., A. Scarpa and G. F. Azzone (1967c) *Biochemistry*, **6**, 3902
Rossi, C. S., N. Siliprandi, E. Carafoli, J. Bielawski and A. L. Lehninger (1967d) *Eur. J. Biochem.*, **2**, 332
Rottenberg, H. and A. K. Solomon (1965) *Biochem. Biophys. Res. Commun.*, **20**, 85
Sanui, H. and N. Pace (1959) *J. Gen. Physiol.*, **42**, 1325
Sanui, H. and N. Pace (1965) *J. Cell Comp. Physiol.*, **65**, 27
Sanui, H. and N. Pace (1967) *J. Cell Physiol.*, **69**, 3

Saris, N.-E. (1963) *Finska Vet. Soc.-Comm. Phys. Math.*, **28**, No. 11
Scarpa, A. and A. Azzi (1968) *Biochim. Biophys. Acta.*, **150**, 473
Scarpa, A. and G. F. Azzone (1968) *J. Biol. Chem.*, **243**, 5132
Schnaitman, C. and J. W. Greenawalt (1968) *J. Cell Biol.*, **38**, 158
Scott, R. L. and J. L. Gamble (1961) *J. Biol. Chem.*, **236**, 570
Settlemire, C. T., G. R. Hunter and G. P. Brierley (1968) *Biochim. Biophys. Acta.*, **162**, 487
Siekevitz, P. and V. R. Potter (1953) *J. Biol. Chem.*, **201**, 1
Slater, E. C. (1953) *Nature*, **172**, 975
Slater, E. C. (1967) *Eur. J. Biochem.*, **1**, 317
Slater, E. C. and K. W. Cleland (1953) *Biochem. J.*, **55**, 566
Slautterback, D. B. (1965) *J. Cell Biol.*, **24**, 1
Stanbury, S. W. and G. H. Mudge (1953) *Proc. Soc. Exptl. Biol. Med.*, **82**, 675
Stein, W. D. (1967) *The Movements of Molecules across Cell Membranes*, Academic Press, London
Swanson, M. A. (1957) *Federation Proc.*, **16**, 258
Tager, J. M., R. D. Veldsema-Currie and E. C. Slater (1966) *Nature*, **212**, 376
Tarr, J. S., Jr. and J. L. Gamble, Jr. (1966) *Amer. J. Physiol.*, **211**, 1187
Tedeschi, H. and D. L. Harris (1955) *Arch. Biochem. Biophys.*, **58**, 52
Thomas, R. S. and J. W. Greenawalt (1968) *J. Cell Biol.*, **39**, 55
Thompson, T. E. and C. Huang (1966) *Ann. N.Y. Acad. Sci.*, **137**, 740
Ulrich, F. (1960) *Amer. J. Physiol.*, **198**, 847
Ussing, H. H. (1949) *Physiol. Revs.*, **29**, 127
Utsumi, K. and L. Packer (1967) *Arch. Biochem. Biophys.*, **121**, 633
Van Dam, K. and C. S. Tsou (1968) *Biochim. Biophys. Acta.*, **162**, 301
Vasington, F. D. (1966) *Biochim. Biophys. Acta.*, **113**, 414
Vasington, F. D. and J. V. Murphy (1962) *J. Biol. Chem.*, **237**, 2670
Vasington, F. D. and J. W. Greenawalt (1964) *Biochem. Biophys. Res. Commun.*, **15**, 133
Veldsema-Currie, R. D. and E. C. Slater (1968) *Biochim. Biophys. Acta.*, **162**, 310
Weinbach, E. C. and T. von Brand (1965) *Biochem. Biophys. Res. Commun.*, **19**, 133
Weinbach, E. C. and J. Garbus (1968a) *Biochem. J.*, **106**, 711
Weinbach, E. C. and J. Garbus (1968b) *Biochim. Biophys. Acta.*, **162**, 500
Wenner, C. E. (1966) *J. Biol. Chem.*, **241**, 2810
Wenner, C. E. and J. H. Hackney (1967) *J. Biol. Chem.*, **242**, 5053
Werkheiser, W. C. and W. Bartley (1957) *Biochem. J.*, **66**, 79
Williams, R. J. P. (1961) *J. Theoret. Biol.*, **1**, 1
Williams, R. J. P. (1962) *J. Theoret. Biol.*, **3**, 209
Winkler, H. H. and A. L. Lehninger (1968) *J. Biol. Chem.*, **243**, 3000
Winkler, H. H., F. L. Bygrave and A. L. Lehninger (1968) *J. Biol. Chem.*, **243**, 20

CHAPTER 9

The movement of inorganic ions and water across the nuclear envelope

D. J. Fry

Medical Professorial Unit,
Saint Bartholomew's Hospital,
London, E.C.1, England.

I. INTRODUCTION

There are several grounds for believing that the nuclear envelope fulfills a role similar to that of the medieval city wall or the settlement stockade, enclosing a special environment and controlling the traffic between this environment and the surroundings. In a number of instances substances have appeared to be able to diffuse freely throughout nucleoplasm and cytoplasm, but be restricted in their passage between them by the nuclear envelope. To quote two, Feldherr (1965) found that ten minutes after injection of colloidal gold into *Amoebae*, the particles were distributed randomly in the cytoplasm, but few had entered the nucleus, whereas after twenty-four hours, the particles were randomly distributed in the nucleoplasm also. Similarly, Stirling and Kinter (1967) noted that labelled galactose taken up by hamster intestine cells spread rapidly throughout the cytoplasm then, after a few minutes' delay, was seen throughout the nucleoplasm too. Evidence for specific transport systems, presumably located in the nuclear membrane, has also been obtained (Allfrey and coworkers, 1961; Lee and Holbrook,

1964; Navon and Lajtha, 1969). The substances investigated (amino acids and nucleosides) could be accumulated against a concentration gradient, and appeared to be in free solution in the nucleus. The presence of an energy-dependent transport in the other direction, from nucleus to cytoplasm, is suggested by the effect of anoxia on the distribution of tritiated nucleotide in nurse cells of housefly larvae (Bier, 1965). In a nitrogenated medium or at low temperature, the initial concentration of label in the nucleus was less than in uncooled oxygenated medium, but the complete abolition of the subsequent movement of label into the cytoplasm was most striking.

Such results argue strongly that the nuclear envelope acts as a barrier to the diffusion of some substances in free solution, and can regulate the composition of the nuclear sap to some extent. Whether it can also govern the movement of water and/or ions is not so easy to determine, and yet is of some importance, since variations in ion concentration have a marked effect on the activity of the chromosomes (Kroeger and Lezzi, 1966; Pogo and coworkers, 1967). In amphibian embryos and chick embryonic myocardium there is a considerable reduction in intracellular sodium at one particular stage in development (Yeh, Hoffman and Spiro, 1966; Kostello and Morrill, 1968). If the nuclear envelope allows ions and water to pass freely, then these changes in total cell ion concentration will be reflected in similar intranuclear changes, and the activity of the plasma membrane ion pump may thus control the type of nuclear activity. On the other hand, a poorly permeable nuclear envelope could protect the nucleoplasm against ion changes in the cytoplasm, and such changes as do occur would be the result rather than the cause of alterations in nuclear behaviour. Similarly, if the nuclear envelope has some control over the movement of ions, the 'non-specific stimulus' to reactivation of the quiescent nucleus in a cell hybrid (Bolund and coworkers, 1969) could be a change in ion permeability. As John Hunter would have pointed out, there is only one way to answer such questions, and the purpose of this chapter is to consider at least some of the relevant experimental evidence.

II. THE STRUCTURE OF THE NUCLEAR ENVELOPE

Under the electron microscope, the nuclear envelope has been resolved into two approximately parallel triple-layered membranes, enclosing a perinuclear space. At several points the membranes appear to be in continuity, leaving gaps in the surface which have been termed 'pores' (Figure 1). The basic structure, with average measurements of the components, is shown diagrammatically in figure 2. In sections tangential to the nuclear surface numerous apparently circular annuli are seen, in roughly hexagonally-spaced arrays: these represent cross-sections through the nuclear pores (Watson,

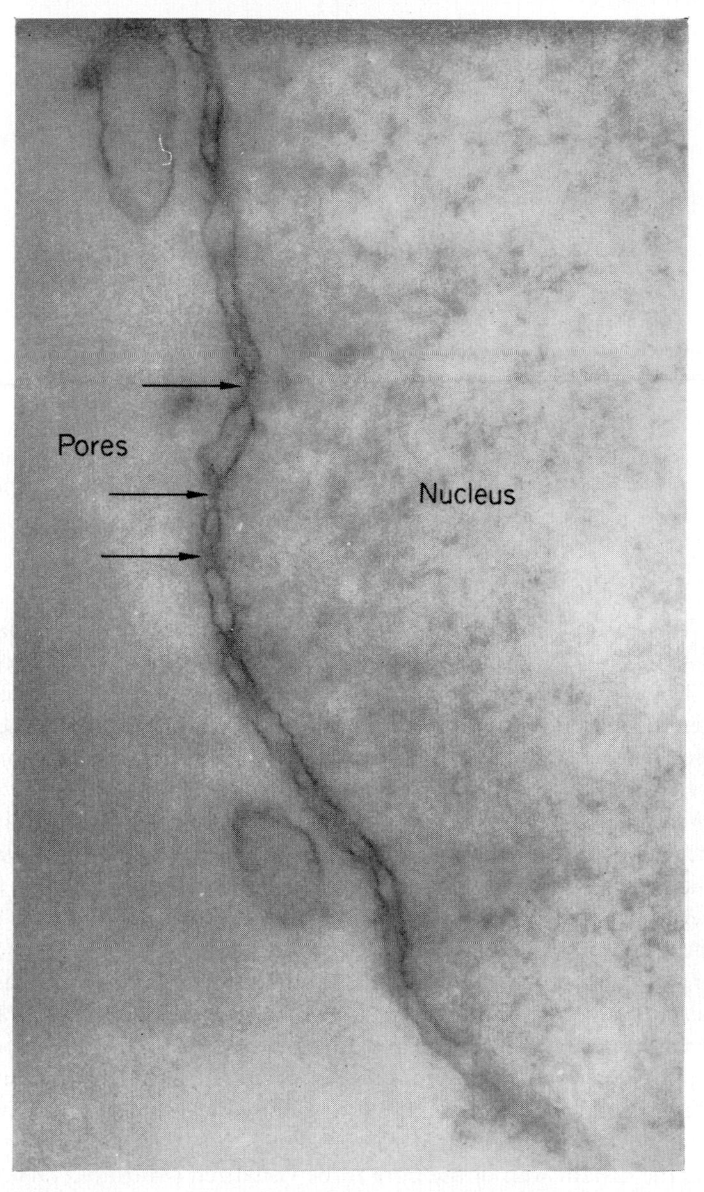

Figure 1. Electron micrograph of the envelope of an isolated toad oocyte nucleus (magnification × 73,000).

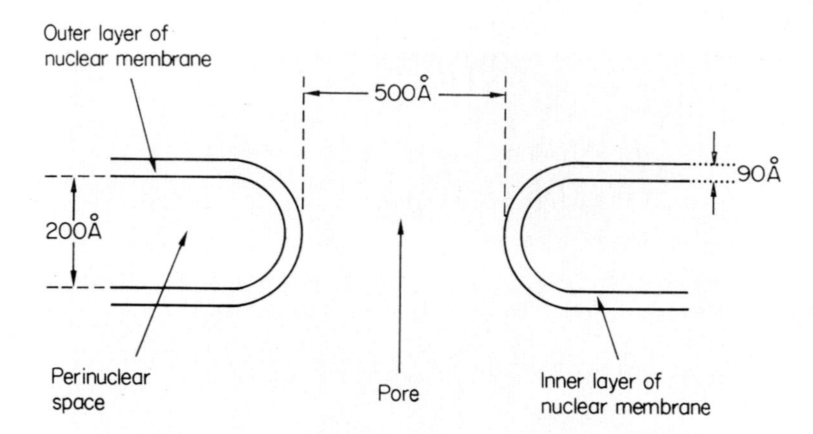

Figure 2. Diagrammatic cross-section through the nuclear envelope showing a pore. Figures given are averages. The range for the perinuclear space is 100–300 Å, that for the nuclear membrane 80–100 Å (Gall, 1964). Pore diameter varies from 300–1000 Å (see text).

1955). This bilamellar porous structure is common to both plant and animal cells, and is seen in neoplastic as well as normal tissue (Wischnitzer, 1960). There is, however, some quantitative variation: for instance, pore diameters from 300 Å (Callan and Tomlin, 1950) to 1000 Å (Franke, 1966) have been reported and the total pore area varies considerably. Watson (1955) estimated it as 5 to 15 per cent of the nuclear surface, but values up to 31 per cent (Wunderlich, 1969) have been given. If Gall's (1967) suggestion that the pores are octagonal rather than circular is accepted, part of this variation can be ascribed to different diameters of the octagon being taken as the true diameter of the pore. Even taking this into account, though, there are marked differences between different nuclear envelopes.

Clearly if these pores allowed direct continuity between nucleoplasm and cytoplasm, the regulatory function of the nuclear envelope would be limited. It is generally agreed, however, that in most cells the pores are blocked in some way. Sometimes the results suggest the presence of a diaphragm across the pore (Merriam 1961; Friend, 1966; Edmonds, 1968; Ward and Ward, 1968), but this may be an artifact caused by the thickness of the section allowing the distant wall of the pore to be visualized (Marinos, 1960; Vivier, 1967). Diffuse material often appears to be filling the pores (Watson, 1955, 1959; Feldherr, 1965; Murray and coworkers, 1965; Stevens and Swift, 1966) but in many cases the results have suggested a definite structure at the periphery of the pore, and a central channel where nucleoplasm and cytoplasm are apparently in continuity (see Vivier, 1967, and Mentré, 1969). It is interesting that where particles have been observed apparently in passage through the pores, they have been located centrally (Beerman, 1964; Feldherr, 1965, Stevens and Swift, 1966; Takamoto, 1966). However, if an open central channel existed, the permeability of nuclear envelopes with pores of similar size and no other detectable anatomical differences might be expected to be related to the pore frequency; this is not the case (Wiener and coworkers, 1965; Feldherr, 1969). In a number of cells a 'fibrous lamina' just deep to the inner layer of the nuclear membrane, and remaining continuous at the pores, has been observed (Fawcett, 1966; Kalifat and coworkers, 1967; Edmonds, 1968). The reconstitution of this structure after cell division is associated with a fall in nuclear envelope permeability (Feldherr, 1968), and it seems possible that in some cells it is this fibrous lamina which 'seals off' the pores.

A connexion between the outer layer of the nuclear membrane and the membranes of the endoplasmic reticulum has been noted in a variety of cells (Watson, 1955; Marinos, 1960; Grasso and coworkers, 1962; Blackburn and Vinijchaikul, 1969). Whether this is more than a transient phenomenon is uncertain; it does mean that there is a possible direct route for substances from the endoplasmic reticular fluid to that in the perinuclear space, which may or may not be exploited. In addition, the occasional observation of continuity between the endoplasmic reticulum membranes and the plasma membrane (Epstein, 1957) indicates the possibility of connection between the perinuclear space and the exterior. The presence of fat globules in the perinuclear space (Palay, 1960), suggests that this route can be functional.

With this complex structure, the exchanges across the nuclear envelope are unlikely to be simple. Five theoretical routes whereby substances could enter or leave the nucleus are indicated in figure 3. They are:

1) Through the pores;

2) Via two layers of nuclear membrane and the perinuclear space;

3) Across one layer of nuclear membrane to the fluid of the endoplasmic reticulum;

4) Across one layer of nuclear membrane and via the endoplasmic reticulum to the exterior;

5) By pinocytosis.

For simplicity this range of possibilities can be reduced to two important exchanges, namely, between the nucleoplasm and the cytoplasm via the pores, and the nucleoplasm and the contents of the perinuclear space via the inner layer of nuclear membrane. Either of these could be the dominant influence under different circumstances.

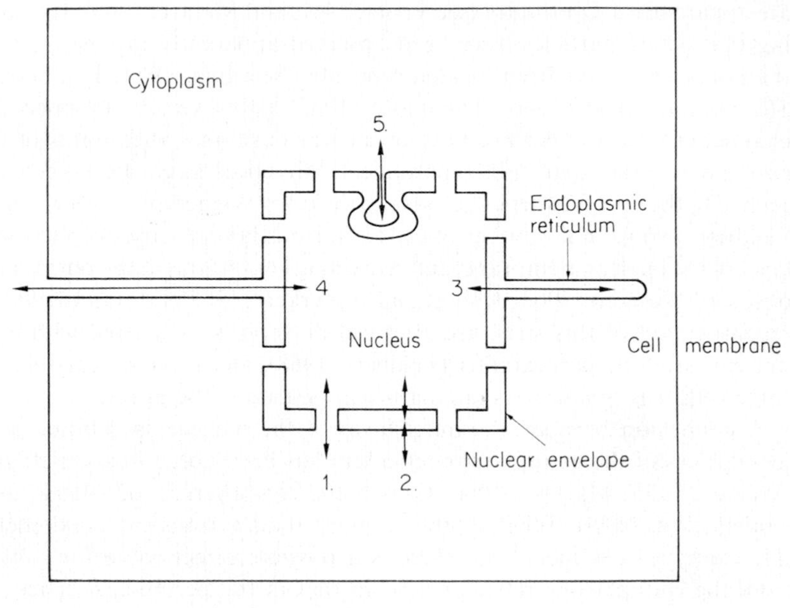

Figure 3. Possible routes leading to exchange across nuclear envelope.

III. THE COMPOSITION OF THE NUCLEAR ENVELOPE

In keeping with the differences in morphology, there are several differences in chemical constitution between the nuclear envelope and the plasma membrane. Whereas the latter is at least 50 per cent lipid, the major component of the nuclear envelope is protein (Callan, 1949). Recently, the isolation of relatively pure nuclear envelopes has enabled them to be thoroughly analysed. Kasper and Kashnig (1969) estimate that 63 per cent of the dry weight is protein, 29 per cent lipid, 5 per cent carbohydrate and 3 per cent RNA. Zbarsky and others (1969), however, give a lower figure for protein, although they used the same tissue (rat liver). They found the nuclear envelopes could be separated into two fractions, the heavy containing 46 to 48 per cent protein, the light 24 to 27 per cent; the figures were similar for a variety of tissues. The lipid content varied from 27 to 50 per cent in both fractions, but the heavy fraction contained more RNA (1·4 to 3·3 per cent) and DNA (0·9 to 1·3 per cent) than the light (1·1 to 2·1 per cent and 0·2 per cent respectively).

The higher content of protein found in the nuclear envelope is probably partly attributable to the presence of the pores. The ground substance of the nuclear pores is digested by proteolytic enzymes and is presumably proteinaceous (Merriam, 1961). The formed elements (fibrils, central granule) of the pore complex are susceptible to RNAse, and the central granules could be dissolved by molar KCl, suggesting that these might be nucleoprotein (Mentré, 1969).

The variety of enzymes found in isolated nuclear membranes emphasizes that control of nucleo-cytoplasmic exchanges is not the only role of the nuclear envelope. Oxidative phosphorylation is probably carried out by the nuclear membrane (Zbarsky and coworkers, 1969) and it may be an important site of DNA synthesis (Milner, 1969). Also, the nuclear envelope is derived from cytoplasmic structures (Barer, Joseph and Meek, 1959), and performs a similar function to the endoplasmic reticulum, with which it is often in continuity (Smith and coworkers, 1969).

IV. THE PERMEABILITY OF THE NUCLEAR ENVELOPE

It seems most unlikely that where the nuclear envelope presents no significant barrier to the diffusion of ions or water, these cross by any other means. Natural selection would not have been kind to such inefficiency! The permeability of the nuclear envelope is therefore a useful guide to the processes involved in ion and water movements into and out of the nucleus. Ideally, the characteristics of each part of the nuclear envelope should be assessed, since the permeabilities of the membrane and the pore material are not necessarily the same. However, it has not proved easy even to determine the permeability of the nuclear envelope as a whole. Much of the data comes from the distribution of marker substances *in vivo*, where binding to cellular constituents or intracellular transport may affect the result, and from experiments on isolated nuclei, which are essentially uncontrolled. There are some documented sources of error: for instance, if the isolation medium permits nuclear swelling to occur, the permeability may increase markedly (Macgregor, 1962). It is also realized that some isolation techniques leave the nuclei in an unsuitable condition for permeability studies: homogenation in non-aqueous solvents is liable to cause complete loss of the nuclear envelope (Siebert and coworkers, 1965a), and methods using citrate or detergent may remove the outer layer (Gurr and coworkers, 1963; Sadowski and Steiner, 1968; Smith and coworkers, 1969). These techniques find their places in, respectively, the isolation of nuclei for analysis and the investigation of the properties of the different layers of the nuclear envelope. The absence of morphological change with other isolation methods, however, is no guarantee that the permeability of the nuclear envelope has not altered.

In the early work, the permeability of the nuclear envelope was deduced from the response of the nucleus to osmotic swelling and shrinking of the cytoplasm. The changes in nuclear volume when various gut epithelia (Hamburger, 1904) and invertebrate eggs (Beck and Shapiro, 1936; Churney, 1942) were exposed to solutions of different tonicity parallelled those of the whole cell, though actually slightly greater in magnitude. Hamburger also noted that isolated nuclei behaved osmotically. The obvious conclusion was that the nuclear envelope was permeable to water but not to some osmotically active substances.

This has since been found to be generally true, but in certain special cells the nuclear envelope appears to retain little, if any, osmotically active material. For example, changes in refraction indicate that macromolecules in the isolation medium diffuse into, and nuclear constituents diffuse out of, nuclei isolated from *Chaetopterus* eggs (Merriam, 1959). In nucleated erythrocytes, the concomitant loss of haem fluorescence from both cytoplasm and nucleoplasm during haemolysis suggests that equilibration across the nuclear envelope occurs very rapidly (Davies, 1961). In these cells and in mammalian erythrocyte precursors, haemoglobin masses extend from cytoplasm to nucleoplasm through the pores (Davies, 1961; Grasso and coworkers, 1962; Hammel and Bessman, 1964; Schjeide and coworkers, 1964; Granick and Levere, 1965). The microscopic appearance suggests the pores allow direct continuity between nucleoplasm and cytoplasm, in which case not only haemoglobin but also most soluble macromolecules would readily pass through them.

In amphibian oocytes, the nuclear envelope is apparently permeable to water and salts but not to substances of high molecular weight. Oocyte nuclei isolated in salt or sucrose solutions swell no matter how concentrated the solute, but if the isolation medium contains sufficient protein or other colloid this swelling can be prevented (Callan, 1949, 1952; Battin, 1959; Dick and Fry, unpublished; Macgregor, 1962). In oocytes, the injection of hypertonic salt solutions into the cytoplasm results in swelling of the nucleus, whereas albumin or polyvinyl pyrrolidone solutions produce swelling or shrinking according to the concentration used (Harding and Feldherr, 1959). The injection of a 4 per cent polyvinyl pyrrolidone solution causes a marked reduction in nuclear volume, but if the whole cell is subsequently immersed in hypotonic solution, the nucleus swells in proportion to the osmotic swelling of the cytoplasm. The important factor in the control of nuclear volume in these cells thus appears to be the colloid osmotic pressure. The corollary that the nuclear envelope is impermeable to large molecules is borne out by the distribution of colloidal gold particles following injection into the cytoplasm of whole oocytes. There appears to be some accumulation of particles along the outside of the membrane but none actually enters the nucleus (Feldherr, 1964). On the other hand, the osmotic studies indicate that ions readily penetrate the nuclear envelope, and this has been confirmed by the finding of an electrical resistance little different from that of nucleoplasm or cytoplasm (Kanno and Loewenstein, 1963). The nuclear envelopes of invertebrate eggs, too, have a low resistance (Kanno and coworkers, 1965) and they may be considered as permeable to ions.

In contrast, the salivary gland cells of *Drosophila* larvae possess nuclear envelopes whose electrical resistance (1·5 ohms.cm^2) is more than 1000 times that of nucleoplasm or cytoplasm, though considerably less than that of the plasma membrane (640 ohms.cm^2). On penetrating the nuclear envelope with a microelectrode, a stable potential difference of some 13 mV can be recorded, which falls to zero if the envelope is punctured (Loewenstein and Kanno, 1963). Such a high resistance indicates that the nuclear envelope presents a formidable barrier to the diffusion of ions. Similar results have been obtained with other Dipteran larvae (Loewenstein, 1964; Ito and Loewenstein, 1965), and there is the possibility that a significant potential difference exists across the nuclear envelope in *Tritonia* giant neurones (Veprintsev and coworkers, 1964).

Amoeba nuclei isolated in salt solutions swell no matter how concentrated the solute (Korohoda and coworkers, 1968) and the nuclear envelopes appear to be freely permeable to ions. Indeed, judging by the distribution of injected ferritin or colloidal gold particles, the nuclear envelopes of *Amoebae* allow particles up to 90 Å to cross (Feldherr, 1962a, 1962b, 1965, 1969)! This may give a false indication of the permeability, however: the particles are accumulated in the nucleus, and one wonders whether their movement across the nuclear envelope may not be just by diffusion. Particles are seen in clusters along the central axes of the pores, but nowhere else in the nuclear envelope, suggesting that some central pore channel provides the route whereby they enter the nucleus (Feldherr, 1965). Whether this indicates that the pore centre is the most permeable part of the nuclear envelope or that the particles are carried by some macromolecule-transporting system located there is still uncertain. Feldherr and Harding (1964) suggest that the particles may adhere to the pore material and be engulfed with that material by a form of pinocytosis.

Systems for the transport of amino acids are apparently present in the nuclear envelopes of several cell types. Chick embryonic myocardium has been extensively studied from this respect (Klein, 1967, 1968; Klein and coworkers, 1968). Nuclei isolated in sucrose solutions show a basic Mg-dependent ATPase activity which can be localized histochemically to the nuclear envelope. In the presence of sodium or potassium, amino acids cause a great increase in ATPase activity, and this enhancement shows saturation kinetics and is inhibited by ouabain. It is still demonstrable after sonic disruption of the nuclei. Isolated nuclei also show a net uptake of amino acids with characteristics paralleling those of the ATPase. Nuclear concentrations of apparently free amino acid up to seven times those of the medium can be obtained. This alone indicates that the nuclear envelope is impermeable to these substances, and the existence of what seems to be an active transport system is good circumstantial evidence that the permeability to amino acids is low *in vivo*. Similar conclusions apply in the case of other tissues such as calf thymus and rat brain, where similar amino acid-transporting systems have been demonstrated in the isolated nuclei, along with a sodium-dependent nucleoside transport (Allfrey and coworkers, 1961; Navon and Lajtha, 1969), and rat liver, which can accumulate free adenine in the nucleus (Lee and Holbrook, 1964). It seems unlikely that such elaborate transport systems are artifacts. However, the fast exchange of labelled amino acids between medium and nuclei in both myocardial (Klein and coworkers, 1968) and thymus nuclei, even at 0°C (Ord and Stocken, 1961), seems to call for some additional postulate, an exchange–diffusion mechanism, for instance. Other apparently conflicting results, for example the penetration of sucrose into isolated thymus nuclei (Kodama and Tedeschi, 1968), may be due to there being greater nuclear damage on isolation with some methods. Also, with substances within this comparatively modest range of molecular weights, other factors than the 'sieve effect' of the nuclear envelope, such as the solubility of the substance in the membrane or in the mysterious pore material, may well be important. Calf thymus nuclei seem to show such a selectivity not based on molecular dimensions. The nuclear concentration of succinate has been reported as twice the cytoplasmic, whereas the nuclear and cytoplasmic concentrations of pyruvate, lactate and α-ketoglutarate were approximately equal (McEwen and coworkers, 1963a). In addition, isolated nuclei failed to take up succinate malonate or citrate from the incubation medium, but readily took up acetate and pyruvate (McEwen and coworkers, 1963b). This discrimination may not be entirely passive, however.

The importance of these studies for the question of ion movements is that they set a rough upper limit to the passive permeability of the nuclear envelope. The nuclear envelopes which restrict the passage of small molecules would seem to be those worth scrutinizing closely for signs of a low ion permeability. In the case of liver cells, Siebert and coworkers (1965b) and Langendorf and coworkers (1964) have shown:

i) that the sodium content of nuclei mass-isolated in non-aqueous media is greatly reduced if the fluid perfusing the liver prior to nuclear isolation is low in sodium, and conversely that with a high-sodium perfusion fluid the nuclear sodium content is increased;

ii) that radioactive sodium injected *in vivo* equilibrates with the nuclear sodium within ten minutes;

iii) that the sodium content and sodium exchange of these nuclei is apparently unaffected by cooling or metabolic inhibitors.

The simplest explanation of these results is that sodium can pass freely across the nuclear envelope. In that case other small ions might be expected to have access to the nucleoplasm, and, in support of this, beryllium has been found to enter the isolated liver nucleus readily, and also to penetrate the nucleus *in vivo*. There is no indication that any active process is involved, and the greater concentration achieved in the nucleus is probably due to binding or complex formation (Witschi and Aldridge, 1968). Ruthenium, on the other hand, enters the nucleus of the intact liver cell only very slowly, with a half time of some 45 hours (Gilbert and Radley, 1964). It is difficult to be certain whether this is because the nuclear envelope discriminates against ions of large crystal radius, or because ruthenium binding or complex formation occurs in the cytoplasm.

In the case of chick embryonic myocardium, there is little to indicate whether or not the nuclear envelope is permeable to ions. The isolated nuclei do not accumulate sodium or potassium, and their ion content is similar to that of the medium (Klein and coworkers, 1968), but both a high and a low permeability could give rise to such results. In calf thymus nuclei, however, there is firm evidence that, after sucrose isolation, the nuclear envelope allows sodium and potassium to pass freely. Nuclei prepared using sucrose solutions do not retain sodium well (Allfrey and coworkers, 1961) and their sodium and potassium contents are approximately one-third those of nuclei isolated in non-aqueous media (Itoh and Schwartz, 1957). The sodium apparently lost on sucrose isolation can be regained on incubation in high-sodium media but the extra sodium is rapidly lost again on washing with solutions low in sodium. These ion movements are little affected by temperature or metabolic inhibitors, and seem to be entirely passive (Itoh and Schwartz, 1957). However, how far these results on isolated nuclei reflect the permeability of the nuclear envelope of the intact thymocyte is difficult to assess.

Even taking into account the various sources of error mentioned at the beginning of this section, there does appear to be a genuinely wide range of nuclear envelope permeabilities. This means that the nuclear envelope permeability has to be worked out separately in each case. The mammalian somatic nuclei so far studied have all fitted near the lower end of the range, showing evidence of a low permeability to substances of low molecular weight, but being apparently freely permeable to small ions. Whether this is true of mammalian somatic nuclei in general cannot yet be stated with certainty. At present the only cells for which there is definite evidence that the nuclear envelope has a low permeability to ions are the salivary gland cells of Dipteran larvae. In these it is possible that some process other than simple diffusion governs the movement of ions between nucleoplasm and cytoplasm.

V. DIFFERENCES IN ION COMPOSITION BETWEEN NUCLEOPLASM AND CYTOPLASM

Nuclear and cytoplasmic ion concentrations might be expected to differ if the nuclear envelope were impermeable to ions and, in particular, if some active transport of ions similar to that described for the plasma membrane took place. There are several instances of such differences in ion concentrations. This is especially well documented for sodium, and the nuclear concentration of this ion may be over ten times the cytoplasmic (Table 1). Such concentration gradients could, however, be due to 'binding'—in the sense of an association of the ion with a large molecule such that the activity of the ion is reduced. Ion binding in this sense to nucleic acids and other cell constituents has been demonstrated *in vitro* (Carr, 1956; Lewis and Saroff, 1957; Baker and Saroff, 1965; Barber and Noble, 1966; Weinstock and coworkers, 1967). The important question, then, is whether there are significant differences in ion activity between nucleoplasm and cytoplasm, and

Table 1. Cells in which the sodium concentration of the nucleus apparently exceeds that of the cytoplasm.

Cell	Estimated N/C Ratio	Method	Reference
Rat liver	13	Mass-isolation in non-aqueous media and Na analysis	Langendorf and coworkers (1961)
Calf thymus, liver and kidney	3	,,	Itoh and Schwartz (1957)
Frog oocytes	2	^{24}Na autoradiography of frozen sections	Abelson and Duryee (1949)
	3	^{22}Na autoradiography of frozen sections	Naora and coworkers (1962)
	3	Analysis of nuclei dissected from single frozen cells	Naora and coworkers (1962)
Tritonia giant neurones	2–3	^{22}Na autoradiography of fixed tissue	Antonov, Kurella and Yaglova (1965b, 1968)
Larus salt-gland	—	Precipitation of Na with pyroantimonate during fixing of the tissue	Komnick and Komnick (1963)
Chick embryonic myocardium	—	,,	Yeh, Hoffman and Spiro (1966)
Rat epithelia and Schwann cells; mouse lymphocytes	—	,,	Spicer, Hardin and Greene (1968)
Frog skin and renal tubule	—	,,	Zadunaisky and coworkers (1968)
Rat kidney	—	,,	Bulger (1969)

$$\text{N/C Ratio} = \frac{\text{Nuclear Na concentration (per litre nuclear water)}}{\text{Cytoplasmic Na concentration (per litre cytoplasmic water)}}$$

unfortunately in none of the cells listed in Table 1 is the nuclear sodium activity known. Some indication of the activity coefficient can, however, be inferred from other data. Thus in frog oocytes there is evidence that the nuclear envelope is freely permeable to ions, and in this case the accumulation of sodium in the nucleus is presumably due to association of the ion with nuclear macromolecules. There is less adequate information on the permeability of the nuclear envelope in the other cells, but a number of observations suggest that the sodium activity coefficient in the nucleoplasm may well be low. Spicer and coworkers (1968) noted the close correlation between the distribution of pyroantimonate precipitate and that of heterochromatin in various cells. The cells with the most heterochromatin had the most nuclear precipitate, and the heaviest precipitation was over the heterochromatin, suggesting that the sodium was 'bound' to one or more of its constituents. Similarly, the nuclear accumulation of sodium found in Antonov and coworkers' experiments (1965b, 1968) is most likely due to 'binding': the giant neurones were fixed before autoradiography, and the persistence of a nuclear–cytoplasmic gradient after such treatment plus the presence of a higher concentration in the nucleoli makes it improbable that the distribution observed is that of 'free' sodium. In thymus nuclei, as mentioned above, the ready loss and uptake of sodium after sucrose isolation indicates that the envelope allows the cation to pass freely (Itoh and Schwartz, 1957), yet even in low-sodium medium approximately one-third of the total nuclear sodium is retained. Thus at least this fraction of the sodium is non-diffusible, and the linear relationship between the nuclear sodium content and the sodium concentration of the medium suggests that almost all the sodium in these nuclei is in fact 'bound'.

The high concentration of sodium in liver nuclei has been extensively investigated (Langendorf and coworkers, 1961, 1964, 1966; Siebert and coworkers, 1965a, b). Sodium appears to penetrate the nuclear envelope rapidly, for equilibration between nucleoplasm and extracellular fluid occurs within ten minutes of injection of radioactive sodium into the whole animal, and there is no indication that the high concentration of sodium in the nucleus is dependent on a virtually impermeable nuclear envelope. As the amount of nuclear sodium is unaffected by cooling or the presence of metabolic inhibitors, it is probably not accumulated actively (the total cell sodium did not change greatly on cooling either, so there is some doubt about this as a test of active transport). The readiness with which ion changes could be brought about by changes in the perfusion medium, on the other hand, suggests an ion exchange mechanism, the high concentration of ion in the nucleus being due to 'trapping' among the nuclear macromolecules. Siebert and coworkers raised the question whether the sodium may not be intranuclear but in the perinuclear space, which was thought to communicate with the exterior. There are two objections to this. Firstly, unless the sodium concentration in the perinuclear space is very much greater than that of the extracellular fluid, the volume of the perinuclear space observed under the electron microscope is far too small to account for a nuclear sodium concentration thirteen times the cytoplasmic. Secondly, the method used removes the whole of the nuclear envelope (Siebert and coworkers, 1965a). Further evidence in favour of a large part of the sodium being associated with nuclear macromolecules is that the concentration of inorganic anions in the nucleus is sufficient to neutralize only 40 per cent of the nuclear cations (Langendorf and coworkers, 1966). In rat liver both potassium and chloride are present as well as sodium in higher concentration in the nucleus than in the cytoplasm, and if all these ions were in free solution, and the nuclear membrane impermeable to them, the crystalloid osmotic pressure difference between nucleoplasm and cytoplasm would be very high, and unlikely to be offset by differences in colloid osmotic pressure. If the perinuclear space contained fluid similar in composition to the extracellular fluid, the osmotic gradient across the nuclear membrane, as distinct from the whole envelope, would be enormous.

In all these cells, then, the nuclear sodium activity coefficient is likely to be low, and the high sodium concentration due to association of the ion with nuclear constituents. The high concentrations of potassium and chloride in liver and thymus nuclei are also probably due to 'binding', and it is tempting to ascribe the apparent high nuclear sodium in the other cells and the high concentration of phosphate in nuclei of maize root tips (Libanati and Tandler, 1969) to the same cause. However, there is little direct evidence for this, although the localized higher concentration of phosphate in maize root nucleoli suggests that at least part of the nuclear phosphate is 'bound'. Another approach follows from the correlation of Na–K ATPase activity with cation transport across the cell membrane (Skou, 1965). ATPase activity at the nuclear envelope has been demonstrated histochemically in several cells (Novikoff and coworkers, 1962; Novikoff, 1967), although this is not universal (Kaye and Pappas, 1965), and the method is liable to artifact (Moses and coworkers, 1966; Rosenthal and coworkers, 1966; Gillis and Page, 1967). In addition, the presence of a Na–ATPase in mammallian brain and thymus nuclei *in vitro*, and in frog and chick embryonic cell nuclei has already been mentioned. One wonders whether at least part of the activity could be accounted for by active ion transport, but an Na–K ATPase similar to that of the plasma membrane has yet to be described. An analysis of the enzyme composition of isolated nuclear membranes has failed to reveal it (Zbarsky and coworkers, 1969). The only Na ATPase to be fully studied is that of chick embryonic myocardium (Klein and coworkers, 1968), and this differs from the Na–K ATPase of the cell membrane in some important respects. Thus the increase in basal activity brought about by sodium is not inhibited by ouabain, and potassium can substitute fully for sodium. A nuclear ion transport system with different characteristics, however, cannot be discounted.

VI. SUMMARY AND CONCLUSIONS

In all the cells studied water appears to cross the nuclear envelope by diffusion. Movements of water between nucleus and cytoplasm are passive and follow osmotic pressure differences brought about by changes in the concentrations of substances to which the nuclear envelope is impermeable. In nucleated red cells, *Amoebae* and the oocytes of several species, there is evidence that ions also pass freely through the nuclear envelope, probably via the nuclear pores. Whether the membrane component of the nuclear envelope is impermeable to ions is not yet clear. Differences in ion composition between nucleoplasm and cytoplasm in these cells have been observed, but presumably are due to the association of the ions with nuclear constituents to which the envelope is impermeable. In a sense, then, the nuclear envelope governs the movement of ions in these cases by regulating the passage of macromolecules. In other cases, in particular in the salivary gland cells of Dipteran larvae, electrical measurements indicate that the nuclear envelope presents a significant barrier to the diffusion of ions. These cells maintain a steady electric potential between nucleoplasm and cytoplasm, and if this is a true membrane potential then some form of ion transport presumably operates to prevent it being dissipated by the inevitable leak. There is no direct evidence for active transport of ions, however.

Several instances of an ion being in greater concentration in the nucleus than the cytoplasm are on record. Unfortunately, the activities of the respective ions in the nucleoplasm are unknown, and in most cases there is little information about the permeability of the nuclear envelope. There is little to suggest that the nuclear envelope is involved in this ion accumulation, whereas there are a number of indications that association of ions with nuclear macromolecules occurs. A system similar to the cation pump of the cell membrane has been looked for without success, and other types of ion transport have not come to light. In some nuclei an ion-dependent accumulation of amino acids and other small molecules has been found but there is no evidence for concomitant ion transport.

It therefore seems likely that in all except a few cells ions are able to move freely across the nuclear envelope by diffusion. This only applies to passage across the envelope as a whole; this probably takes place by the nuclear pores. No conclusions can yet be drawn about the movement of substances across the nuclear membrane, i.e. between nucleoplasm or cytoplasm and perinuclear space.

REFERENCES

Abelson, P. H. and W. R. Duryee (1949) *Biol. Bull.*, **96**, 205
Afzelius, B. A. (1955) *Expl. Cell. Res.*, **8**, 147
Allfrey, V. G., R. Meudt, J. W. Hopkins and A. E. Mirsky (1961) *Proc. Nat. Acad. Sci. U.S.*, **47**, 907
Antonov, V. F., G. A. Kurella, L. F. Meshchisen and V. Ben'-tse (1965a) *Dokl. Akad. Nauk. S.S.S.R., Biol.*, **161**, 691
Antonov, V. F., G. A. Kurella and L. G. Yaglova (1965b) *Biophysics*, **10**, 1200
Antonov, V. F., G. A. Kurella and L. G. Yaglova (1968) *Biophysics*, **13**, 398
Baker, H. P. and H. A. Saroff (1965) *Biochemistry, Wash.*, **4**, 1670
Barber, R. and M. Noble (1966) *Biochim. Biophys. Acta*, **123**, 205
Barer, R., S. Joseph and G. A. Meek (1959) *Exp. Cell Res.*, **18**, 179
Battin, W. T. (1959) *Exp. Cell Res.*, **17**, 59
Beck, L. V. and H. Shapiro (1936) *Proc. Soc. Exp. Biol. Med.*, **34**, 170
Beerman, W. (1964) *J. Exp. Zool.*, **157**, 49
Beritashvili, D. R., I. Sh. Kvavilashvili and C. A. Kafiani (1969) *Exp. Cell Res.*, **56**, 113
Bier, K. (1965) *Chromosoma*, **16**, 58
Blackburn, W. R. and K. Vinijchaikul (1969) *Lab. Invest.*, **20**, 305
Bolund, L., N. R. Ringertz and H. Harris (1969) *J. Cell Sci.*, **4**, 71
Boyd, J. B., H. D. Berendes and H. Boyd. (1968) *J. Cell Biol.*, **38**, 369
Bulger, R. E. (1969) *J. Cell Biol.*, **40**, 79
Callan, H. G. (1949) *Hereditas*, **35**, suppl. 547
Callan, H. G. (1952) *Symp. Soc. Exp. Biol.*, **6**, 243
Callan, H. G. and S. G. Tomlin (1950) *Proc. Roy. Soc. (London), Ser. B*, **137**, 367
Carr, C. W. (1956) *Arch. Biochem. Biophys.*, **62**, 476
Churney, L. (1942) *Biol. Bull.*, **82**, 52
Clever, V. and C. G. Rombull (1966) *Proc. Nat. Acad. Sci. U.S.*, **56**, 1470
Davies, H. G. (1961) *J. Biophys. Biochem. Cytol.*, **9**, 671
Edmonds, R. H. (1968) *J. Ultrastruct. Res.*, **24**, 295
Ellgaard, E. G. and R. G. Kessel (1966) *Exp. Cell Res.*, **42**, 302
Epstein, M. A. (1957) *J. Biophys. Biochem. Cytol.*, **3**, 851
Fawcett, D. W. (1966) *Am. J. Anat.*, **119**, 129
Feldherr, C. M. (1962a) *J. Cell Biol.*, **12**, 159
Feldherr, C. M. (1962b) *J. Cell Biol.*, **14**, 65
Feldherr, C. M. (1964) *J. Cell Biol.*, **20**, 188
Feldherr, C. M. (1965) *J. Cell Biol.*, **25**, 43
Feldherr, C. M. (1968) *J. Cell Biol.*, **39**, 49
Feldherr, C. M. (1969) *J. Cell Biol.*, **42**, 841
Feldherr, C. M. and A. B. Feldherr (1960) *Nature*, **185**, 250
Feldherr, C. M. and C. V. Harding (1964) *Protoplasmatologia*, **5**, (2), 35
Franke, W. W. (1966) *J. Cell Biol.*, **31**, 619
Franke, W. and J. Kartenbeck (1969) *Experientia*, **25**, 396
Friend, D. S. (1966) *J. Cell Biol.*, **29**, 317
Gall, J. G. (1964) *Protoplasmatologia*, **5** (2), 4
Gall, J. G. (1967) *J. Cell Biol.*, **32**, 391
Gilbert, I. G. F. and J. M. Radley (1964) *Biochim. Biophys. Acta*, **79**, 568
Gillis, J. M. and S. G. Page (1967) *J. Cell Sci.*, **2**, 113
Granick, S. and R. D. Levere (1965) *J. Cell Biol.*, **26**, 167
Grasso, J. A., H. Swift and G. A. Ackerman (1962) *J. Cell. Biol.*, **14**, 235
Gurr, M. I., J. B. Finean and J. N. Hawthorne (1963) *Biochim. Biophys. Acta.*, **70**, 406
Hamburger, J. (1904) *Osmotischer Druck und Ionenlehre in dem medicinischen Wissenschaften, Band III*, J. F. Bergmann, Wiesbaden
Hammell, C. L. and S. P. Bessman (1964) *J. Biol. Chem.*, **239**, 2228
Harding, C. V. and C. Feldherr (1959) *J. Gen. Physiol.*, **42**, 1155
Ito, S. and W. R. Loewenstein (1965) *Science, N.Y.*, **150**, 909

Itoh, S. and I. L. Schwartz (1957) *Am. J. Physiol.*, **188**, 490
Kalifat, S. R., M. Bouteille and J. Delarve (1967) *J. Microscop.*, **6**, 1019
Kanno, Y. and W. R. Loewenstein (1963) *Exp. Cell Res.*, **31**, 149
Kanno, Y., R. F. Ashman and W. R. Loewenstein (1965) *Exp. Cell Res.*, **39**, 184
Kasper, C. B. and D. M. Kashnig (1969) *Federation Proc.*, **28**, 404
Kaye, G. I. and G. D. Pappas (1965) *J. Microscop.*, **4**, 497
Klein, R. L. (1967) *Proc. Soc. Exp. Biol. Med.*, **124**, 1258
Klein, R. L. (1968) *Exp. Cell Res.*, **49**, 69
Klein, R. L., C. R. Horton and A. Thureson-Klein (1968) *European J. Biochem.*, **6**, 514
Kodama, R. M. and H. Tedeschi (1968) *J. Cell Biol.*, **37** ,747
Komnick, H. and V. Komnick (1963) *Z. Zell for sch.*, **60**, 163
Korohoda, W., J. A. Forrester, K. G. Moreman and E. J. Ambrose (1968) *Nature*, **217**, 615
Kostello, A. B. and G. A. Morrill (1968) *Exp. Cell Res.*, **50**, 679
Kroeger, H. (1963) *Nature*, **200**, 1234
Kroeger, H. (1964) Cited in Loewenstein, W. R. (1964), *Protoplasmatologia*, **5** (2), 26
Kroeger, H. (1967) *Mem. Soc. Endocrinol.*, **15**, 55
Kroeger, H. and M. Lezzi (1966) *Ann. Rev. Entomol.*, **11**, 1
Lane, B. P. and E. Martin (1969) *J. Histochem. Cytochem.*, **17**, 102
Langendorf, H., G. Siebert, K. Kesselring and R. Hannover (1966) *Nature*, **209**, 1130
Langendorf, H., G. Siebert, I. Lorenz, R. Hannover and R. Beyer (1961) *Biochem. Z.*, **335**, 273
Langendorf, H., G. Siebert and D. Nitz-Litzow (1964) *Nature*, **204**, 888
Lee, H-J. and D. J. Holbrook (1964) *Arch. Biochem. Biophys.*, **108**, 275
Lewis, M. S. and H. A. Saroff (1957) *J. Am. Chem. Soc.*, **79**, 2112
Lezzi, M. and H. Kroeger (1966) *Z. Naturf.*, **21**B, 274
Libanati, C. M. and C. J. Tandler (1969) *J. Cell Biol.*, **42**, 754
Loewenstein, W. R. (1964) *Protoplasmatologia*, **5** (2), 26
Loewenstein, W. R. (1966) *Ann. N.Y. Acad. Sci.*, **137**, 441
Loewenstein, W. R. and Y. Kanno (1963) *J. Cell Biol.*, **16**, 421
Loewenstein, W. R., Y. Kanno and S. Ito (1966) *Ann. N.Y. Acad. Sci.*, **137**, 708
Macgregor, H. C. (1962) *Exp. Cell Res.*, **26**, 520
McEwen, B. S., V. G. Allfrey and A. E. Mirsky (1963a) *J. Biol. Chem.*, **238**, 2571
McEwen, B. S., V. G. Allfrey and A. E. Mirsky (1963b) *J. Biol. Chem.*, **238**, 2579
Marinos, N. G. (1960) *J. Ultrastruct. Res.*, **3**, 328
Mentré, P. (1969) *J. Microscop.*, **8**, 51
Merriam, R. W. (1959) *J. Biophys. Biochem. Cytol.*, **6**, 353
Merriam, R. W. (1961) *J. Biophys. Biochem. Cytol.*, **11**, 559
Milner, G. R. (1969) *J. Cell Sci.*, **4**, 569
Mityushin, V. M. (1968) *Biophysics*, **13**, 446
Morrill, G. A. and D. E. Watson (1966) *J. Cell Physiol.*, **67**, 85
Moses, H. L., A. S. Rosenthal, D. L. Beaver and S. S. Schuffman (1966) *J. Histochem. Cytochem.*, **14**, 702
Murray, R. G. A. Murray and A. Pizzo (1965) *Anat. Rec.*, **151**, 17
Naora, H., H. Naora, M. Izawa, V. G. Allfrey and A. E. Mirsky (1962) *Proc. Nat. Acad. Sci. U.S.*, **48**, 853
Navon, S. and A. Lajtha (1969) *Biochim. Biophys. Acta*, **173**, 518
Novikoff, A. B. (1967) *J. Histochem. Cytochem.*, **15**, 353
Novikoff, A. B., E. Essner, S. Goldfischer and M. Heus (1962) *Symp. Internat. Soc. Cell Biol.*, **1**, 149
Ord, M. G. and L. A. Stocken (1961) *Biochem. J.*, **81**, 1
Palay, S. L. (1960) *J. Biophys. Biochem. Cytol.*, **7**, 391
Pogo, A. O., V. C. Littau, V. G. Allfrey and A. E. Mirsky (1967) *Proc. Nat. Acad. Sci., U.S.*, **57**, 743
Riemann, W., C. Muir and H. C. Macgregor (1969) *J. Cell Sci.*, **4**, 299
Ristow, H. and S. Arends (1968) *Biochim. Biophys. Acta..* **157**, 178

Rosenthal, A. S., H. L. Moses, D. L. Beaver and S. S. Schuffman (1966) *J. Histochem. Cytochem.*, **14**, 698
Sadowski, P. D. and J. W. Steiner (1968) *J. Cell Biol.*, **37**, 147
Schjeide, O. A., R. G. McCandless and R. J. Munn (1964) *Growth*, **28**, 29
Siebert, G., G. B. Humphrey, H. Themann and W. Kersten (1965a) *Hoppe-Seyler's Z. Physiol. Chem.*, **340**, 51
Siebert, G., H. Langendorf, R. Hannover, D. Nitz-Litzow, B. C. Pressman and C. Moore (1965b) *Hoppe-Seyler's Z. Physiol. Chem.*, **343**, 101
Skou, J. C. (1965) *Physiol. Rev.*, **45**, 596
Smith, S. J., H. R. Adams, K. Smetana and H. Busch (1969) *Exp. Cell Res.*, **55**, 185
Spicer, S. S., J. H. Hardin and W. B. Greene (1968) *J. Cell Biol.*, **39**, 216
Stevens, B. J. (1964) *J. Ultrastruct. Res.*, **11**, 329
Stevens, B. J. and H. Swift (1966) *J. Cell Biol.*, **31**, 55
Stirling, C. E. and W. B. Kinter (1967) *J. Cell Biol.*, **35**, 585
Takamoto, T. (1966) *Nature*, **211**, 772
Veprintsev, B. N., I. V. Krasts, and D. A. Sakharov (1964) *Biophysics*, **9**, 351
Vivier, E. (1967) *J. Microsc.*, **6**, 371
Ward, R. T. and E. Ward (1968) *J. Cell Biol.*, **39**, 139A
Watson, M. L. (1955) *J. Biophys. Biochem. Cytol.*, **1**, 257
Watson, M. L. (1959) *J. Biophys. Biochem. Cytol.*, **6**, 147
Weinstock, A., P. C. King and R. E. Wuthier (1967) *Biochem. J.*, **102**, 983
Wiener, J., D. Spiro and W. R. Loewenstein (1965) *J. Cell Biol.*, **27**, 107
Wischnitzer, S. (1960) *Int. Rev. Cytol.*, **10**, 137
Witschi, H. P. and W. N. Aldridge (1968) *Biochem. J.*, **106**, 811
Wunderlich, F. (1969) *Exp. Cell Res.*, **56**, 369
Yasuzumi, G., Y. Nakai, I. Tsubo, M. Yasuda and T. Sugioka (1967) *Exp. Cell Res.*, **45**, 261
Yeh, B. K. and B. F. Hoffman (1967) *Federation Proc.*, **26**, 382
Yeh, B. K., B. F. Hoffman and D. Spiro (1966) *The Physiologist*, **2**, 324
Zadunaisky, J. A., J. F. Gennaro, N. Bashirelahi and M. Hilton (1968) *J. Gen. Physiol.*, **51**, 290s.
Zbarsky, I. B., K. A. Perevoshchikova, L. N. Delektorskaya and V. V. Delektorsky (1969) *Nature*, **221**, 257

Index